Synthesis of Metal-Organic Frameworks via Water-Based Routes

Synthesis of Metal-Organic Frameworks via Water-Based Routes

A Green and Sustainable Approach

Edited by

Yasser Azim
Assistant Professor, Department of Applied Chemistry, Zakir Husain College of Engineering & Technology, Faculty of Engineering & Technology, Aligarh Muslim University, Aligarh, Uttar Pradesh, India

Sami-Ullah Rather
Department of Chemical and Materials Engineering, King Abdulaziz University, Jeddah, Saudi Arabia

Showkat Ahamd Bhawani
Faculty of Resource Science and Technology, Universiti Malaysia Sarawak, Kota Samarahan, Sarawak, Malaysia

Prashant M. Bhatt
Staff Scientist, Advanced Membranes and Porous Materials Centre, King Abdullah University of Science and Technology (KAUST), Thuwal, Saudi Arabia

ELSEVIER

Elsevier
Radarweg 29, PO Box 211, 1000 AE Amsterdam, Netherlands
The Boulevard, Langford Lane, Kidlington, Oxford OX5 1GB, United Kingdom
50 Hampshire Street, 5th Floor, Cambridge, MA 02139, United States

Notices
Knowledge and best practice in this field are constantly changing. As new research and experience broaden our understanding, changes in research methods, professional practices, or medical treatment may become necessary.

Practitioners and researchers must always rely on their own experience and knowledge in evaluating and using any information, methods, compounds, or experiments described herein. In using such information or methods they should be mindful of their own safety and the safety of others, including parties for whom they have a professional responsibility.

To the fullest extent of the law, neither the Publisher nor the authors, contributors, or editors, assume any liability for any injury and/or damage to persons or property as a matter of products liability, negligence or otherwise, or from any use or operation of any methods, products, instructions, or ideas contained in the material herein.

ISBN: 978-0-323-95939-1

For Information on all Elsevier publications visit our website at https://www.elsevier.com/books-and-journals

Publisher: Candice Janco
Acquisitions Editor: Gabriela Capille
Editorial Project Manager: Kathrine Esten
Production Project Manager: Rashmi Manoharan
Cover Designer: Miles Hitchen

Typeset by Aptara, New Delhi, India

Dedication

Sir Syed Ahmad Khan for his enormous efforts in promoting modern education and the establishment of Muhammadan Anglo-Oriental College, which later evolved into Aligarh Muslim University.

Contents

Contributors

Mohd. Sufian Abbasi, Department of Civil Engineering, Integral University, Lucknow, Uttar Pradesh, India

Musheer Ahmad, Department of Applied Chemistry, ZHCET, Faculty of Engineering and Technology, Aligarh Muslim University, Aligarh, Uttar Pradesh, India

Naseem Ahmad, Department of Chemistry, Integral University, Lucknow, Uttar Pradesh, India

Akhtaruzzaman, Department of Chemistry, Aliah University, New Town, Kolkata, West Bengal, India

Shaikh Arfa Akmal, Functional Inorganic Materials Lab (FIML), Department of Chemistry, Aligarh Muslim University, Aligarh, Uttar Pradesh, India

Seikh Mafiz Alam, Department of Chemistry, Aliah University, Kolkata, West Bengal, India

Arif Ali, Department of Applied Chemistry, ZHCET, Faculty of Engineering and Technology, Aligarh Muslim University, Aligarh, Uttar Pradesh, India

Farrukh Aqil, UofL Health-Brown Cancer Center and Department of Medicine, University of Louisville, Louisville, KY, United States

Nargis Akhter Ashashi, Department of Chemistry, University of Jammu, Jammu and Kashmir, India

Mohd. Asif, Department of Chemistry, Integral University, Lucknow, Uttar Pradesh, India

Yasser Azim, Department of Applied Chemistry, Zakir Husain College of Engineering & Technology, Faculty of Engineering & Technology, Aligarh Muslim University, Aligarh, Uttar Pradesh, India

Taposi Chatterjee, Department of Chemistry, Aliah University, Kolkata, West Bengal, India; Department of Basic Science & Humanities, Techno International New Town, Kolkata, West Bengal, India

Basudeb Dutta, Department of Chemistry, Aliah University, New Town, Kolkata, West Bengal, India

Utsav Garg, Department of Applied Chemistry, Zakir Husain College of Engineering & Technology, Faculty of Engineering & Technology, Aligarh Muslim University, Aligarh, Uttar Pradesh, India

Atif Husain, Department of Chemistry, Integral University, Lucknow, Uttar Pradesh, India

Saima Kamaal, Department of Applied Chemistry, ZHCET, Faculty of Engineering and Technology, Aligarh Muslim University, Aligarh, Uttar Pradesh, India

Samrah Kamal, Functional Inorganic Materials Lab (FIML), Department of Chemistry, Aligarh Muslim University, Aligarh, Uttar Pradesh, India

Mohd Khalid, Functional Inorganic Materials Lab (FIML), Department of Chemistry, Aligarh Muslim University, Aligarh, Uttar Pradesh, India

Abdul Rahman Khan, Department of Chemistry, Integral University, Lucknow, Uttar Pradesh, India

Samim Khan, Department of Chemistry, Aliah University, New Town, Kolkata, West Bengal, India

Mohammad Hedayetullah Mir, Department of Chemistry, Aliah University, New Town, Kolkata, West Bengal, India

Mohd Muslim, Department of Applied Chemistry, ZHCET, Faculty of Engineering and Technology, Aligarh Muslim University, Aligarh, Uttar Pradesh, India

Malik Nasibullah, Department of Chemistry, Integral University, Lucknow, Uttar Pradesh, India

Bhaskar Nath, A. M. School of Education, Assam University, Silchar, Assam, India

Zaib ul Nisa, Department of Chemistry, University of Jammu, Jammu and Kashmir, India

SK Khalid Rahaman, Department of Chemistry, Aliah University, Kolkata, West Bengal, India

Sami-Ullah Rather, Department of Chemical and Materials Engineering, King Abdulaziz University, Jeddah, Saudi Arabia

M. Shahid, Functional Inorganic Materials Lab (FIML), Department of Chemistry, Aligarh Muslim University, Aligarh, Uttar Pradesh, India

Haq Nawaz Sheikh, Department of Chemistry, University of Jammu, Jammu and Kashmir, India

Benjamin Siddiqui, Department of Chemistry, Integral University, Lucknow, Uttar Pradesh, India

Kafeel Ahmad Siddiqui, Department of Chemistry, National Institute of Technology Raipur, Raipur, Chhattisgarh, India

Priyanka Singh, Department of Chemistry, National Institute of Technology Raipur, Raipur, Chhattisgarh, India

Farhat Vakil, Functional Inorganic Materials Lab (FIML), Department of Chemistry, Aligarh Muslim University, Aligarh, Uttar Pradesh, India

Preface

This book offers solid, quantitative descriptions and reliable guidelines, reflecting the maturation and demand of the field and the development of metal-organic frame (MOF) works. It summarizes the fundamental approaches and principles to prepare MOF works. The book particularly emphasizes the exciting preparation and applications of zeolitic imidazolate frameworks (ZIFs), isoreticular metal-organic frameworks (IRMOFs), coordination pillared-layer (CPL), and more. This book will be interesting for researchers working in the fields of MOF works, composite materials, material science, applied science, organic chemistry, and environmental chemistry. Additionally, the book will be useful for the scientists working on the preparation of MOF works from water-based systems. Furthermore, it will be equally helpful for the students in the development MOF works as well as graduates in material science, chemical engineering, environmental engineering, organic synthesis, and environmental chemistry. The book will serve as a reference book for industries working on the design and manufacturing process of MOF works.

The introductory chapter begins by covering basic information about MOFs and their applications. In the second chapter, the fundamentals of MOFs are discussed. The third chapter covers the kind and role of linkers for MOFs. Chapters four to eight describe different ways of synthesis of MOFs. Chapter nine provides valuable information about the solubility and thermodynamic stability of MOFs. Chapters 10 to 13 concentrate on the preparation and applications of essential water-based MOFs. Chapter 14 provides information about the applications of MOFs for wastewater treatment. Finally, the last chapter is dedicated to the industrial aspects of water-based MOFs.

Finally, we assure the readers that the information provided in this book can serve as a very important tool for anyone working on the MOF works. We are grateful to all the authors who contributed chapters to this book and who helped to turn our thoughts into reality. Lastly, we are grateful to the Elsevier team for their continuous support at every stage to make it possible to publish on time.

Chapter 1

Introduction to metal–organic frameworks

Utsav Garg and Yasser Azim
Department of Applied Chemistry, Zakir Husain College of Engineering & Technology, Faculty of Engineering & Technology, Aligarh Muslim University, Aligarh, Uttar Pradesh, India

1.1 Historical background

Porous materials have humongous applications in various life aspects and industries such as filters, masks, sponges, adsorbents, foams, and catalysts. The low density of porous materials, due to high void spaces, is the key feature that helps in designing countless functionalities for desired applications. In addition, long-range ordered structures (i.e., crystalline porous materials) enable versatile and straightforward control over their networks and attributes [1]. Metal–organic frameworks (MOFs), also known as porous coordination polymers (PCPs), comprise inorganic building blocks (metal ions or clusters) and organic ligands. They are the most advanced branch of coordination polymers [2–5]. Over the past two decades, more than 83,000 publications have been reported in the "metal–organic frameworks subset" and more than 23,000 publications have been reported in the SciFinder^{n} categories for "metal–organic frameworks" and "green synthesis" combined, and the number of publications relating to MOF-related research has been steadily rising (Fig. 1.1).

Synthetic zeolites consisting exclusively of inorganic silicates and aluminates had been intensively explored since the 1940s; nonetheless, the first crystalline porous materials with pore sizes of 1 nm or 2 nm were described in the early 1990s [6,7]. In 1995, Yaghi created MOFs, a novel category of crystalline porous materials with persistent porosity and large surface areas (MOFs) [8,9]. The unique feature of MOFs, in contrast to zeolites, was the strong bonds between charged organic ligands and metal ions. The discovery of MOFs with highly tunable organic and inorganic building units revolutionized the applications and design of porous materials. Contemporary, Kitagawa reported gas adsorption isotherms at room temperature and high pressures in which metal–organic polymers acted as hosts to take up gas-phase guest molecules [10]. Yaghi reported the first MOF with nitrogen adsorption/desorption isotherms at

Synthesis of Metal–Organic Frameworks via Water-Based Routes: A Green and Sustainable Approach.
DOI: https://doi.org/10.1016/B978-0-323-95939-1.00013-7

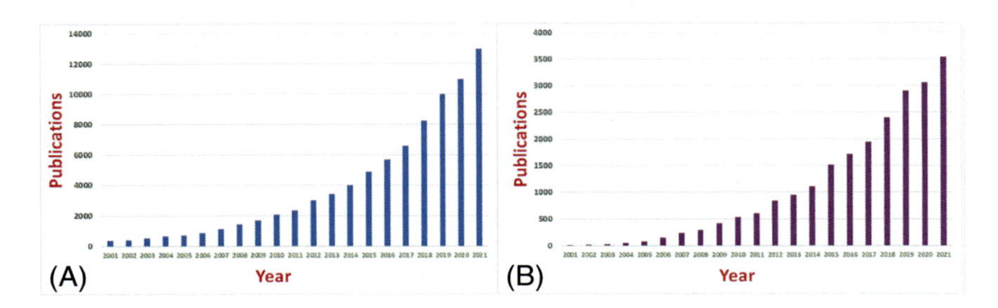

FIGURE 1.1 The number of metal–organic framework publications interactionsin last two decades, from 2001 to 2021. (A) The term "metal–organic frameworks" (B) the terms of "metal–organic frameworks" and "green synthesis" were used to search in SciFinder^n.

77 K and low pressure in 1998, resulting in the first values of pore volume and apparent surface area in a MOF [11] (Fig. 1.2). The previous works that support the presence of porosity were only focussed on guest exchange/removal in MOFs; however, gas adsorption measurements on MOF in the de-solvated condition lay the groundwork for creating permanent porosity [10]. Since then, the interest in MOFs has increased substantially due to their attractive physicochemical properties, including their flexible chemical structure, permanent porosity, large surface area, and great active sites [3,12–16]. The high porosity of MOFs is the paragon to make their utilization in opulent fields including catalysis [17–19], drug delivery [20,21], gas storage [22,23], sensing [24,25], small-molecule separations [26–28], conductivities [29,30], biomolecule encapsulation [31–33], magnetism [34–36], and many more.

1.2 Porosity of MOFs

1.2.1 Classification

Based on pore sizes, the International Union of Pure and Applied Chemistry (IUPAC) broadly classified porous materials (such as MOFs, activated carbon, and zeolites) into three categories, (1) for pore width <2 nm, materials are microporous, (2) for pore width 2 nm to <50 nm, materials are mesoporous, and (3) for pore width >50 nm, materials are termed as macroporous [37,38]. The porosity of MOFs has been studied generally by collecting adsorption isotherms of probe molecules (N_2 or Ar) at their standard boiling points. Ar is preferred and recommended by IUPAC over N_2 for calculating the porosity of materials with polar functionalities as N_2 has a high quadrupole moment and can interact with functional surfaces. However, due to the easy accessibility and low cost of N_2, most research groups use N_2 for adsorption experiments. CO_2 has much more quadrupole moment than N_2, hence not recommended by IUPAC for analyzing porous materials with polar surfaces. Researchers, however, utilized CO_2 to

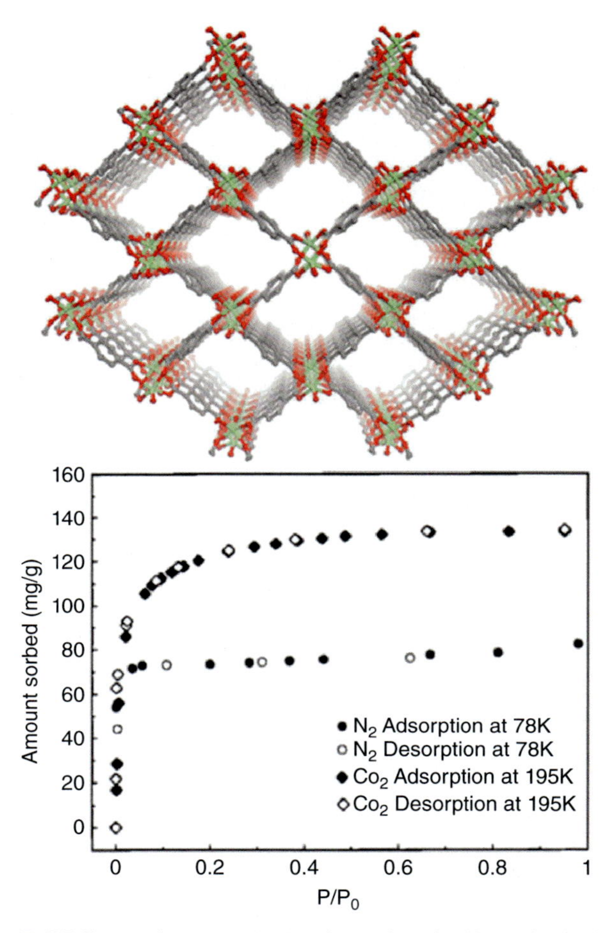

FIGURE 1.2 Zn(BDC) crystal structure (*top*) and gas adsorption/desorption isotherms (*bottom*) demonstrating permanent porosity. (*Reproduced with permission from ref. [11]. Copyright 1998, American Chemical Society*).

analyze petite pore size MOFs (<0.45 nm) as Ar and N_2 are hard to diffuse into such a small pore space [39].

Physisorption isotherms are generally plotted for adsorption studies and presented as the quantity of gas adsorbed versus relative pressure. IUPAC categorized six main types of physisorption isotherms viz. type I to type VI as representative isotherms such that the rigid MOFs often show the type I (microporous) or type IV (mesoporous) isotherm [37,39]. In contrast, the flexible MOFs have structural transformation throughout the adsorption process, and display shapes different from the above representative isotherms [40–45]. Moreover, the other adsorbates like water also show various shapes of adsorption isotherm and are dependent on MOF adsorbents [46–49]. The interpretation of isotherm shapes of flexible MOFs is complicated, and thus the development of novel approaches is in dire need to categorize new types of physisorption isotherms [50].

1.2.2 Surface area, pore volume, and pore size distribution

The collection of physisorption isotherms (N_2 or Ar adsorption isotherm) is the routinely used characterization of MOFs after discovering permanent porosity [11]. The interpretation of these isotherms helps to calculate surface area, pore size distribution, and pore volume. The Brunauer–Emmett–Teller (BET) method, based on the multilayer adsorption model, is the favored method for evaluating the surface area of porous MOFs [51]. The type II and type IV isotherms of porous materials (with pore width greater than 4 nm) are generally examined using the BET equation. However, IUPAC recommends applying the BET method with extreme care when micropores are present, which is the general case of most MOFs [39]. The differentiation between micropore filling, monolayer adsorption, and multilayer adsorption is difficult in microporous MOFs and problematic to locate a linear range of BET calculations. Hence, the recommended terms, "estimated BET area" or "apparent BET surface area," are widely used.

The four BET consistency criteria were developed by Rouquerol et al. to establish the linear range and calculate the projected BET area [52]. It is interesting that many reported BET area values only met the first two BET consistency requirements; nevertheless, for extremely porous MOFs with large surface areas, all four BET consistency criteria must be applied. Apart from their limitations, the BET area can be utilized as the fingerprint of MOF adsorbents and to compare the surface area of different MOFs [53].

The porosity of MOFs is also evaluated by measuring their pores volume. Pores are the empty spaces that are formed due to removal of guest molecules within any porous materials (Fig. 1.3). Ideally, a virtual horizontal plateau must be observed in the isotherm of a MOF that does not contain large pores and has a minimal external surface area, e.g., an adsorption isotherm plateau is seen in the case of type I isotherms for a microporous MOF and type IV isotherms for a mesoporous MOF when the saturation point is reached for the molecules that are being adsorbed.

The adsorbed capacity can be used to compute the experimental pore volume by applying the Gurvich rule under the assumption that the pores of MOFs are filled with condensed adsorbate in the liquid state [37,55]. Usually, uptake at a pressure close to unity is used to figure out the total pore volume. MOFs can have theoretical pore volumes calculated from their crystal structures that serve as a good benchmark for pore volumes measured experimentally. It is important to consider both pore geometries (e.g., cylinder, slit, and spherical) and kernels when interpreting the pore size distribution from isotherms. There

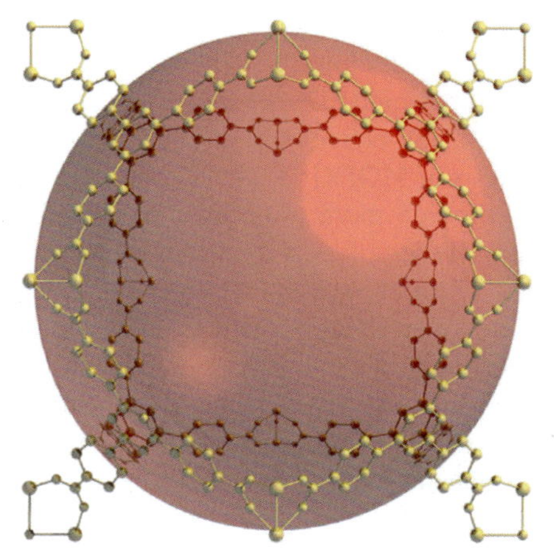

FIGURE 1.3 PCN-9 pore (*red sphere*) viewed through ligand-created window. (*Reproduced with permission from ref. [54]. Copyright 2009, Elsevier*).

can be significant variations in pore size distributions depending on the chosen shape models and kernels. Different methods, mainly density functional theory (DFT), Horvath–Kawazoe (HK), and Monte Carlo simulation (MC) for deriving pore size distribution, have been developed over the years [56,57]. Although the assumption of a molecularly smooth surface places theoretical restrictions on the DFT pore size distribution, the majority of porous MOFs can still be reliably assessed. Therefore, improving pore size distribution models is urgently needed, especially for flexible MOFs.

1.3 Green methods of synthesizing MOFs

The scientific literature provides a variety of approaches to the synthesis of MOFs, all of which are based on the 12 principles of green chemistry [58]. The fundamentals involve renewable feedstock, energy efficiency, non-hazardous solvents/safer reagents, waste elimination, and biodegradable final products. Also, the green and scalable synthesis of MOFs can be judged by (1) using less energy, (2) using theoretical predictions to design the performance of MOFs, (3) using continuous production methods, (4) using building blocks that are compatible with living things, and (5) using safe reaction media (such as water and supercritical solvents) [14,59–61]. Some of the chief methods for the green production of MOFs are briefly discussed.

1.3.1 Solvent-based synthesis

In MOF synthesis, solvent selection is crucial, as solvents primarily govern the coordination environment. The solvent can either coordinate with metal ions or act as a structure-directing agent. Consequently, the solvent selection is crucial to the final lattice structure [62]. The acid–base chemistry of organic solvents like dimethylformamide (DMF) and N, N′-diethylformamide (DEF) makes them attractive for non-aqueous media reactions and are generally used in the traditional synthesis of MOFs [63,64]. Moreover, due to their high boiling points, these solvents provide more prolonged periods for the reaction. However, heating or burning these solvents can produce hazardous amines in large quantities [65]. To ensure process safety and minimize environmental impacts, developing alternatives to solvents is critical. A green solvent can facilitate chemical processes that would otherwise be unfeasible, but their economic impact must also be considered. Bio-degradability, health impacts, and ease of extraction from natural or renewable resources are used to evaluate solvents' environmental impact. Hazards and toxicity can be reduced by using solvents derived from reusable and recyclable waste [66,67]. Several eco-friendly options have been suggested to replace environmentally unfriendly solvents, including water, lactone-based solvents (like c-butyrolactone and valerolactone), glycerol derivatives (like triacetin), dimethyl sulfoxide (DMSO), ethyl or methyl lactate, and ionic liquids (ILs) [68–71].

1.3.1.1 Water

Water is the most abundant, safest, and cheapest solvent. A water-based MOF synthesis is ideal since it allows easy purification and recycling [72]. Organic solvents cannot be used for industrial manufacturing since they must be managed carefully. Using water as a solvent led to new structural MOFs [73]. Most current research on MOF syntheses such as zeolitic imidazolate frameworks (ZIFs) and Materials Institute Lavoisier (MILs) uses water-based processes at atmospheric pressure and ambient temperature [74]. However, a few limitations, viz. phase transition, loss in crystallinity, and limited stability of the final product, are also associated with water-based MOF synthesis (Fig. 1.4). The plausible reason is the interfacial interaction of hydrophilic moieties of the MOF with water molecules [75,76]. Moreover, the latent heat and high specific heat of water results in increased energy costs for vaporization. The insolubility of most organic salts and linkers in water poses another challenge in the water-based synthesis of coordination compounds [77]. The solution to overcome these limitations will require (1) a detailed understanding of the nature of linkers (like N-linked, fluorinated, and chiral ligands) and metal salts (like sulfates and oxides) and (2) the use of modulators (like formic acid and polyvinylpyrrolidone [PVP]) and hydrophobic organic moieties (like organosilicon polymers and alkyl phosphonic acids) [78–80].

1.3.1.2 Supercritical liquids

The use of supercritical liquids (mainly $scCO_2$) as a medium is another intriguing option for eco-friendly synthesis of MOF [81]. Green synthesis techniques

FIGURE 1.4 Important considerations in choosing water as the solvent for metal–organic framework synthesis. (*Reproduced with permission from ref. [61]. Copyright 2020, Elsevier*).

can be used to produce $scCO_2$ and water due to their tunable characteristics (viscosity, polarity, surface tension, phase, etc.) depending on the conditions under which they are synthesized. These tunable green solvents allow low energy demands, avoid distillation, and for recycling [82]. CO_2 is nearly limitless, noncombustible, and low-impact due to its low critical temperature (31.1°C) and gentle pressure (73.84 bar). Temperature and pressure can change the phase, density, and consistency of $scCO_2$. Because of its altering properties, $scCO_2$ and CO_2-expanded liquid (CXL) as a solvent have garnered much research. The supercritical fluid was initially used for post-synthesis pore cleaning of entrapped solvents, but its utility has been shown in creating new and sustainable procedures for highly porous MOFs [83]. Supercritical CO_2 solvents have been used to make 1D, 2D, and 3D coordination frameworks by reactive crystallisation. Moreover, the water-insoluble organic compounds can be dissolved in $scCO_2$ for various applications, viz., catalytic reactions, polymerization, extraction, biomedicine, and synthetic methods. In the past two decades, many highly porous MOFs have been activated by $scCO_2$ drying, shown in Table 1.1.

1.3.1.3 Ionic liquids

Ionic liquids are organic and inorganic cation/anion liquids below 100°C. In ILs, the cationic portion can be pyridinium, imidazolium, or quaternary ammonium/phosphonium, whereas anionic species can be halogens, hexafluorophosphate, triflate, and tetrachloroalumanuide [102,103]. They have a lower melting point than water and act as promising sustainable and environmentally friendly solvents. However, a few exceptions of ILs like perfluorinated anions, 1-butyl-3-methylimidazolium, and some imidazoles are toxic [104,105]. In ILs, the

TABLE 1.1 Selected highly porous metal–organic frameworks reported in last two decades.

Materials	Activation method[a]	BET area[b] (m^2/g)	Pore volume[b] (cm^3/g)	Year	References
MOF-5	Thermal	3800	1.55	1999, 2007	[15,84]
MIL-101c	Chemical and thermal	4230	2.15	2005, 2008	[85,86]
MOF-177	Thermal	4750	1.89	2007	[87]
UMCM-2	Room temperature vaccum	5200	2.32	2009	[88]
MOF-205/DUT-6	Thermal	4460	2.16	2009, 2010	[89,90]
MOF-200	Supercritical CO_2	4530	3.59	2010	[90]
MOF-210	Supercritical CO_2	6240	3.60	2010	[90]
PCN-68	Thermal	5110	2.13	2010	[91]
NU-100	Supercritical CO_2	6140	2.82	2010	[92]
bio-MOF-100	Supercritical CO_2	4300	4.3	2012	[93]
DUT-49	Supercritical CO_2	5480	2.91	2012	[94]
NU-110	Supercritical CO_2	7140	4.40	2012	[95]
DUT-32	Supercritical CO_2	6410	3.16	2014	[96]
Al-soc-MOF-1	Thermal	5590	2.3	2015	[97]
DUT-76	Supercritical CO_2	6340	3.25	2015	[98]
NU-1103	Supercritical CO_2	6550	2.91	2015	[99]
DUT-60[c]	Supercritical CO_2	7840	5.02	2018	[100]
NU-1501-Al[d]	Supercritical CO_2	7310	2.91	2020	[101]
NU-1501-Al[c]	Supercritical CO_2	9140	2.91	2020	[101]

[a] *Thermal activation.*
[b] *Brunauer–Emmett–Teller (BET) area and experimental pore volume.*
[c] *BET area calculated after satisfying first two BET consistency criteria.*
[d] *BET area calculated after satisfying all four BET consistency criteria.*

strong electrostatic interactions are the only interactions that hold the pure salts in liquid form at ambient conditions and thus determine the fundamental properties. The lower volatility and flammability of ILs are the reasons for seeking attention as alternate green solvents [106]. By changing the ions, there are a huge number of possible ionic combinations, thus called designer solvents. They direct MOF precursors by serving as structure-coordinating agents and modulating the porosity of structures [107]. A variety of MOF structures like 3D ferroelectric MOF, polyoxometalate-based MOF, and 1,4-ndc-based MOF have been synthesized using ILs [108]. Apart from flexibility in design, ILs also have a number of other unique properties, such as being able to dissolve both organic and inorganic substances and having very high stability, low volatility, and the ability to be recycled.

1.3.1.4 Bio-derived solvents

Biomass utilization by people has a long history. It is possible to classify the catalytic conversion of biomass into biochemicals and useable energy as sustainable renewable feedstock. Biomass has a number of intriguing qualities that could make it useful as a high-value chemical raw material, including chirality, oxygenation pattern, and diversity [109]. There are numerous techniques commercialized from research benches [110]. However, commercialized aprotic solvents like NMP (N-methyl-2-pyrrolidone) and DMF cannot be substituted immediately in synthetic chemistry [111]. Cyrene (IUPAC: 6,8-dioxabicyclo [3.2.1] octanone) is a popularly used bio-derived solvent. Cyrene is essential to MOF synthesis as it provides crystallinity and facilitates the manufacture of MOFs with a large surface area, like $Zn_2(BDC)_2(DABCO)$, UiO-66, HKUST-1, ZIF-8, and MOF-74 [112]. Levoglucosenone (LGO)/-Cyrene (2H-LGO) is a dipolar aprotic solvent whose biomass resource is cellulose obtained from corn cobs, poplar wood, bagasse, larch logs, bilberry press cake, and crude waste softwoods. The conversion of biomass into 2H-LGO is a two-step process: primarily the conversion of solid biomass into LGO followed by hydrogenation (reduction) of LGO into 2H-LGO. Water is the only product released to the environment in this energy-neutral conversion [113,114]. Pyrolysis is the most common process for converting cellulose into LGO then cyrene. There have been various reports on developing novel methods for directly converting cellulosic biomass to 2H-LGO without pre-treatment [111].

1.3.2 Solvent-free synthesis

Over the past decades, there has been an intense focus on developing solvent-free approaches to ligand–metal coordination polymerization. Mechano-synthesis has proven powerful for the clean crystallization of MOFs [115,116] as well as in harnessing the different crystalline solid forms [117–120]. In addition, mechano-synthesis have quantitative yields, short reaction times, and eco-friendly nature [121]. Water is the only reaction's by-product in the mechano-synthesis of MOFs [122]. Different methods for mechanochemical synthesis of MOFs have been reported in the literature, such as (1) neat grinding (no solvent); (2) liquid-assisted grinding(LAG); (3) ion-and-liquid assisted grinding (ILAG); and (4) extrusion and compression for pilot-scale production of MOFs [123,124]. In 2006, the first mesoporous MOF was reported through neat grinding method [125] in which isonicotinic acid and copper(II) acetate monohydrate were grounded for 10 min in a shaker mill to achieve coordination polymerization. Nowadays, several archetypal MOFs are being synthesized by this process [126,127]. Liquid assisted grinding (LAG) method, in which a few drops of a solvent are added during grinding, can also be used to enhance the crystallization and reaction rate of MOFs [128]. According to recent research, the mechano-chemically synthesized MOFs outperform analogs made by solvent-based methods [129]. By using mechano-synthesis, low-cost metal precursors (oxides and carbonates) can be used instead of metal nitrates and chlorides, and the waste generated can be reduced. The commercially available MOFs, like HKUST-1, having high crystallinity and large surface area, can be formed by different mechano-chemical approaches (ND, LAG, and ILAG) [130].

MOF synthesis has also been explored using microwave (MW) and sonication approaches [131]. MW-assisted uniform heating induces fast interfacial self assembly of metal salts and linkers. Almeida Paz and co-workers used MWs irradiation to synthesize yttrium-based MOFs having sustainable processing, high stability, and protonic conductivity at 98% relative humidity (2.58×10^{-2} S/cm) [132]. The MW-assisted synthesis leads to more efficient MOFs (large surface area and high yield) as the MOF synthesis takes place in aqueous solutions and has been extended to the heating of solvent-free organic reactions [133,134]. MW irradiation has been utilized to melt ligands and metal sources in a reaction mixture for MOF synthesis (e.g., ZIF-67, MOF-199, MIL-100(Fe/Cl), and MIL-100(Fe/NO_3)) [135]. The solvent-free microwave heating has a poor surface area, and lower crystallinity, but similar reaction yields in comparison with conventional oven heating or other solvent-free processes [135].

1.3.3 Sustainable metal precursors

Green chemistry in MOF crystallization involves choosing metal salts with the fewest risks. The environmental implications of counter ions associated with metal ions should be considered, as harmful by-products can impair human health and eco-systems in general. Nitrates, perchlorates, and chloride salts are commonly used as precursors for coordination complexes, and they are the primary focus of our attention because of their high solubility and low interfacial contact strength. Alternatively, the metal precursors like carbonates, oxides, acetates, and hydroxides can be used to fabricate MOFs in combination with linker molecules and generate water only as a by-product [58]. The sulfate anion-based metal salts are relatively soluble in water and can serve as templates for MOF synthesis when the circumstances are ambient. However, the strong interfacial interactions of sulfate anions inside the framework topology can have an effect on the rigidity, porosity, and crystalline structure of MOFs [136,137]. The sulfonated linkers coordinate with metal sites to produce Brønsted acidity in the framework and provide improved catalytic performance [138]. Several industrial methods have adopted metal acetylacetonate salts as eco-friendly reagents for the hydrothermal synthesis of MOFs. Researchers are seeking greener options for MOF synthesis by exploring different salts. A significant concern with the industrialization of MOF synthesis is the disposal of the frameworks, as their decomposition leads to metal ion pollution such that metal species of almost all types are widely used in framework crystallization. Despite this, only a few reports have examined MOFs' toxicity. In recent years, rare oxalate minerals like stepanovite and zhemchuzhnikovite have been reported to contain naturally occurring hexagonal aluminum–iron frameworks [139]. The presence of such framework architecture in organic minerals confirms the suitability of framework architecture in geological environments.

1.3.4 Eco-friendly alternatives for linkers

The choice of polytopic linkers also influences MOF topologies in addition to metal ions. The polytopic carboxylic acids, like trimesic and terephthalic acids, are the conventional and popularly used linkers in MOF synthesis due to the coordination reaction of metal nodes with them. Therefore, linkers play a vital role in MOFs' structural and functional properties. However, the type of linker affects the industrial cost of MOF production as all the commercially available linkers are generally costly [140]. Furthermore, using linker salts to prepare MOFs for industrial application can add complexity to the reagent preparation and limit their mass production. The petrochemical origins of linkers contribute significantly to their high cost and environmental pollution. Ideally, the linker molecules should be non-toxic for industrial production and sustainability, allowing the generation of non-toxic intermediates or by-products. Initial MOF synthesis relied on protonated linkers (e.g., imidazoles or carboxylic acids), which led to a stream of toxic by-products [77]. However, linkers in their anionic forms (such as sodium salts) can reduce the release of these toxic chemicals. Recently, research efforts have focused on developing renewable, cheaper, and non-toxic linkers derived from waste or biomass. Likewise, Badische Anilin- und Sodafabrik (BASF, largest chemical producer of North America) commercializes Basolite A520, a natural organic linker-based aluminum fumarate framework with a superior safety profile [141]. This greener approach to MOF synthesis at room temperature is faster, more innovative, and more widespread than hydrothermal synthesis.

1.4 MOF applications for a sustainable future

Porous materials show plethora of applications viz. in gas storage, shape/size-selective catalysis, adsorption based gas separation, and drug storage and delivery. Porous materials are traditionally organic or inorganic; perhaps activated carbon is the most common organic porous material but the structures of inorganic porous frameworks are highly organized (e.g., zeolites). To develop powerful interactions within the inorganic framework, a synthesis frequently needs an inorganic or organic template. One strategy to benefit from these traits is to create porous hybrids, commonly referred to as MOFs, which have both the characteristics of inorganic and organic porous materials.

1.4.1 Methane and hydrogen storage in MOFs

Due to the demand to minimize the world's reliance on fossil fuels, methane and hydrogen are becoming increasingly significant for gas storage applications. Gases can be kept at a substantially lower pressure in a tank with a porous adsorbent than in a tank with no adsorbent of the same size. In this way, multi-stage compressors and high-pressure tanks can be avoided, resulting in a safer and more efficient way to store gases. Activated carbon, carbon nanotubes, and zeolites are examples of porous adsorbents that have been the subject of several investigations [142]. MOFs are increasingly being thought of as such adsorbents because of their flexible frameworks and customizable pore geometries. MOFs are better than other porous materials because some of them can be made to make unsaturated metal centres (UMCs) by removing coordinated solvent molecules in a vacuum [143]. The interaction between hydrogen and UMCs is significantly stronger than the interaction between hydrogen and pure carbon materials, and the isosteric heat of adsorption can sometimes exceed 12–13 kJ/mol [144], which is quite near to the projection of 15.1 kJ/mol [145].

1.4.2 Selective gas adsorption in MOFs

A growing desire for more environmentally friendly and cost-effective industrial processes has prompted researchers to look at new ways to adsorb and separate gases. Cryogenic, membrane, and adsorption-based approaches are being used in the majority of industrial operations for selective gas adsorption. Molecular sieves, zeolites, silica gel, aluminosilicates, and carbon nanotubes are among the adsorbents typically employed in adsorption-based separation [146,147]. The two main criteria for selecting gas adsorption materials are: (1) adsorbent selectivity for an adsorbate, and (2) adsorption capacity of the adsorbent [148]. These parameters are determined during adsorption by the adsorbent's temperature, equilibrium pressure, chemical composition, and structure. The selective adsorption of gas by MOFs is very promising, leading to gas separation. Mesh-adjustable molecular sieve (MAMS) is a novel form of MOF distinguished by exceptionally high selectivity and configurable characteristics [149].

1.4.3 Catalysis in MOFs

MOFs may prove useful in catalysis as porous materials. It is theoretically possible to customize the pores of MOFs in a systematic way to optimize them for specific catalytic applications. MOFs are known for their high metal content. One of the greatest advantage is their highly crystalline nature, which leads to rarely different active sites. However, there have only been a few examples of their use in catalysis to date. These MOFs are capable of size- and shape-selective catalysis based on their porosity and catalytically active metal centres.

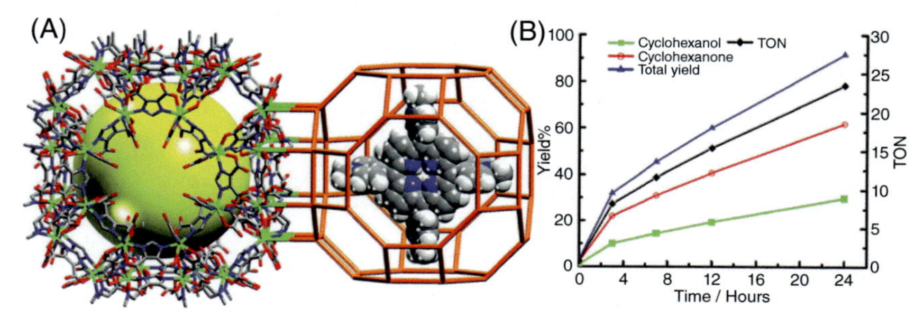

FIGURE 1.5 (A) Crystal structure of rho-ZMOF (*left*) and schematic presentation of $[H_2TMPyP]^{4+}$ porphyrin ring enclosed in rho-ZMOF R-cage (*right*, drawn to scale). (B) Cyclohexane catalytic oxidation using Mn-RTMPyP as a catalyst at 65°C. *(Reproduced with permission from ref. [150])*.

Alkordi and co-workers have demonstrated the effectiveness of *rho*-ZMOFs based on In-HImDC as hosts for large catalytically active molecules, specifically metalloporphyrins [150] (Fig. 1.5A). A cyclohexane oxidation experiment was carried out in the presence of Mn-RTMPyP (5,10,15,20-tetrakis(1-methyl-4-pyridinio) porphyrin tetra(p-toluene sulfonate encapsulated in rho-ZMOF) to assess its catalytic activity (Fig. 1.5B).

1.4.4 Magnetic properties of MOFs

Magnets are an extremely important class of materials that are finding use in an increasing variety of contexts. Consequently, enhancing the properties of magnets and exploring novel functionalities, particularly in conjunction with other valuable phenomena, are primary objectives of magnetic materials research. Polymetallic systems have properties that are ferromagnetic, anti-ferromagnetic, and ferrimagnetic. This is because the paramagnetic metal ions or organic radicals interact with each other through diamagnetic bridging entities. Thus, their magnetic characteristics depend on both the metal and the organic ligand, as well as the metal–ligand coordination relationship. Therefore, ligand design is of the utmost significance for the purpose of pursuing the magnetism of MOFs. This is true for both the organisation of the paramagnetic metal ions in an ideal topology and for the transmission of exchange contacts amongst the metal ions in an effective manner. There must be an appropriate framework to allow the parallel or antiparallel coupling of non-zero spin carriers in a bulk material.

In general, there is an anti-parallel tendency to coupling of spins because a low-spin state is more stable than a high-spin state [151]. Studies of magnetic MOFs fall under the purview of molecular magnets, magnetic sensors, low-dimensional magnetic materials, and multi-functional materials. As a result, organic ligands used in MOFs typically exhibit only weak magnetic interactions.

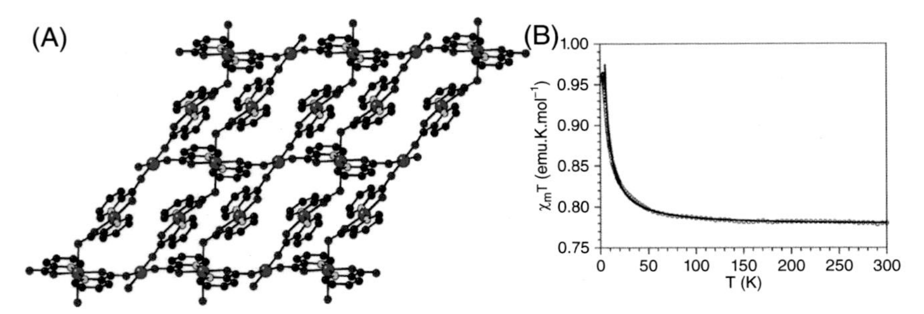

FIGURE 1.6 (A) 2D layer [110] of the Cu(II)-picolinate-bridged structure in $\{[Cu_2(pic)_3(H_2O)]ClO_4\}_n$. (B) Thermal variation of the product of the molar magnetic susceptibility and temperature (χ_mT) for $\{[Cu_2(pic)_3(H_2O)]ClO_4\}_n$. *(Reproduced with permission from ref. [157]).*

Short oxo, cyano, or azido bridges need to be built between the metal centres for a strong coupling [152–154]. Thus, MOFs with adjustable magnetic properties can be designed by crystal engineering or chemical coordination techniques [155,156].

A magnetized MOF material, $[Cu_2(pic)_3(H_2O)]ClO_4]_n$ (pic = 2-picolinate), was prepared by Biswas et al. based on "fish backbone" chains synthesized by syn-anti carboxylate groups [157]. Chains of syn-anti carboxylates link together to form a rectangular grid-like 2D network and the ClO_4^- anions were located between the 2D cationic sheets, shown in Fig. 1.6A. Magnetic susceptibility measurements indicated that this MOF fitted with the appropriate model has weak ferromagnetic coupling (Fig. 1.6B).

1.4.5 Drug storage and delivery in MOFs

The absence of a controlled release of typical pharmaceuticals that are taken orally has prompted a significant amount of interest and research into alternative means of medication delivery. There have been developed systems based on polymers, liposomes, mesoporous silicon, microporous zeolites, and other mesoporous materials [158–160]. Both organic and inorganic delivery systems can be thought of as broad categories for the many distribution methods. However, organic systems do not have a regulated release mechanism, despite their high biocompatibility for the uptake of a variety of medicines [161,162]. Due to their ordered porosity network, inorganic delivery materials are able to transport adsorbates at a controlled rate, although they have a lower loading capacity [163,164]. MOFs have shown that they could be good candidates for drug delivery, but more research is needed to find out how well they could work. Mesoporous MOFs, which are increasingly synthesized, are showing promising potential for drug delivery.

1.5 Future outlook and challenges

Over the course of the past few decades, there has been a proliferation of MOF research aimed at generating novel materials and making use of these highly programmable materials in a wide variety of applications. Synthesis of MOFs as well as applications of MOFs in academic and industry research are both subjects of current research. MOF synthesis has moved from the lab to mass production to the point where several companies, such as NuMat Technologies, novoMOF, Prof-MOF, MOF apps, BASF, etc., are making these materials to develop new applications. Commercial use of these MOFs is confined to environmental, energy, and catalysis applications; nevertheless, industry research has already made attempts to improve sustainable and clean methods of MOF manufacturing. The combination of green chemistry and sustainability may be able to meet the needs of the industrial sector. MOF applications are expected to overgrow with greener synthesis approaches shortly. Even avoiding harmful solvents for only a part of the synthesis procedure is a significant breakthrough owing to the high cost and environmental impact attached to producing, recycling, or treating solvents. In general, simply eliminating toxic solvents for a portion of the synthesis procedure represents significant progress at the practical synthesis level, because the manufacturing, recycling, or treatment of solvents is connected with substantial costs and environmental degradation. As part of this book, we discuss recent developments in water-based synthesis strategies of MOFs and their green production. Water-based synthesis is a highly efficient and environmentally friendly method of producing MOFs. Their simplicity, versatility, and ability to post-treat solvent make them much easier to scale up and produce fewer harmful by-products. However, the realization of large-scale industrial production of MOFs still faces some challenges and limitations as the MOFs synthesized in aqueous solutions generally have a poor crystalline structure, low yields, and decreased porosity. The synthesis of MOFs pre-dominantly occurs in solvothermal or hydrothermal conditions, typically involving high temperatures and pressure with prolonged reaction times due to the poor solubility of organic ligands in water. These severe reaction conditions consume a lot of energy and reduce the production rate. The ability to synthesize MOF materials cost-effectively in large quantities is another crucial pre-requisite for green applications in the real world.

A large-scale MOF production process requires low costs and addresses the related safety risks and environmental impacts. MOFs are therefore primarily produced and used commercially by applying "green synthesis" principles. Despite numerous advances in the preparation of MOF materials with water in recent years, a method capable of "green synthesis" has rarely been reported. The idea of "green synthesis of MOFs" is still in its early stage because other factors, like high yields, room temperature, and normal pressure, have not yet been taken into account. Because of this, it is very important for the industry that makes MOFs to look into more advanced ways to make them so that they can reach the level of "green synthesis." When academia and business realize this, it will be a big step toward putting MOFs on the market.

Acknowledgment

UG and YA thank the Department of Applied Chemistry, Aligarh Muslim University, for providing research facilities. UG is also thankful to the CSIR-SRF (Grant No: 09/112(0633)/2019-EMR-I) for providing financial assistance.

References

[1] M.E. Davis, Ordered porous materials for emerging applications, Nature 417 (2002) 813–821.

[2] S. Abednatanzi, P. Gohari Derakhshandeh, H. Depauw, F.-X. Coudert, H. Vrielinck, P. Van Der Voort, et al., Mixed-metal metal–organic frameworks, Chem. Soc. Rev. 48 (2019) 2535–2565.

[3] M. Ding, R.W. Flaig, H.-L. Jiang, O.M. Yaghi, Carbon capture and conversion using metal–organic frameworks and MOF-based materials, Chem. Soc. Rev. 48 (2019) 2783–2828.

[4] S. Kamaal, M. Gupta, R. Mishra, A. Ali, A. Alarifi, M. Afzal, et al., A three-dimensional pentanuclear Co(II) coordination polymer: structural topology, hirshfeld surface analysis and magnetic properties, ChemistrySelect 5 (2020) 13732–13737.

[5] A. Ahmed, A. Ali, M. Ahmed, K.N. Parida, M. Ahmad, A. Ahmad, Construction and topological studies of a three dimensional (3D) coordination polymer showing selective adsorption of aromatic hazardous dyes, Sep. Purif. Technol. 265 (2021) 118482.

[6] M.E. Davis, C. Saldarriaga, C. Montes, J. Garces, C. Crowdert, A molecular sieve with eighteen-membered rings, Nature 331 (1988) 698–699.

[7] C.T. Kresge, M.E. Leonowicz, W.J. Roth, J.C. Vartuli, J.S. Beck, Ordered mesoporous molecular sieves synthesized by a liquid-crystal template mechanism, Nature 359 (1992) 710–712.

[8] O.M. Yaghi, H. Li, Hydrothermal synthesis of a metal-organic framework containing large rectangular channels, J. Am. Chem. Soc. 117 (1995) 10401–10402.

[9] O.M. Yaghi, G. Li, H. Li, Selective binding and removal of guests in a microporous metal–organic framework, Nature 378 (1995) 703–706.

[10] M. Kondo, T. Yoshitomi, H. Matsuzaka, S. Kitagawa, K. Seki, Three-dimensional framework with channeling cavities for small molecules:{[M2(4, 4′-bpy)$_3$(NO$_3$)$_4$]·xH$_2$O}$_n$(M = Co, Ni, Zn), Angew. Chemie Int. Ed. English 36 (1997) 1725–1727.

[11] H. Li, M. Eddaoudi, T.L. Groy, O.M. Yaghi, Establishing microporosity in open metal—organic frameworks: gas sorption isotherms for Zn(BDC) (BDC = 1,4-benzenedicarboxylate), J. Am. Chem. Soc. 120 (1998) 8571–8572.

[12] K. Shen, L. Zhang, X. Chen, L. Liu, D. Zhang, Y. Han, et al., Ordered macro-microporous metal-organic framework single crystals, Science 359 (2018) 206–210.

[13] A. Dhakshinamoorthy, Z. Li, H. Garcia, Catalysis and photocatalysis by metal organic frameworks, Chem. Soc. Rev. 47 (2018) 8134–8172.

[14] S. Wang, C. Serre, Toward green production of water-stable metal–organic frameworks based on high-valence metals with low toxicities, ACS Sustain. Chem. Eng. 7 (2019) 11911–11927.

[15] H. Li, M. Eddaoudi, M. O'Keeffe, O.M. Yaghi, Design and synthesis of an exceptionally stable and highly porous metal-organic framework, Nature 402 (1999) 276–279.

[16] Y. Bai, Y. Dou, L.-H. Xie, W. Rutledge, J.-R. Li, H.-C. Zhou, Zr-based metal–organic frameworks: design, synthesis, structure, and applications, Chem. Soc. Rev. 45 (2016) 2327–2367.

[17] J. Lee, O.K. Farha, J. Roberts, K.A. Scheidt, S.T. Nguyen, J.T. Hupp, Metal–organic framework materials as catalysts, Chem. Soc. Rev. 38 (2009) 1450.

[18] X. Zhang, Z. Huang, M. Ferrandon, D. Yang, L. Robison, P. Li, et al., Catalytic chemoselective functionalization of methane in a metal−organic framework, Nat. Catal. 1 (2018) 356–362.

[19] L. Jiao, H.-L. Jiang, Metal-organic-framework-based single-atom catalysts for energy applications, Chem. 5 (2019) 786–804.

[20] M.-X. Wu, Y.-W. Yang, Metal-organic framework (MOF)-based drug/cargo delivery and cancer therapy, Adv. Mater. 29 (2017) 1606134.

[21] Y. Chen, P. Li, J.A. Modica, R.J. Drout, O.K. Farha, Acid-resistant mesoporous metal–organic framework toward oral insulin delivery: protein encapsulation, protection, and release, J. Am. Chem. Soc. 140 (2018) 5678–5681.

[22] M. Eddaoudi, J. Kim, N. Rosi, D. Vodak, J. Wachter, M. O'Keeffe, et al., Systematic design of pore size and functionality in isoreticular MOFs and their application in methane storage, Science 295 (2002) 469–472.

[23] C. Gropp, S. Canossa, S. Wuttke, F. Gándara, Q. Li, L. Gagliardi, et al., Standard practices of reticular chemistry, ACS Cent. Sci. 6 (2020) 1255–1273.

[24] B. Chen, Y. Yang, F. Zapata, G. Lin, G. Qian, E.B. Lobkovsky, Luminescent open metal sites within a metal–organic framework for sensing small molecules, Adv. Mater. 19 (2007) 1693–1696.

[25] L.E. Kreno, K. Leong, O.K. Farha, M. Allendorf, R.P. Van Duyne, J.T. Hupp, Metal–organic framework materials as chemical sensors, Chem. Rev. 112 (2012) 1105–1125.

[26] X. Cui, K. Chen, H. Xing, Q. Yang, R. Krishna, Z. Bao, et al., Pore chemistry and size control in hybrid porous materials for acetylene capture from ethylene, Science 353 (2016) 141–144.

[27] J.-R. Li, J. Sculley, H.-C. Zhou, Metal–organic frameworks for separations, Chem. Rev. 112 (2012) 869–932.

[28] B. Chen, C. Liang, J. Yang, D.S. Contreras, Y.L. Clancy, E.B. Lobkovsky, et al., A microporous metal–organic framework for gas-chromatographic separation of alkanes, Angew. Chemie Int. Ed. 45 (2006) 1390–1393.

[29] W. Xu, X. Pei, C.S. Diercks, H. Lyu, Z. Ji, O.M. Yaghi, A metal–organic framework of organic vertices and polyoxometalate linkers as a solid-state electrolyte, J. Am. Chem. Soc. 141 (2019) 17522–17526.

[30] L. Sun, M.G. Campbell, M. Dincă, Electrically conductive porous metal-organic frameworks, Angew. Chemie Int. Ed. 55 (2016) 3566–3579.

[31] Y. Chen, P. Li, H. Noh, C. Kung, C.T. Buru, X. Wang, et al., Stabilization of formate dehydrogenase in a metal–organic framework for bioelectrocatalytic reduction of CO_2, Angew. Chemie Int. Ed. 58 (2019) 7682–7686.

[32] X. Lian, Y. Fang, E. Joseph, Q. Wang, J. Li, S. Banerjee, et al., Enzyme–MOF (metal–organic framework) composites, Chem. Soc. Rev. 46 (2017) 3386–3401.

[33] Y. Chen, P. Li, J. Zhou, C.T. Buru, L. Đorđević, P. Li, et al., Integration of enzymes and photosensitizers in a hierarchical mesoporous metal–organic framework for light-driven CO_2 reduction, J. Am. Chem. Soc. 142 (2020) 1768–1773.

[34] X. Zhang, M.R. Saber, A.P. Prosvirin, J.H. Reibenspies, L. Sun, M. Ballesteros-Rivas, et al., Magnetic ordering in TCNQ-based metal–organic frameworks with host–guest interactions, Inorg. Chem. Front. 2 (2015) 904–911.

[35] G. Mínguez Espallargas, E. Coronado, Magnetic functionalities in MOFs: from the framework to the pore, Chem. Soc. Rev. 47 (2018) 533–557.

[36] D. Aulakh, H. Xie, Z. Shen, A. Harley, X. Zhang, A.A. Yakovenko, et al., Systematic investigation of controlled nanostructuring of Mn 12 single-molecule magnets templated by metal–organic frameworks, Inorg. Chem. 56 (2017) 6965–6972.

[37] K.S.W. Sing, D.H. Everett, R.A.W. Haul, L. Moscou, R.A. Pierotti, J. Rouquerol, et al., Reporting physisorption data for gas/solid systems with special reference to the determination of surface area and porosity, Pure Appl. Chem. 54 (1985) 2201–2218.

[38] J. Rouquerol, D. Avnir, C.W. Fairbridge, D.H. Everett, J.M. Haynes, N. Pernicone, et al., Recommendations for the characterization of porous solids, Pure Appl. Chem. 66 (1994) 1739–1758.

[39] M. Thommes, K. Kaneko, A.V. Neimark, J.P. Olivier, F. Rodriguez-Reinoso, J. Rouquerol, et al., Physisorption of gases, with special reference to the evaluation of surface area and pore size distribution (IUPAC technical report), Pure Appl. Chem. 87 (2015) 1051–1069.

[40] X. Zhang, Z. Chen, X. Liu, S.L. Hanna, X. Wang, R. Taheri-Ledari, et al., A historical overview of the activation and porosity of metal–organic frameworks, Chem. Soc. Rev. 49 (2020) 7406–7427.

[41] R. Kitaura, K. Seki, G. Akiyama, S. Kitagawa, Porous coordination-polymer crystals with gated channels specific for supercritical gases, Angew. Chemie Int. Ed. 42 (2003) 428–431.

[42] J.H. Lee, S. Jeoung, Y.G. Chung, H.R. Moon, Elucidation of flexible metal-organic frameworks: research progresses and recent developments, Coord. Chem. Rev. 389 (2019) 161–188.

[43] A. Schneemann, V. Bon, I. Schwedler, I. Senkovska, S. Kaskel, R.A. Fischer, Flexible metal–organic frameworks, Chem. Soc. Rev. 43 (2014) 6062–6096.

[44] Z.-J. Lin, J. Lü, M. Hong, R. Cao, Metal–organic frameworks based on flexible ligands (FL-MOFs): structures and applications, Chem. Soc. Rev. 43 (2014) 5867–5895.

[45] K. Uemura, R. Matsuda, S. Kitagawa, Flexible microporous coordination polymers, J. Solid State Chem. 178 (2005) 2420–2429.

[46] N. Hanikel, M.S. Prévot, O.M. Yaghi, MOF water harvesters, Nat. Nanotechnol. 15 (2020) 348–355.

[47] F.-X. Coudert, Water adsorption in soft and heterogeneous nanopores, Acc. Chem. Res. 53 (2020) 1342–1350.

[48] H. Furukawa, F. Gándara, Y.-B. Zhang, J. Jiang, W.L. Queen, M.R. Hudson, et al., Water adsorption in porous metal–organic frameworks and related materials, J. Am. Chem. Soc. 136 (2014) 4369–13681.

[49] J. Canivet, A. Fateeva, Y. Guo, B. Coasne, D. Farrusseng, Water adsorption in MOFs: fundamentals and applications, Chem. Soc. Rev. 43 (2014) 5594–5617.

[50] K.A. Cychosz, M. Thommes, Progress in the physisorption characterization of nanoporous gas storage materials, Engineering 4 (2018) 559–566.

[51] P. Llewellyn, G. Maurin, J. Rouquerol, Adsorption by metal-organic frameworks, in: Adsorption by Powders and Porous Solids, Elsevier, 2014, pp. 565–610.

[52] J. Rouquerol, F. Rouquerol, P. Llewellyn, G. Maurin, K. Sing, Adsorption by Powders and Porous Solids, 2nd ed., Elsevier, 2013.

[53] Z. Chen, P. Li, X. Wang, K.-I. Otake, X. Zhang, L. Robison, et al., Ligand-directed reticular synthesis of catalytically active missing zirconium-based metal–organic frameworks, J. Am. Chem. Soc. 141 (2019) 12229–12235.

[54] R.J. Kuppler, D.J. Timmons, Q.-R. Fang, J.-R. Li, T.A. Makal, M.D. Young, et al., Potential applications of metal-organic frameworks, Coord. Chem. Rev. 253 (2009) 3042–3066.

[55] S. Lowell, J.E. Shields, M.A. Thomas, M. Thommes, Characterization of Porous Solids and Powders: Surface Area, Pore Size and Density, Springer, 2012.
[56] J.S.S. Lowell, M.A. Tomas, M. Thommes, Characterization of Porous Solids and Powders: Surface Area, Pore Size and Density, Springer, 2004.
[57] M. Thommes, K.A. Cychosz, Physical adsorption characterization of nanoporous materials: progress and challenges, Adsorption 20 (2014) 233–250.
[58] P.A. Julien, C. Mottillo, T. Friščić, Metal–organic frameworks meet scalable and sustainable synthesis, Green Chem. 19 (2017) 2729–2747.
[59] H. Reinsch, S. Waitschat, S.M. Chavan, K.P. Lillerud, N. Stock, A facile "green" route for scalable batch production and continuous synthesis of zirconium MOFs, Eur. J. Inorg. Chem. 2016 (2016) 4490–4498.
[60] P. Rocío-Bautista, I. Taima-Mancera, J. Pasán, V. Pino, Metal-organic frameworks in green analytical chemistry, Separations 6 (2019) 33.
[61] S. Kumar, S. Jain, M. Nehra, N. Dilbaghi, G. Marrazza, K.-H. Kim, Green synthesis of metal–organic frameworks: a state-of-the-art review of potential environmental and medical applications, Coord. Chem. Rev. 420 (2020) 213407.
[62] B. Krūkle-Bērziņa, S. Mishnev, Stability and phase transitions of nontoxic γ-cyclodextrin-K^+ metal-organic framework in various solvents, Crystals 10 (2020) 37.
[63] A.J. Howarth, Y. Liu, P. Li, Z. Li, T.C. Wang, J.T. Hupp, et al., Chemical, thermal and mechanical stabilities of metal–organic frameworks, Nat. Rev. Mater. 1 (2016) 15018.
[64] D. Yang, B.C. Gates, Catalysis by metal organic frameworks: perspective and suggestions for future research, ACS Catal. 9 (2019) 1779–1798.
[65] M.J. Raymond, C.S. Slater, M.J. Savelski, LCA approach to the analysis of solvent waste issues in the pharmaceutical industry, Green Chem. 12 (2010) 1826.
[66] Y. Tao, G. Huang, H. Li, M.R. Hill, Magnetic metal–organic framework composites: solvent-free synthesis and regeneration driven by localized magnetic induction heat, ACS Sustain. Chem. Eng. 7 (2019) 13627–13632.
[67] L. Lomba, E. Zuriaga, B. Giner, Solvents derived from biomass and their potential as green solvents, Curr. Opin. Green Sustain. Chem. 18 (2019) 51–56.
[68] M. Tobiszewski, J. Namieśnik, F. Pena-Pereira, Environmental risk-based ranking of solvents using the combination of a multimedia model and multi-criteria decision analysis, Green Chem. 19 (2017) 1034–1042.
[69] P. Sánchez-Camargo A del, M. Bueno, F. Parada-Alfonso, A. Cifuentes, E. Ibáñez, Hansen solubility parameters for selection of green extraction solvents, Trends Analyt. Chem. 118 (2019) 227–237.
[70] M.A. Rasool, I.F.J. Vankelecom, Use of γ-valerolactone and glycerol derivatives as bio-based renewable solvents for membrane preparation, Green Chem. 21 (2019) 1054–1064.
[71] M. Asakawa, A. Shrotri, H. Kobayashi, A. Fukuoka, Solvent basicity controlled deformylation for the formation of furfural from glucose and fructose, Green Chem. 21 (2019) 6146–6153.
[72] E.V. Ramos-Fernandez, A. Grau-Atienza, D. Farrusseng, S. Aguado, A water-based room temperature synthesis of ZIF-93 for CO_2 adsorption, J. Mater. Chem. 6 (2018) 5598–5602.
[73] J. Jacobsen, B. Achenbach, H. Reinsch, S. Smolders, F.-D. Lange, G. Friedrichs, et al., The first water-based synthesis of Ce(IV)-MOFs with saturated chiral and achiral C_4-dicarboxylate linkers, Dalt. Trans. 48 (2019) 8433–8441.
[74] Z. Hu, I. Castano, S. Wang, Y. Wang, Y. Peng, Y. Qian, et al., Modulator effects on the water-based synthesis of Zr/Hf Metal–organic frameworks: quantitative relationship studies

between modulator, synthetic condition, and performance, Cryst. Growth Des. 16 (2016) 2295–2301.

[75] H. Chevreau, A. Permyakova, F. Nouar, P. Fabry, C. Livage, F. Ragon, et al., Synthesis of the biocompatible and highly stable MIL-127(Fe): from large scale synthesis to particle size control, CrystEngComm 18 (2016) 4094–4101.

[76] B. Zhang, J. Zhang, C. Liu, X. Sang, L. Peng, X. Ma, et al., Solvent determines the formation and properties of metal–organic frameworks, RSC Adv. 5 (2015) 37691–37696.

[77] M. Sánchez-Sánchez, N. Getachew, K. Díaz, M. Díaz-García, Y. Chebude, I. Díaz, Synthesis of metal–organic frameworks in water at room temperature: salts as linker sources, Green Chem. 17 (2015) 1500–1509.

[78] X. Qian, F. Sun, J. Sun, H. Wu, F. Xiao, X. Wu, et al., Imparting surface hydrophobicity to metal–organic frameworks using a facile solution-immersion process to enhance water stability for CO_2 capture, Nanoscale 9 (2017) 2003–2008.

[79] Y. Sun, Q. Sun, H. Huang, B. Aguila, Z. Niu, J.A. Perman, et al., A molecular-level superhydrophobic external surface to improve the stability of metal–organic frameworks, J. Mater. Chem. 5 (2017) 18770–18776.

[80] M. Joharian, A. Morsali, A. Azhdari Tehrani, L. Carlucci, D.M. Proserpio, Water-stable fluorinated metal–organic frameworks (F-MOFs) with hydrophobic properties as efficient and highly active heterogeneous catalysts in aqueous solution, Green Chem. 20 (2018) 5336–5345.

[81] H.-Y. Guan, R.J. LeBlanc, S.-Y. Xie, Y. Yue, Recent progress in the syntheses of mesoporous metal–organic framework materials, Coord. Chem. Rev. 369 (2018) 76–90.

[82] S. Abou-Shehada, J.H. Clark, G. Paggiola, J. Sherwood, Tunable solvents: shades of green, Chem. Eng. Process. Process Intensif. 99 (2016) 88–96.

[83] K. Matsuyama, M. Motomura, T. Kato, T. Okuyama, H. Muto, Catalytically active Pt nanoparticles immobilized inside the pores of metal organic framework using supercritical CO_2 solutions, Microporous Mesoporous Mater. 225 (2016) 26–32.

[84] S.S. Kaye, A. Dailly, O.M. Yaghi, J.R. Long, Impact of Preparation and handling on the hydrogen storage properties of Zn_4O(1,4-benzenedicarboxylate)$_3$ (MOF-5), J. Am. Chem. Soc. 129 (2007) 14176–14177.

[85] P.L. Llewellyn, S. Bourrelly, C. Serre, A. Vimont, M. Daturi, L. Hamon, et al., High uptakes of CO_2 and CH_4 in mesoporous metal—organic frameworks MIL-100 and MIL-101, Langmuir 24 (2008) 7245–7250.

[86] G. Férey, C. Mellot-Draznieks, C. Serre, F. Millange, J. Dutour, S. Surblé, et al., A chromium terephthalate-based solid with unusually large pore volumes and surface area, Science 309 (2005) 2040–2042.

[87] H. Furukawa, M.A. Miller, O.M. Yaghi, Independent verification of the saturation hydrogen uptake in MOF-177 and establishment of a benchmark for hydrogen adsorption in metal–organic frameworks, J. Mater. Chem. 17 (2007) 3197.

[88] K. Koh, A.G. Wong-Foy, A.J. Matzger, A porous coordination copolymer with over 5000 m^2/g BET surface area, J. Am. Chem. Soc. 131 (2009) 4184–4185.

[89] N. Klein, I. Senkovska, K. Gedrich, U. Stoeck, A. Henschel, U. Mueller, et al., A mesoporous metal–organic framework, Angew. Chemie Int. Ed. 48 (2009) 9954–9957.

[90] H. Furukawa, N. Ko, Y.B. Go, N. Aratani, S.B. Choi, E. Choi, et al., Ultrahigh porosity in metal-organic frameworks, Science 329 (2010) 424–428.

[91] D. Yuan, D. Zhao, D. Sun, H.-C. Zhou, An isoreticular series of metal-organic frameworks with dendritic hexacarboxylate ligands and exceptionally high gas-uptake capacity, Angew. Chemie Int. Ed. 49 (2010) 5357–5361.

[92] O.K. Farha, A. Özgür Yazaydın, I. Eryazici, C.D. Malliakas, B.G. Hauser, M.G. Kanatzidis, et al., De novo synthesis of a metal–organic framework material featuring ultrahigh surface area and gas storage capacities, Nat. Chem. 2 (2010) 944–948.
[93] J. An, O.K. Farha, J.T. Hupp, E. Pohl, J.I. Yeh, N.L. Rosi, Metal-adeninate vertices for the construction of an exceptionally porous metal-organic framework, Nat. Commun. 3 (2012) 604.
[94] U. Stoeck, S. Krause, V. Bon, I. Senkovska, S. Kaskel, A highly porous metal–organic framework, constructed from a cuboctahedral super-molecular building block, with exceptionally high methane uptake, Chem. 48 (2012) 10841.
[95] O.K. Farha, I. Eryazici, N.C. Jeong, B.G. Hauser, C.E. Wilmer, A.A. Sarjeant, et al., Metal–organic framework materials with ultrahigh surface areas: is the sky the limit? J. Am. Chem. Soc. 134 (2012) 15016–15021.
[96] R. Grünker, V. Bon, P. Müller, U. Stoeck, S. Krause, U. Mueller, et al., A new metal–organic framework with ultra-high surface area, Chem. Commun. 50 (2014) 3450.
[97] D. Alezi, Y. Belmabkhout, M. Suyetin, P.M. Bhatt, Ł.J. Weseliński, V. Solovyeva, et al., MOF crystal chemistry paving the way to gas storage needs: aluminum-based soc -MOF for CH_4, O_2, and CO_2 storage, J. Am. Chem. Soc. 137 (2015) 13308–13318.
[98] U. Stoeck, I. Senkovska, V. Bon, S. Krause, S. Kaskel, Assembly of metal–organic polyhedra into highly porous frameworks for ethene delivery, Chem. 51 (2015) 1046–1049.
[99] T.C. Wang, W. Bury, D.A. Gómez-Gualdrón, N.A. Vermeulen, J.E. Mondloch, P. Deria, et al., Ultrahigh surface area zirconium MOFs and insights into the applicability of the BET theory, J. Am. Chem. Soc. 137 (2015) 3585–3591.
[100] I.M. Hönicke, I. Senkovska, V. Bon, I.A. Baburin, N. Bönisch, S. Raschke, et al., Balancing mechanical stability and ultrahigh porosity in crystalline framework materials, Angew. Chemie Int. Ed. 57 (2018) 13780–13783.
[101] Y. Chen, X. Zhang, M.R. Mian, F.A. Son, K. Zhang, R. Cao, et al., Structural diversity of zirconium metal–organic frameworks and effect on adsorption of toxic chemicals, J. Am. Chem. Soc. 142 (2020) 21428–21438.
[102] M. Dašić, I. Stanković, K. Gkagkas, Molecular dynamics investigation of the influence of the shape of the cation on the structure and lubrication properties of ionic liquids, Phys. Chem. Chem. Phys. 21 (2019) 4375–4786.
[103] A.S. Hanamertani, R.M. Pilus, A.K. Idris, S. Irawan, I.M. Tan, Ionic liquids as a potential additive for reducing surfactant adsorption onto crushed Berea sandstone, J. Pet. Sci. Eng. 162 (2018) 480–490.
[104] A. Figoli, T. Marino, S. Simone, E. Di Nicolò, X.-M. Li, T. He, et al., Towards non-toxic solvents for membrane preparation: a review, Green Chem. 16 (2014) 4034.
[105] P. Isosaari, V. Srivastava, M. Sillanpää, Ionic liquid-based water treatment technologies for organic pollutants: current status and future prospects of ionic liquid mediated technologies, Sci. Total Environ. 690 (2019) 604–619.
[106] T. Erdmenger, C. Guerrero-Sanchez, J. Vitz, R. Hoogenboom, U.S. Schubert, Recent developments in the utilization of green solvents in polymer chemistry, Chem. Soc. Rev. 39 (2010) 3317.
[107] H.-C. Oh, S. Jung, I.-J. Ko, E.-Y. Choi, Ionothermal synthesis of metal-organic framework, in: Recent Advancements in the Metallurgical Engineering and Electrodeposition, IntechOpen, 2020.
[108] B. Zhang, J. Zhang, B. Han, Assembling metal-organic frameworks in ionic liquids and supercritical CO_2, Chem. Asian J. 11 (2016) 2610–2619.

[109] M.B. Comba, Y. Tsai, A.M. Sarotti, M.I. Mangione, A.G. Suárez, R.A. Spanevello, Levoglucosenone and its new applications: valorization of cellulose residues, Eur. J. Org. Chem. 2018 (2018) 590–604.
[110] M. Sharma, D. Mondal, N. Singh, K. Prasad, Biomass derived solvents for the scalable production of single layered graphene from graphite, Chem. Commun. 52 (2016) 9074–9077.
[111] J.E. Camp, Bio-available solvent cyrene: synthesis, derivatization, and applications, ChemSusChem 11 (2018) 3048–3055.
[112] J. Zhang, G.B. White, M.D. Ryan, A.J. Hunt, M.J. Katz, Dihydrolevoglucosenone (cyrene) as a green alternative to N,N -dimethylformamide (DMF) in MOF synthesis, ACS Sustain. Chem. Eng. 4 (2016) 7186–7192.
[113] X. Huang, S. Kudo, J. Hayashi, Two-step conversion of cellulose to levoglucosenone using updraft fixed bed pyrolyzer and catalytic reformer, Fuel Process. Technol. 191 (2019) 29–35.
[114] T.W. Bousfield, K.P.R. Pearce, S.B. Nyamini, A. Angelis-Dimakis, J.E. Camp, Synthesis of amides from acid chlorides and amines in the bio-based solvent cyrene^{tm}, Green Chem. 21 (2019) 3675–3681.
[115] L.S. Germann, A.D. Katsenis, I. Huskić, P.A. Julien, K. Užarević, M. Etter, et al., Real-time in situ monitoring of particle and structure evolution in the mechanochemical synthesis of UiO-66 metal–organic frameworks, Cryst. Growth Des. 20 (2020) 49–54.
[116] G. Ayoub, B. Karadeniz, A.J. Howarth, O.K. Farha, I. Đilović, L.S. Germann, et al., Rational synthesis of mixed-metal microporous metal–organic frameworks with controlled composition using mechanochemistry, Chem. Mater. 31 (2019) 5494–5501.
[117] U. Garg, Y. Azim, Challenges and opportunities of pharmaceutical cocrystals: a focused review on non-steroidal anti-inflammatory drugs, RSC Med. Chem. 12 (2021) 705–721.
[118] U. Garg, Y. Azim, A. Kar, C.P. Pradeep, Cocrystals/salt of 1-naphthaleneacetic acid and utilizing Hirshfeld surface calculations for acid–aminopyrimidine synthons, CrystEngComm 22 (2020) 2978–2989.
[119] U. Garg, Y. Azim, M. Alam, In acid-aminopyrimidine continuum: experimental and computational studies of furan tetracarboxylate-2-aminopyrimidinium salt, RSC Adv. 11 (2021) 21463–21474.
[120] U. Garg, Y. Azim, M. Alam, A. Kar, C.P. Pradeep, Extensive analyses on expanding the scope of acid–aminopyrimidine synthons for the design of molecular solids, Cryst. Growth Des. 22 (2022) 4316–4331.
[121] T. Friščić, C. Mottillo, H.M. Titi, Mechanochemistry for synthesis, Angew. Chemie 132 (2020) 1030–1041.
[122] D. Chen, J. Zhao, P. Zhang, S. Dai, Mechanochemical synthesis of metal–organic frameworks, Polyhedron 162 (2019) 59–64.
[123] J.-L. Do, T. Friščić, Mechanochemistry: a force of synthesis, ACS Cent. Sci. 3 (2017) 13–19.
[124] D. Tan, F. García, Main group mechanochemistry: from curiosity to established protocols, Chem. Soc. Rev. 48 (2019) 2274–2292.
[125] A. Pichon, A. Lazuen-Garay, S.L. James, Solvent-free synthesis of a microporous metal–organic framework, CrystEngComm 8 (2006) 211.
[126] B. Szczęśniak, S. Borysiuk, J. Choma, M. Jaroniec, Mechanochemical synthesis of highly porous materials, Mater. Horiz. 7 (2020) 1457–1473.
[127] I. Brekalo, W. Yuan, C. Mottillo, Y. Lu, Y. Zhang, J. Casaban, et al., Manometric real-time studies of the mechanochemical synthesis of zeolitic imidazolate frameworks, Chem. Sci. 11 (2020) 2141–2147.

[128] M. Rubio-Martinez, C. Avci-Camur, A.W. Thornton, I. Imaz, D. Maspoch, M.R. Hill, New synthetic routes towards MOF production at scale, Chem. Soc. Rev. 46 (2017) 3453–3480.

[129] D. Prochowicz, J. Nawrocki, M. Terlecki, W. Marynowski, J. Lewiński, Facile mechanosynthesis of the archetypal Zn-based metal–organic frameworks, Inorg. Chem. 57 (2018) 13437–13442.

[130] T. Stolar, L. Batzdorf, S. Lukin, D. Žilić, C. Motillo, T. Friščić, et al., In situ monitoring of the mechanosynthesis of the archetypal metal–organic framework HKUST-1: effect of liquid additives on the milling reactivity, Inorg. Chem. 56 (2017) 6599–6608.

[131] V.I. Isaeva, L.M. Kustov, Microwave activation as an alternative production of metal-organic frameworks, Russ. Chem. Bull. 65 (2016) 2103–2114.

[132] A.D.G. Firmino, R.F. Mendes, M.M. Antunes, P.C. Barbosa, S.M.F. Vilela, A.A. Valente, et al., Robust multifunctional yttrium-based metal–organic frameworks with breathing effect, Inorg. Chem. 56 (2017) 1193–1208.

[133] A.C. Dreischarf, M. Lammert, N. Stock, H. Reinsch, Green synthesis of Zr-CAU-28: structure and properties of the first Zr-MOF based on 2,5-furandicarboxylic acid, Inorg. Chem. 56 (2017) 2270–2277.

[134] Y. Liu, A. Hori, S. Kusaka, N. Hosono, M. Li, A. Guo, et al., Microwave-assisted hydrothermal synthesis of [Al(OH)(1,4-NDC)] membranes with superior separation performances, Chem. Asian J. 14 (2019) 2072–2076.

[135] M. Lanchas, S. Arcediano, A.T. Aguayo, G. Beobide, O. Castillo, J. Cepeda, et al., Two appealing alternatives for MOFs synthesis: solvent-free oven heating vs. microwave heating, RSC Adv. 4 (2014) 60409–60412.

[136] H. Reinsch, B. Bueken, F. Vermoortele, I. Stassen, A. Lieb, K.-P. Lillerud, et al., Green synthesis of zirconium-MOFs, CrystEngComm 17 (2015) 4070–4074.

[137] H. Reinsch, "Green" synthesis of metal-organic frameworks, Eur. J. Inorg. Chem. 2016 (2016) 4290–4299.

[138] J. Jiang, F. Gándara, Y.-B. Zhang, K. Na, O.M. Yaghi, W.G. Klemperer, Superacidity in sulfated metal–organic framework-808, J. Am. Chem. Soc. 136 (2014) 12844–12847.

[139] I. Huskić, I.V. Pekov, S.V. Krivovichev, T. Friščić, Minerals with metal-organic framework structures, Sci. Adv. 2 (2016) e1600621–e1600621.

[140] D. DeSantis, J.A. Mason, B.D. James, C. Houchins, J.R. Long, M. Veenstra, Techno-economic analysis of metal–organic frameworks for hydrogen and natural gas storage, Energy Fuels 31 (2017) 2024–2032.

[141] N. Tannert, C. Jansen, S. Nießing, C. Janiak, Robust synthesis routes and porosity of the Al-based metal–organic frameworks Al-fumarate, CAU-10-H and MIL-160, Dalt. Trans. 48 (2019) 2967–2976.

[142] R.E. Morris, P.S. Wheatley, Gas storage in nanoporous materials, Angew. Chemie Int. Ed. 47 (2008) 4966–4981.

[143] M. Dincă, J.R. Long, Hydrogen storage in microporous metal-organic frameworks with exposed metal sites, Angew. Chemie Int. Ed. 47 (2008) 6766–6779.

[144] B. Chen, X. Zhao, A. Putkham, K. Hong, E.B. Lobkovsky, E.J. Hurtado, et al., Surface interactions and quantum kinetic molecular sieving for H_2 and D_2 adsorption on a mixed metal−organic framework material, J. Am. Chem. Soc. 130 (2008) 6411–6423.

[145] S.K. Bhatia, A.L. Myers, Optimum conditions for adsorptive storage, Langmuir 22 (2006) 1688–1700.

[146] R. Xu, W. Pang, J. Yu, Q. Huo, J. Chen, Chemistry of Zeolites and Related Porous Materials: Synthesis and Structure, John Wiley & Sons, 2007.

[147] J. Rouquerol, F. Rouquerol, P. Llewellyn, G. Maurin, K. Sing, Adsorption by Powders and Porous Solids, second ed., John Wiley & Sons, Hoboken, NJ, 2013.
[148] J.-R. Li, R.J. Kuppler, H.-C. Zhou, Selective gas adsorption and separation in metal–organic frameworks, Chem. Soc. Rev. 38 (2009) 1477.
[149] S. Ma, D. Sun, X.-S. Wang, H-C. Zhou, A mesh-adjustable molecular sieve for general use in gas separation, Angew. Chemie Int. Ed. 46 (2007) 2458–2462.
[150] M.H. Alkordi, Y. Liu, R.W. Larsen, J.F. Eubank, M. Eddaoudi, Zeolite- like metal−organic frameworks as platforms for applications: on metalloporphyrin-based catalysts, J. Am. Chem. Soc. 130 (2008) 12639–12641.
[151] M. Kurmoo, Magnetic metal–organic frameworks, Chem. Soc. Rev. 38 (2009) 1353.
[152] F.A. Mautner, C. Berger, B. Sudi, R.C. Fischer, A. Escuer, R. Vicente, et al., Azido-bridging cobalt(II) systems: crystal structures and magnetic properties, Acta Crystallogr. 69 (2013) s624 –s624.
[153] X.-Y. Wang, Z.-M. Wang, S. Gao, Constructing magnetic molecular solids by employing three-atom ligands as bridges, Chem. Commun. (2008) 281–294. https://doi.org/10.1039/B708122G.
[154] L.M.C. Beltran, J.R. Long, Directed assembly of metal−cyanide cluster magnets, Acc. Chem. Res. 38 (2005) 325–334.
[155] E. Pardo, R. Ruiz-García, J. Cano, X. Ottenwaelder, R. Lescouëzec, Y. Journaux, et al., Ligand design for multidimensional magnetic materials: a metallosupramolecular perspective, Dalt. Trans. 47 (2008) 2780.
[156] D. Maspoch, D. Ruiz-Molina, J. Veciana, Magnetic nanoporous coordination polymers, J. Mater. Chem. 14 (2004) 2713.
[157] C. Biswas, P. Mukherjee, M.G.B. Drew, C.J. Gómez-García, J.M. Clemente-Juan, A. Ghosh, Anion-directed synthesis of metal−organic frameworks based on 2-picolinate Cu(II) complexes: a ferromagnetic alternating chain and two unprecedented ferromagnetic fish backbone chains, Inorg. Chem. 46 (2007) 10771–10780.
[158] M.G. Rimoli, M.R. Rabaioli, D. Melisi, A. Curcio, S. Mondello, R. Mirabelli, et al., Synthetic zeolites as a new tool for drug delivery, J. Biomed. Mater. Res. Part A 87A (2008) 156–164.
[159] J. Salonen, A.M. Kaukonen, J. Hirvonen, V.-P. Lehto, Mesoporous silicon in drug delivery applications, J. Pharm. Sci. 97 (2008) 632–653.
[160] M. Vallet-Regí, F. Balas, D. Arcos, Mesoporous materials for drug delivery, Angew. Chemie Int. Ed. 46 (2007) 7548–7558.
[161] K.S. Soppimath, T.M. Aminabhavi, A.R. Kulkarni, W.E. Rudzinski, Biodegradable polymeric nanoparticles as drug delivery devices, J. Control. Release 70 (2001) 1–20.
[162] S. Freiberg, X.X. Zhu, Polymer microspheres for controlled drug release, Int. J. Pharm. 282 (2004) 1–18.
[163] C.-Y. Lai, B.G. Trewyn, D.M. Jeftinija, K. Jeftinija, S. Xu, S. Jeftinija, et al., A mesoporous silica nanosphere-based carrier system with chemically removable CdS nanoparticle caps for stimuli-responsive controlled release of neurotransmitters and drug molecules, J. Am. Chem. Soc. 125 (2003) 4451 –1459.
[164] P. Horcajada, C. Serre, G. Maurin, N.A. Ramsahye, F. Balas, M. Vallet-Regí, et al., Flexible porous metal-organic frameworks for a controlled drug delivery, J. Am. Chem. Soc. 130 (2008) 6774–6780.

Chapter 2

Fundamentals of metal–organic frameworks

Atif Husain and Malik Nasibullah
Department of Chemistry, Integral University, Lucknow, Uttar Pradesh, India

2.1 Introduction

Metal–organic frameworks (MOFs) have seen extraordinary attention from the scientific audience over the past three decades due to their fascinating properties playing a key role in their potential applications. These MOFs are porous crystalline solids or porous coordination polymers (PCPs) which are constructed by metal sites and organic or inorganic building blocks (metal ions or clusters) as their basic representatives of coordination polymers. During this course of time in the past few decades, they have received tremendous research attention from people of different domains due to their high surface area, permanent porosity, controllable morphology, tunable chemical properties, and flexible chemical structure [1]. They are structurally made up of metal ions or metal clusters which are further linked by organic ligands or other similar structures into cage-like network structures often exhibiting permanent porosity [2] and also imparting new bulk properties [3]. Their tendency to exhibit an unprecedented degree of tunability has given them a superior selection where other materials have failed. This has led to newer advances and further developments of new structures with improved features paving a brighter way for new emerging potential applications as selective catalysts in organic synthesis particularly heterogenous catalysis by overcoming the non-neglectable issues arising from contemporary organic synthesis [4].

Apart from the established synthetic procedures which may or may not be oriented towards a greener approach, an elated technique needs to be developed to overcome the disadvantages of the traditional techniques including reduction in the environmental costs, energy, and the need for toxic organic solvents which consequently reduce the production cost. The development of these greener and safer pathways attaining the goal of a sustainable approach is the need of the hour. In the current scenario, many directional efforts have been made to follow alternate pathways that lead to the minimal generation of hazardous organic solvents in the synthesis of MOFs through solvent-free methods, aerosol

Synthesis of Metal–Organic Frameworks via Water-Based Routes: A Green and Sustainable Approach.
DOI: https://doi.org/10.1016/B978-0-323-95939-1.00012-5

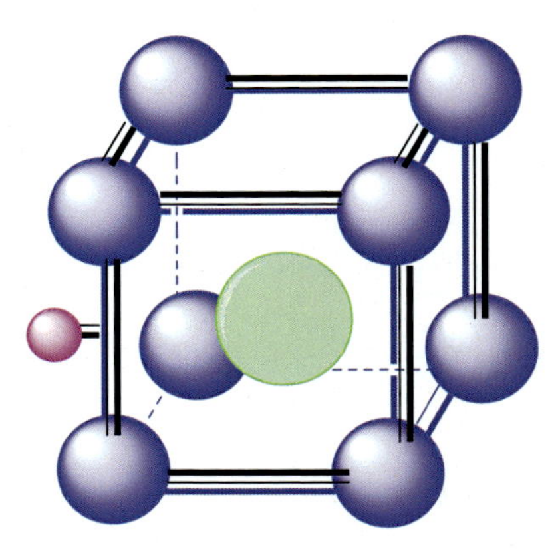

FIGURE 2.1 An illustration of a typical metal–organic framework showing the building unit of the framework (*blue*), linker molecule (*pink*), and external unit (*green*).

routes, microwave radiation techniques, etc. [5]. The possibility of structural modifications leading to dynamic and corresponding changes in the potential porosity and other structural changes has led to a major scope in the further advanced developments in the same field attracting the interest of the chemical industry on a greener scale (Fig. 2.1) [6].

This chapter presents a concise perspective on the basic fundamentals of MOF, the structural MOF chemistry as an important element of inorganic chemistry, and the selective properties of these privileged materials making them core materials for the upcoming research associated with them. It also presents the established and emerging synthetic procedures for an efficient synthesis pathway in accordance with green chemistry leading to the emerging potential applications keeping in mind the developments made in the chemistry of MOF during the past three decades, MOF studies are shifting from the designing and synthetic part to the application and commercialization of these MOF and their derivatives.

The fundamentals of MOF upon which the synthesis and commercialization of these promising materials are based are shown with appropriate illustrations and the synthetic approaches adopted in the literature of the past several years approaching greener and sustainable pathways.

2.2 Background of metal–organic frameworks

Analyzing the rapid increase in the available literature on the chemistry of MOF during the last few years and the abruptly increasing advances in the same field one might think that the studies on these privileged materials are being made over the past many centuries, albeit the systematic studies on the chemistry of MOF and their applications started only about three decades ago. The orientation of the scientific community towards the field of MOF started in the late 1990s henceforth many around 13 seminal works and 7 works have been documented by Hoskins and Robson in the successive years establishing the foundation for the future of MOFs [7,8]. Their work creates a foundation for today's researchers by laying down the facts about structural features such as crystalline nature, microporous, stable solids, with applications such as ion exchange, gas sorption, and catalytic behavior thus allowing various groups in their structure by post-synthetic modifications [9].

Since then moving forward with a brighter vision of the creation of such selective materials with modifiable structural, electrical, magnetic, and catalytic properties by appropriate selection of metal ions and organic ligands became a driving factor for many new scientists to further explore this field and efficiently exploit its applications. These properties like crystalline and microporous nature have led to the development of a new array of applications. The very first 3D MOF was reported in 1997 which displays salutary properties towards ambient conditions [10]. Various synthetic procedures were initially developed by using the same established methods which lead to the development of important materials such as zeolite by varying their morphology and elemental compositions [11]. These synthetic procedures employed in the development of these MOFs are very important for the purpose of tuning the MOF characteristics for functional applications. There are various methods adopted for the synthesis of MOF by modifying the synthetic approach and many strategies. These MOFs can also be designed using hydrothermal, solvothermal, sonochemical, microwave-assisted, and many other methods. Some advanced methods apart from the traditional methods were also developed like ultrasound- or microwave-assisted synthesis which effectively alter the crystallization behavior of MOFs [12]. This chapter will focus on the basic fundamentals of MOF ranging from its historical background to the brief MOF chemistry and the possibility of potential applications emerging by the selective selection of organic ligands and post-synthetic modifications.

During the phase of developments made to further advance the chemistry of MOFs, certain important conclusions made were: (1) The geometric principle for the construction of them where metal–containing units were kept in rigid shapes. This approach not only led to the identification of a small number of preferred topologies that could be targeted in designed synthesis but was also the core area to achieve a permanent porosity. (2) The use of the isoreticular principle where the size and the nature of a structure change without changing its topology led to MOFs with ultrahigh porosity and unusually large pore openings. (3) Post-synthetic modification of MOFs increased their functionality by reacting organic units and metal–organic complexes with linkers. (4) Multifunctional MOFs incorporated multiple functionalities in a single framework [13].

2.3 Metal–organic framework chemistry and field of inorganic chemistry

The present elaboration of the term "MOF" is made as "metal," "organic," and "framework" and here all three words can be understood more or less by any

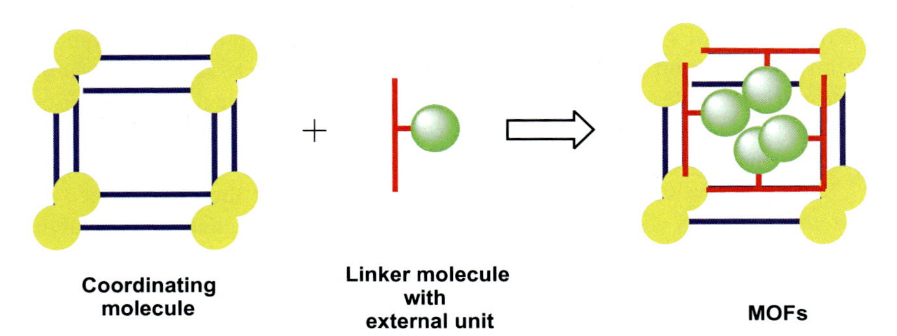

FIGURE 2.2 A schematic illustration of the binding of a coordinating molecule with an external linker.

non-specific audience. The term "organic" is significant outside the inorganic chemistry community albeit the role and dominance of inorganic chemistry in the chemistry of MOF is inalienable. On a fundamental basis, the connection between MOF chemistry and the field of inorganic chemistry is very strong. In multiple ways, MOF chemistry enables inorganic chemists to establish a link between existing coordination complexes the emerging new structures. The coordination bonds which are the building blocks of these complexes are mainly molecular interactions between the metal ions and organic linkers, whereas H–bonding and π–π interactions are present as intermolecular interactions. This implies that the basic understanding of the tool used in the design of MOFs is very crucial to control the reaction [14].

However, the probability of the formation of the specific geometry of new MOFs is strictly governed by the formation of different coordination environments by different central metal atoms. A variety of metal ions including the first-row transition metal ions and lanthanides are extensively studied for different geometries, oxidation states, and topologies offering unusual synthetic and structural integrity. In the designing process of the desired MOFs, the flexibility and the role of the rigid organic ligand is very crucial as the flexible linkers increase the degree of freedom concerning the rigid ones, which leads to unusual crystal formation [15].

The initial detailed study on MOF chemistry was performed by chemists from the fields of coordination and solid-state especially zeolite chemistry; although this has also enclosed the keen attention of material chemists. This has led to the introduction of functionalities and imparting them into MOFs similar to the earlier made attempts by chemists of the post-synthetic modification of MOFs [16]. Another interesting category of materials having the possibility of forming conjugated interactions between them and benzene rings of target molecules are metal–organic frameworks derived from nanoporous carbons. These specialized materials are used in some advanced applications as useful adsorbents (Fig. 2.2) [17].

2.4 Morphologically modifiable structure of MOFs

One of the extraordinary features of these special classes of materials is their capacity to modify their structure according to the reaction conditions and building units. This provides a scope for a large number of morphologically different structures differing in their properties and their topology ultimately acting as substrates for different potential applications. According to the Cambridge Structural Database presently there are over 80,000 MOFs with a wide diversity of topological differences [18]. Among the huge library of structures available the MOFs which have been synthesized in water are quite less, some of them like zeolitic imidazolate frameworks (ZIFs), isoreticular MOFs (IRMOFs), MIL, UiO, coordination pillared-layer (CPL), and porous coordination network (PCN) finds a better place in the chemistry of MOF [19].

The advantage of the adjustable nature of these MOFs has played a key role in the development of new applications in the past few decades. This adjustment in the morphological behavior of MOF is applied in the detection of target analytes to further analyze and understand the mechanism of their detecting behavior [20]. This strategy was employed to identify new categories of MOFs exhibiting new topologies, open structures, and the largest possible surface areas to fit in the possible requirement of potential applications. The development of these featured structures has provided a successful press to the researchers and the scientific community in concepts such as post-synthesis functionalization [21], isoreticular frameworks [22], protection/deprotection [23], etc. Crediting the availability of a large number of different metal nodes and organic linkers available, different MOFs with a variety of topologies can be accessed showing differing chemical and physical properties [24].

Post-synthetic modification is one such technique that has gained much attention from the scientific community over the past several years for tuning the molecular building block of MOF. Some of the practical approaches of post-synthetic which are widely accepted in today's scenario are (1) by using various organic ligands with the different functionalized sites during the pre-synthetic process, (2) PSM of monofunctional ligands by tandem grafting different groups, and (3) the PSM using the separate reactions of bifunctional ligand [25].

2.5 The selective properties of metal–organic frameworks making them promising futuristic materials

The very promising properties of MOFs have established a strong foundation of MOF chemistry providing a key role in generation of various selective materials for different sectors. One such property of high surface area has led to the selection of these MOFs as ideal materials to be used as adsorbents to meet various separation needs for different compounds including organic or metal compounds from a wide range of matrices like food samples, drinking water etc. MOF-199, MOF-5(Zn), ZIF-8 and MIL-53(Al) are some typical examples

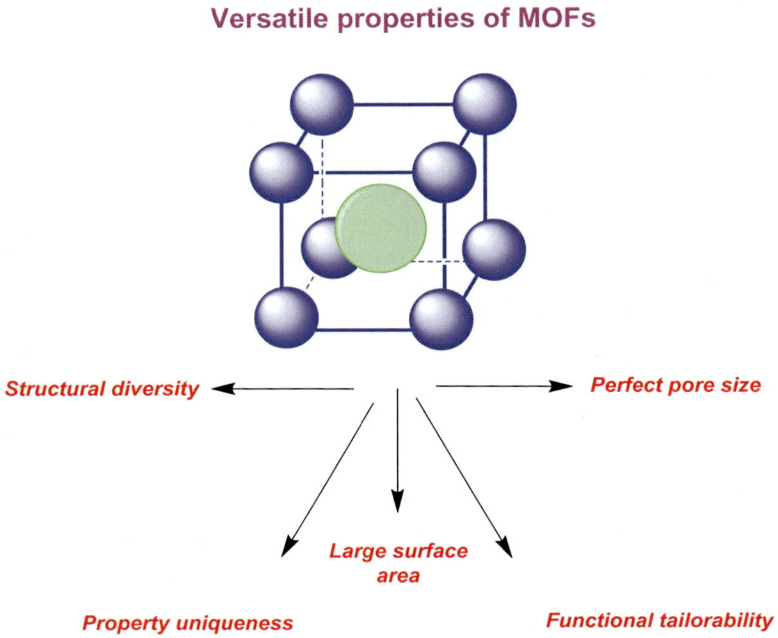

Versatile properties of MOFs

Structural diversity
Perfect pore size
Large surface area
Property uniqueness
Functional tailorability

FIGURE 2.3 Versatile properties of metal–organic frameworks make them futuristic materials.

of MOF materials that have been used as adsorbents for sample preparation [26]. Presence of large surface area, perfect pore size and thermal stability is exclusively seen in MOFs due to the presence of strong bonds such as C–H, C–C, C–O, and M–O bonds are ultimately associated with properties and applications. Keeping in view the potential applications in the field of green chemistry the versatile properties of MOFs like structural diversity, property uniqueness, and functional tailorability have attracted the enormous attention of scientists and researchers especially green chemists over the past decades [27].

The selection of these properties has made rapid growth leading to the broad exploration of the applications of MOF apart from the growth in the design, synthesis, and property exploration of them. Different strategies have been employed to manipulate the synthesis of these MOFs or atomically precise hybrid materials to offer potential application as a unique solid adsorbent designing platform due to their synthetic diversity in some advanced modern-day applications like acting as chemical warfare agents. A possibility in the modification of pore sizes has resulted in the enhanced selection of target gases by incorporation of functional groups on organic linkers thus enabling the tuning of physical properties, including hydrophobicity and the electron donating or accepting ability, as well as to engender selective reactivity toward targeted gases (Fig. 2.3) [28].

As the global energy demand is ever-increasing with the depletion of resources and increasing consumption, an urgent need for promising futuristic MOFs is the need of the hour. The development of greener, clean, safe, and sustainable synthetic procedures leading to the formation of selective MOFs is very important in today's scenario. A sustainable approach towards the synthesis of MOFs will not only satisfy the adverse impacts arising from the traditional methods but will also provide a newer and better strategy for many new upcoming researches and projects.

2.6 Existing and emerging synthetic procedures in compliance with green chemistry

Over the past few years, more focus has been made on the synthesis and applications of MOF in compliance with green chemistry. This includes using solvent-free or mild solvent conditions, reduction in the number of byproducts, use of innocuous solvents, and use of water as a reaction media [29]. Some parameters like choosing cheaper/safer units, decreasing energy input, easy activation, etc. are considered while following synthetic procedures under greener and sustainable conditions. Some developed synthetic procedures allow coordination bonds to be formed and reformed during crystallization. One of the widely adopted methods of this category is solvothermal synthesis in which a metal species and a multitopic organic ligand are mixed with a high-boiling-point solvent [e.g., dimethylformamide (DMF), dimethylacetamide (DMA), or dimethyl sulfoxide (DMSO)], and then heated by analyzing and varying the reaction parameters like pH, solvent, temperature, etc. [30]. Another powerful tool to enhance the formation of new MOFs is High-throughput methods which modify the reaction conditions and is closely related to the concepts of miniaturization and automation [31].

These rewards conferred to the environment and also to the industrial and scientific community by constructing MOF through green synthetic procedures are worth deserving, hence a rapid increase is seen in new emerging applications over the past several years. The rate of some selective applications on the parameters of a green and sustainable pathway is ever-increasing in some of the applications like energy-related applications, separation and adsorption applications, drug delivery, catalysis, etc. Nevertheless, some challenges are always associated including electrochemical, mechanochemical, microwave-assisted synthesis, and continuous flow production needs to cope with new emerging applications. In the industrial preparation of HKUST-1, BASF was the first to demonstrate the scale-up synthesis of MOFs via an electrochemical process by using metal electrodes directly as a metal source in order to exclude metal anions [32]. In gas storage applications a desired MOF adsorbent is developed to act as a fuel storage material for vehicular applications which should fulfill the task of providing a high deliverable capacity apart from having a high fuel uptake capacity [33]. Some strategies like tuning pore chemistry and linker or node functionalization are also investigated to further improve the fuel storage capacity of MOFs. In the case of adsorption-based applications the intrinsic properties of MOFs specifically their high tailorable pore capacity and surface chemistry are extensively worked upon to overcome the need for high-performing separation materials facilitate the sustainability of the separation process [34].

Predicting on the basis of current trends the emerging applications will primarily focus on the green pathways towards a sustainable approach using a facile, efficient, economic approach in bulk to meet the requirements of present-day industries. In this chapter, we have given a glance at some of the major and selectively important applications and many novel emerging applications that have been advancing in recent years and similar ones will be able to compete in the upcoming future with established materials.

2.7 Conclusion and outlook

Metal–organic frameworks are categorized as a novel class of crystalline materials comprising coordination bonds between metal clusters, metal atoms, and polydentate organic linkers that may contain some donor atoms ultimately forming a compact three-dimensional structure. The properties of both the units i.e., coordination molecule and linker unit determine the morphology, appearance, and structural behavior, as well as the physical and chemical properties ultimately deciding the nature of potential applications. These potential applications have played a role as driving factor for synthetic procedures associated with them. On approaching the available synthetic pathways which are posing a threat to the environment and resources a need for greener, safer, environmentally, and economically sustainable is a prerequisite for any upcoming synthetic pathway. Thus, a green synthetic pathway will efficiently reduce the risk of environmental impacts, energy losses, higher costs, complicated scale-up procedures, waste byproduct generation, and reduction in hazardous and toxic chemicals and will lead to a sustainable approach, i.e., green, sustainable, and water-based routes.

In this chapter, we have presented an introductory approach to the synthesis of these privileged structures proceeding through green synthetic water-based routes. The essential fundamentals range from their structural chemistry, the implication of inorganic chemistry on MOF chemistry, the scope in the morphology of MOF making them selective materials for new emerging potential applications, the characteristic properties of MOFs making them effective, the established and emerging synthetic procedures leading to new applications are discussed. Over the past several years multiple efforts have been done by different research groups with the motto of approaching a sustainable pathway whether it be improving solvent conditions or employing water as a reaction media or as a substitute for organic solvents.

Apart from the outstanding advances which have been made in the field of MOF chemistry, especially the green synthetic procedures there still lies a gap in the development of more efficient and effective synthetic methodologies that could be of better interest by further extending the scope of MOFs. The special properties of MOFs need to be further selectively explored in the development of green production of MOF apart from the conventional methods by also inculcating the knowledge of materials, solid chemistry, physics, and engineering into MOF chemistry.

References

[1] P. Szuromi, Mesoporous metal-organic frameworks, Science 359 (2018) 172.9–173.

[2] S. Kitagawa, R. Kitaura, S.-I. Noro, Functional porous coordination polymers, Angew. Chem. Int. Ed. 43 (2004) 2334–2375.

[3] D. Zacher, R. Schmid, C. Wöll, R.A. Fischer, Surface chemistry of metal-organic frameworks at the liquid-solid interface, Angew. Chem. Int. Ed. 50 (2011) 176–199.

[4] V. Pascanu, G.G. Miera, A.K. Inge, B. Martín-Matute, Metal-organic frameworks as catalysts for organic synthesis: a critical perspective, J. Am. Chem. Soc. 2019, 141, 7223–7234.

[5] H. Bux, F. Liang, Y. Li, et al., Zeolitic imidazolate framework membrane with molecular sieving properties by microwave-assisted solvothermal synthesis, J. Am. Chem. Soc. 131 (2009) 16000–16001.

[6] M. Jacoby, Multifunctional metal-organic frameworks: From academia to industrial applications, In Chem. Eng. News 86 (2008) 13.

[7] B.F. Hoskins, R. Robson, Infinite polymeric frameworks consisting of three dimensionally linked rod-like segments, J. Am. Chem. Soc. 111 (1989) 5962.

[8] B.F. Hoskins, R. Robson, Design and construction of a new class of scaffolding-like materials comprising infinite polymeric frameworks of 3D-linked molecular rods. A reappraisal of the zinc cyanide and cadmium cyanide structures and the synthesis and structure of the diamond-related frameworks $[N(CH_3)_4][CuIZnII(CN)_4]$ and CuI[4,4′,4″,4‴-tetracyanotetraphenylmethane]$BF_4 \cdot xC_6H_5NO_2$, J. Am. Chem. Soc. 112 (1990) 1546.

[9] D. Venkataraman, G.B. Gardner, S. Lee, J.S. Moore, Zeolite-like behavior of a coordination network, J. Am. Chem. Soc. 117 (1995) 11600.

[10] M. Kondo, T. Yoshitomi, K. Seki, H. Matsuzaka, S. Kitagawa, Three-dimensional framework with channeling cavities for small molecules: $\{[M_2(4,4'\text{-bpy})_3(NO_3)_4]\cdot xH_2O\}_n$ (M = Co, Ni, Zn), Angew. Chem. Int. Ed. Engl. 36 (1997) 1725.

[11] H. Frost, T. Düren, R.Q. Snurr, Effects of surface area, free volume, and heat of adsorption on hydrogen uptake in metal-organic frameworks, J. Phys. Chem. B 110 (2006) 9565.

[12] Z.H. Xiang, D.P. Cao, X.H. Shao, W.C. Wang, J.W. Zhang, W.Z. Wu, CO_2 adsorption of metal organic framework material Cu-BTC via different preparation routes, Chem. Eng. Sci. 65 (2010) 3140–3146.

[13] V. Gitis, G. Rothenberg (2020). Gitis, V., Rothenberg, G. (Eds.). Handbook of Porous Materials. Vol. 4. Singapore: World Scientific. pp. 110–111.

[14] I. Mantasha, H.A.M. Saleh, K.M.A. Qasem, M. Shahid, M. Mehtab, M. Ahmad, Efficient and selective adsorption and separation of methylene blue (MB) from mixture of dyes in aqueous environment employing a Cu (II) based metal organic framework, Inorg. Chim. Acta 511 (2020) 119787.

[15] M.N. Ahamad, M. Shahid, M. Ahmad, F. Sama, Cu(II) MOFs based on bipyridyls: Topology, magnetism, and exploring sensing ability toward multiple nitroaromatic explosives, ACS Omega 4 (2019) 7738.

[16] I. Ahmed, S. Jhung, Composites of metal–organic frameworks: Preparation and application in adsorption, Mater. Today 17 (2014) 136.

[17] W. Chaikittisilp, K. Ariga, Y. Yamauchi, A new family of carbon materials: synthesis of MOF-derived nanoporous carbons and their promising applications, J. Mater. Chem. A 1 (2013) 14–19.

[18] P.Z. Moghadam, A. Li, S.B. Wiggin, et al., Development of a Cambridge structural database subset: a collection of metal-organic frameworks for past, present, and future, Chem. Mater. 29 (2017) 2618–2625.

[19] H. Reinsch, "Green" synthesis of metal-organic frameworks, Eur. J. Inorg. Chem. 2016 (27) (2016) 4290–4299.

[20] E.A. Dolgopolova, A.M. Rice, C.R. Martin, N.B. Shustova, Photochemistry and photophysics of MOFs: steps towards MOF-based sensing enhancements, Chem. Soc. Rev. 47 (2018) 4710−4728.

[21] Z. Wang, S.M. Cohen, Postsynthetic modification of metal–organic frameworks, Chem. Soc. Rev. 38 (2009) 1315.

[22] M. Eddaoudi, J. Kim, N.L. Rosi, D. Vodak, J. Wachter, M. O'Keeffe, O.M. Yaghi, Systematic design of pore size and functionality in isoreticular MOFs and their application in methane storage, Science 295 (2002) 469.

[23] T. Yamada, H. Kitagawa, Protection and deprotection approach for the introduction of functional groups into metal−organic frameworks, J. Am. Chem. Soc. 131 (2009) 6312.

[24] Z. Chen, S.L. Hanna, L.R. Redfern, D. Alezi, T. Islamoglu, O.K. Farha, Reticular chemistry in the rational synthesis of functional zirconium cluster-based MOFs, Coord. Chem. Rev. 386 (2019) 32−49.

[25] M.N. Ahamad, M.S. Khan, M. Shahid, M. Ahmad, Metal organic frameworks decorated with free carboxylic acid groups: Topology, metal capture and dye adsorption properties, Dalton Trans. 49 (2020) 14690.

[26] Y. Chen, W. Zhang, Y. Zhang, Z. Deng, W. Zhao, H. Du, X. Ma, D. Yin, F. Xie, W. Chen, et al., In situ preparation of core-shell magnetic porous aromatic framework nanoparticles for mixed-mode solid-phase extraction of trace multitarget analytes, J. Chromatogr. A 1556 (2018) 1–9.

[27] H. Furukawa, K.E. Cordova, M. O'Keeffe, O.M. Yaghi, The chemistry and applications of metal-organic frameworks, Science 341 (6149) (2013) 1230444.

[28] T. Islamoglu, Z. Chen, M.C. Wasson, C.T. Buru, K.O. Kirlikovali, U. Afrin, M.R. Mian, O.K. Farha, Metalâ organic frameworks against toxic chemicals, Chem. Rev. (2020). https://doi.org/10.1021/acs.chemrev.9b00828.

[29] X.-J. Kong, J.-R. Li, An overview of metal-organic frameworks for green chemical engineering, Engineering (2021). https://doi.org/10.1016/j.eng.2021.07.001.

[30] N. Stock, S. Biswas, Synthesis of metal-organic frameworks (MOFs): routes to various MOF topologies, morphologies, and composites, Chem. Rev. 112 (2) (2012) 933–969.

[31] Y.G. Chung, J. Camp, M. Haranczyk, B.J. Sikora, W. Bury, V. Krungleviciute, et al., Computation-ready, experimental metal–organic frameworks: a tool to enable high-throughput screening of nanoporous crystals, Chem. Mater. 26 (21) (2014) 6185–6192.

[32] P. Silva, S.M.F. Vilela, J.P.C. Tomé, F.A. Almeida Paz, Multifunctional metal–organic frameworks: from academia to industrial applications, Chem. Soc. Rev. 44 (19) (2015) 6774–6803.

[33] B. Wang, L.H. Xie, X. Wang, X.M. Liu, J. Li, J.R. Li, Applications of metal–organic frameworks for green energy and environment: new advances in adsorptive gas separation, storage and removal, Green Energy Environ. 3 (3) (2018) 191–228.

[34] X. Zhao, Y. Wang, D.S. Li, X. Bu, P. Feng, Metal–organic frameworks for separation, Adv. Mater. 30 (37) (2018) e1705189.

Chapter 3

Kind and role of linkers for metal–organic frameworks

Arif Ali, Mohd Muslim, Saima Kamaal and Musheer Ahmad
Department of Applied Chemistry, ZHCET, Faculty of Engineering and Technology, Aligarh Muslim University, Aligarh, Uttar Pradesh, India

3.1 Introduction

Metal–organic frameworks (MOFs) are inorganic–organic hybrid materials made by a combination of metal ions and organic molecular building blocks/ linkers with highly crystalline and porous coordination networks [1]. More than 90,000 MOFs architecture are synthesized and 500,000 are predicted yet and thousands of MOFs are commercialized [2,3]. Organic ligands (linkers) and inorganic secondary building blocks (SBUs) as connectors are used as precursors to developing molecular building architectures *via* coordination or strong bond [4]. The nuclearity of metal clusters or complexes is depended on the binding affinity or open valance sides of coordination of metal nodes as well as the functionality of organic linker. The binding and coordination sides are built with definite geometry and shape for periodically arranged 3D networks [4–7] (Fig. 3.1). Different types of organic linkers are used to develop good porous architecture such as neutral or anionic terminal linkers are constructed the mono, bicarboxylate, pyridyl ligands, and polycarboxylates. MOF-5 [$Zn_4O(bdc)_3$, bdc = terephthalate] and HKUST-1 ($Cu_3(btc)_2$, btc = 1,3,5-benzenetricarboxylate) were prepared and investigated with a baseline in reticular MOF chemical reactions throughout 1999, demonstrating positive porous structure at low-pressure gas sorption study by crystal structure [8,9]. In 2005, Cr(III) based MOF (MIL-101) was synthesized with high chemical stability by utilizing a terephthalate linker. The MIL-101 has been shown high surface area with 4100 m^2/g and 5900 m^2/g by using Brunauer-Emmett-Teller (BET) and Langmuir surface areas, respectively [10]. In that year, [Zn_2(dhbdc), dhbdc = 2,5-dihydroxy-1,4-benzenedicarboxylate] (MOF-74) has been synthesized to outperform physiological materials for CO_2 absorption at low-pressure [11,12]. Another way, the replacement of neutral or anionic linker with new functionalized or coordination mode provided new well decorate inner space

Synthesis of Metal–Organic Frameworks via Water-Based Routes: A Green and Sustainable Approach.
DOI: https://doi.org/10.1016/B978-0-323-95939-1.00007-1

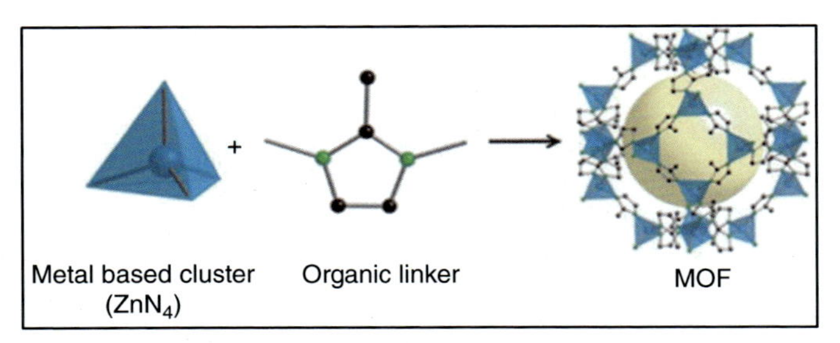

FIGURE 3.1 A typical schematic representation of multi–organic framework structure consists of a metal-based cluster and an organic linker; yellow sphere shows open space [21].

for framework without change in underlying net [13,14] (Fig. 3.2). The high crystallinity of MOFs is confirmed the inner arrangement of SBUs with appropriate distance and dimension of rigid architecture [15]. The orientation and geometrical distance of neighboring terminal ligands could be substituted by multidentate or multitopic linker, this substitution is called linker installation in MOFs [16]. On the basis of the dissolution and recrystallization process linkers could be installed by single crystal to single crystal (SC–SC) transformation, i.e., called post-synthetic modification for the connector of MOF [17,18]. The conventional modification in organic linker is changed the underlying net connectivity and topology of the framework. Yaghi and co-workers proposed this method to trigger the systematic development of different utilities of MOFs with cage matrices [19]. The ***dia*** (diamond) like net, for example, can be built from tetrahedral clusters and bidentate organic linkers, whereas the cylindrical net could be built from 8-connected and bidentate sequential linkers [20].

3.2 Types of organic linkers

3.2.1 Anionic organic linker for metal–organic framework

The six zirconium (Zr) atoms containing SBUs are the most well-known examples of linker installation in MOFs. With the utility of SBUs node in an extended network, the Zr_6 unit does have the unique ability to control the extension number from 4-,6-,8-,10-,11-, and complete 12-connected under diverse synthetic variables [23,24]. The numbers of connection parameter of Zr_6 units with different organic linkers give a well-defined topological connection. The metal nodes with lower coordination networks are a distinct structural subunit of the 12-connected node that can be served as parent MOF with available linker vacancies for installation. On removing of terminal groups and inserting another linker could be changed the topological architecture [25].

Mainly, di, tri, tetra, and hexa-carboxylate organic ligands are used as anionic linkers. Non-symmetric anion exchange binding affinities (generally described as pyridine-X-COO-(X= spacer)) containing coordination polymers, such as the 1,4-dihydroxy-2,5-benzoquinone linker and its derived products used as

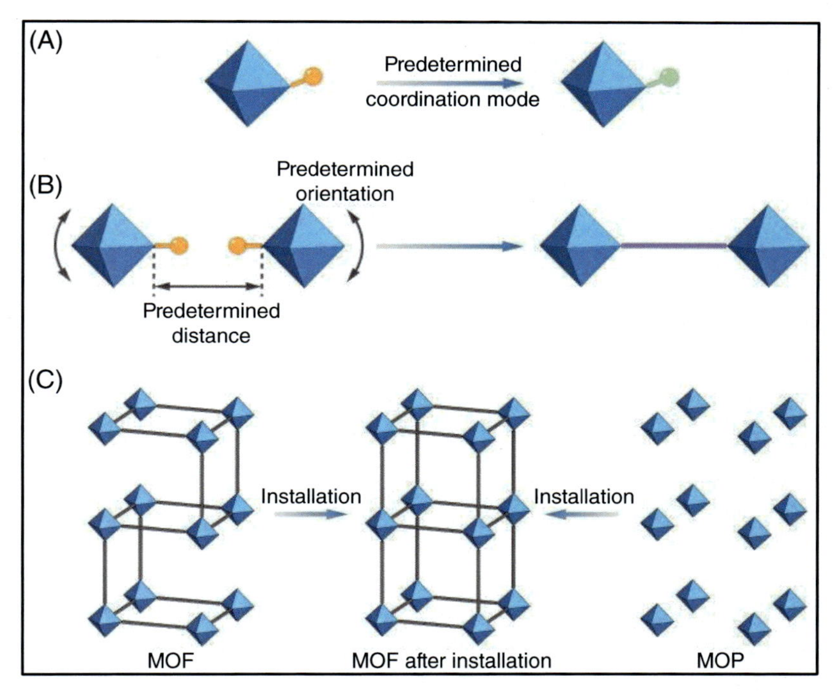

FIGURE 3.2 Schematic representation of installation of organic linker in metal–organic frameworks, (A) functionality of original terminal ligand replaced by predetermined coordination mode of new ligand, (B) elongated framework of metal–organic frameworks and metal–organic polymers (MOPs) by using multitopic linker to construct multicomponent, and (C) metal–organic frameworks by simply replacing several multitopic command prompt ligands from adjacent secondary building blocks with predefined cross distance and orientation of linker [22].

linear linkers to develop a variety of frameworks, have been studied extensively. Several anionic linkers have been synthesized (Fig. 3.3) to fabricate the different MOFs with different topologies.

3.2.1.1 *Ditopic carboxylate-based linker*

In situ synthesis of MOFs is dependent on the condition of the reaction. MOFs are developed on reactions between the SBUs and organic ligands that lead the different crystal structures. For example, the reported MOF-5 was synthesized by using $Zn(NO_3)_2$ and terephthalate (bdc) under the solvothermal method. Another method is slow vapor diffusion which was used to fabricate pyramid-like crystals of Zn(bdc)(DMF)(H_2O) at room temperature. Zn(bdc)(DMF)(H_2O) was created by combining triethylamine and toluene in an N,N-dimethylformamide (DMF) solution that contains $Zn(NO_3)_2$. Rather than the octahedral $Zn_4O(CO_2)_6$ constellation in MOF-5, this structure displayed a 2D simple cubic dizinc propeller unit [26].

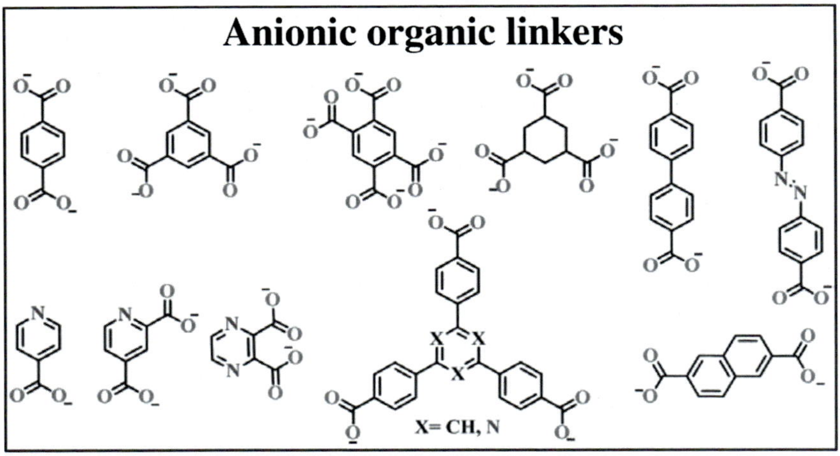

Anionic organic linkers

X= CH, N

FIGURE 3.3 Schematic representation of well-known anionic organic linkers.

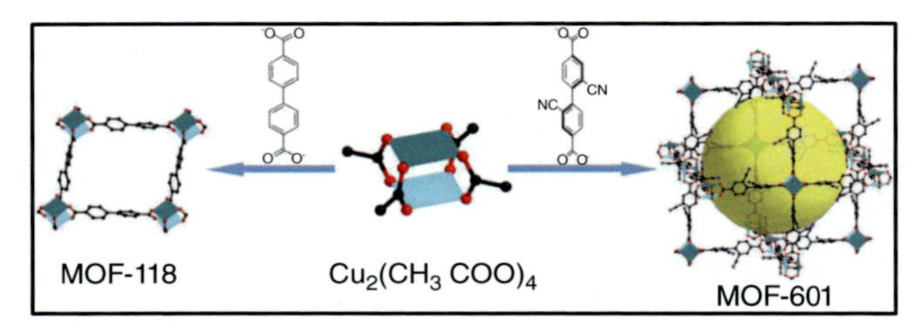

FIGURE 3.4 Same coordination functionality of secondary building blocks lead the 2D MOF-118, 3D MOF-601 with ***sql*** and ***nbo*** topology, respectively [26].

Typically, zinc and copper acetate salts could be used as a propeller structure with a solvent to accomplish the coordination realm. In MOF-118, for example, 4,4′-biphenyldicarboxylate (bpdc) contains a carboxylate group of two phenyl rings that are coplanar with SBUs and form a two-dimensional sheet to SQL configuration. The 2,2′-dicyano-4,4′-biphenyldicarboxylate (cnbpdc) has two carboxylate functional groups that really are parallel to the direction of one another and are compelled by two CN groups, resulting in a three-dimensional network (***nbo*** topology) (MOF-601) (Fig. 3.4). In general, the position of linear dicarboxylates is played an important role in the fabrication of MOFs that controlled the dimensionality and topology of resultant architecture [26].

3.2.1.2 Tritopic carboxylate-based linker

In response to tetradentate organic ligands, the $Zn_4O(CO_2)_6$ showed a 6-connected tetrahedral system, which could produce a presents series of structures of various qom topologies (Fig. 3.5). The green synthesis MOF-177 introduced

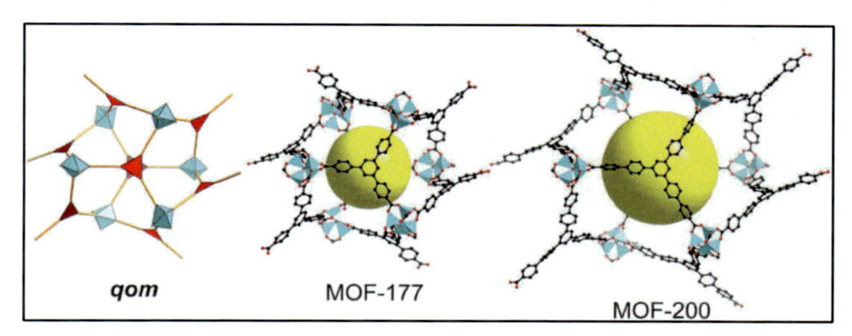

FIGURE 3.5 The MOF-177 and MOF-180 represent ***qom*** net augmented form [28].

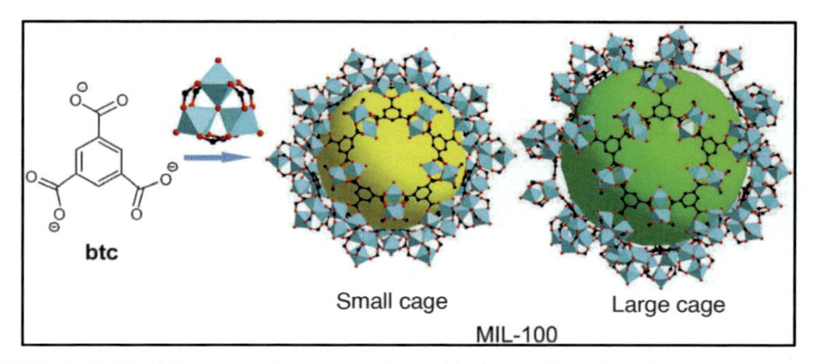

FIGURE 3.6 MIL-100 structural representation, with the btc ligand serving as a spacer and the tetrahedrally cluster serving as secondary building blocks; Cr^{3+}, V^{3+}, Fe^{3+}, and Al^{3+} [29].

a novel underpinning 6-connected and three points net, with the $Zn_4O(CO_2)_6$ group having to serve as the 6-connection site and the btb crosslinks trying to serve as the 3-connection site. MOF-180 and MOF-200 have been produced using 4,4′,4″(benzene-1,3,5-triyltris(ethyne-2,1-diyl))tribenzoate (bte) as fusions and bbc binding sites as comparators, respectively. Because of the ***qom*** topology, the MOF-177 framework is quasi. MOF-180 and -200 exhibit pore volume (89% and 90%, respectively) and super-duper BET surface. The linkers managed to gain additional flexibility by substituting the central benzene ring with a nitrogen atom, which is used in the assembly of the same octahedral $Zn_4O(CO_2)_6$ constellation and 6,6′,6″-nitrilotris(2-naphthoic acid) (ntn) preprocessor; developed a novel porous medium (SNU-150) with new PdF_2-type net configuration [27].

MIL-100(Cr) was produced using the $Cr_3O(CO_2)_6$ groupings and the hexadentate spacer benzene tricarboxylic acid (btc) [29]. The chromium trimers are linked together by the btc spacer, which helps connect four chromium trimers as nodes and four btc as faces to form the zeolite Mobil Thirty-Nine (MTN) network. MIL-100 framework has disclosed two types of highly porous cages with radii of 25 and 29 (Fig. 3.6). Tiny cages in this case are pentagon shaped. On the other hand, the larger cages are made up of polygonal and hexagon shaped windows with radii 4.8 and 8.6, respectively. Following the manufacturing of

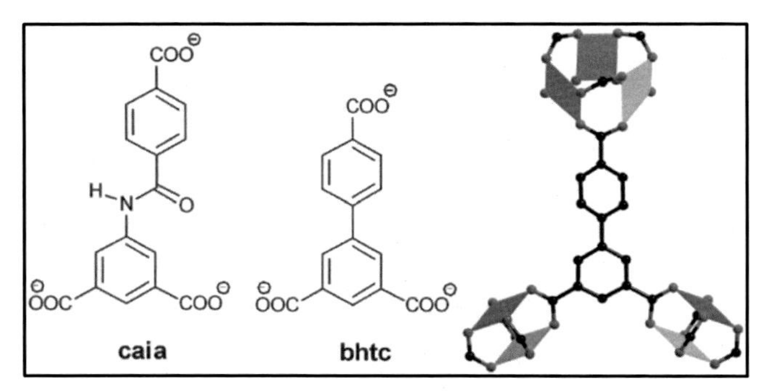

FIGURE 3.7 The coordination mode of with copper secondary building blocks is depicted by tritopic carboxylate (caia and bhtc) linkers; UMCM-150 exhibits dicopper paddle wheel and trinuclear copper clusters [34].

MIL-100 (Cr), the isomorphs sequence of MIL-100 (M) (M = Fe^{3+}, Al^{3+}, and V^{3+}) was fabricated by changing metal ions in the synthetic strategic business units with improved hydrothermal stability and used for a variety of applications including fuel (gas, vapor, and liquid), chemical processing, and therapeutic agents [30–32].

Moreover, the symmetry of btc linker exhibited quite a versatility with multiple SBUs in MOFs. The symmetry of linker may drive the new coordination sphere assembly to satisfy the geometry of the linker. For example, A C_2v symmetrical spacer, biphenyl-3,4′,5-tricarboxylate (bhtc), with three carboxyl groups, connecting two SBUs. Not all carboxylate groups coordinate in the same way [33]. The 3 and 5 positions of the linker's isophathalic moiety are organized with binuclear copper-paddle wheel clusters, and another 4′ position alkoxy group (benzoate moiety) forms a tri-nuclear clump form with an innovative SBU in MOF. Correspondingly, the UMCM-150 (the University of Michigan crystalline material) disclosed an extraordinary three, four, and six-connected underlying net. The isoreticular MOF (NJU-Bai_3) produced by the same SBU with a good space complier 5-(4-carboxybenzoylamino)-isophthalate (caia) demonstrated a network interface spherically [34] (Fig. 3.7). In NJU-Bai_3, an amide group was inserted to decorate the novel linkers, which improved CO_2 binding ability and selectivity over other gas molecules on the pore surface of MOF. NJU-Bai_3 has a higher CO_2 gas adsorption with 6.21 mmol/g at room temperature and 1 bar than UMCM-150 (4.68 mmol/g) [35].

3.2.1.3 Hexatopic carboxylate-based linker

Carboxylate-based hexatopic linkers have been widely used in the production of functional highly permeable structures with high surface and gas storage capacity, such as rht-MOFs [36], NOTTs [37], PMOFs [38], and others [39–41]. The UTSA-20 is made up of the ***rht*** topographic net, which is made up of 3,3′,3″,5,5′,5″-benzene-1,3,5-triyl-hexabenzoate (bhb) and dicopper

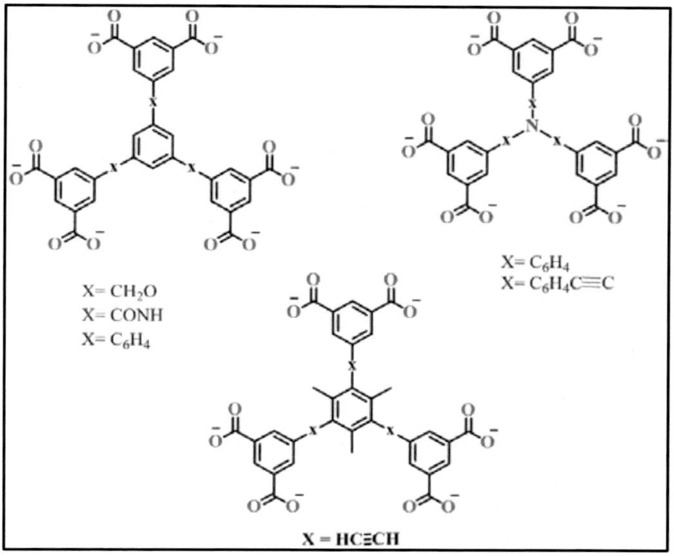

FIGURE 3.8 Schematic representation of hexatopic linker with isophthalate moieties.

propeller. SBU displays a different appearance in hexatopic fusions with C_3 symmetrical. Because of the compound between the peripheral phenyl groups, the bhb linker includes 3 isophthalate organizations that are not perfectly straight. As a result, the structure is a (3,3,4)-c trinodal with a new ***zyp*** topology. The collinear shape of six carboxyl groups with a unique (3,24)-connected ***rht*** net was formed by the bhb complier MOF structures. The Department of Energy methane was used to fully utilize the pore spaces and open sites of copper SBU of UTSA-20 for methane storage at 180 cm^3/cm at ambient temperature and 35 bar [42]. When the three methyl groups are placed on the center phenyl group of the bhb spacer, the resulting steric hindrance causes the isophthalate cluster to be set up parallel to the central phenyl ring. Two new (3,24)-connected ***rht*** topographic networks for SDU-1154 [40] and NPC-5155 [41] have been crafted using the DNA synthesis linker (tmbhb). Two new (3,24)-connected rht topological systems for SDU-1154 [40] and NPC-5155 [41] have been crafted using the recombinant linker (tmbhb). The SDU-1154 and NPC-5155 are made up of 3,4-connected nucleophilic propeller SBUs (Fig. 3.8).

3.2.1.4 Octatopic carboxylate-based linker

Due to the synthetic challenge, octatopic carboxylate-based metal organic frameworks are still rarely synthesized. Octatopic linkers have long arms to lead the formation of interpenetrated structure (Fig. 3.9). On the basis of previous reported octacarboxylate-based MOFs prevent the interpenetration due to high coordination sphere. For example, UTSA-33a synthesized by

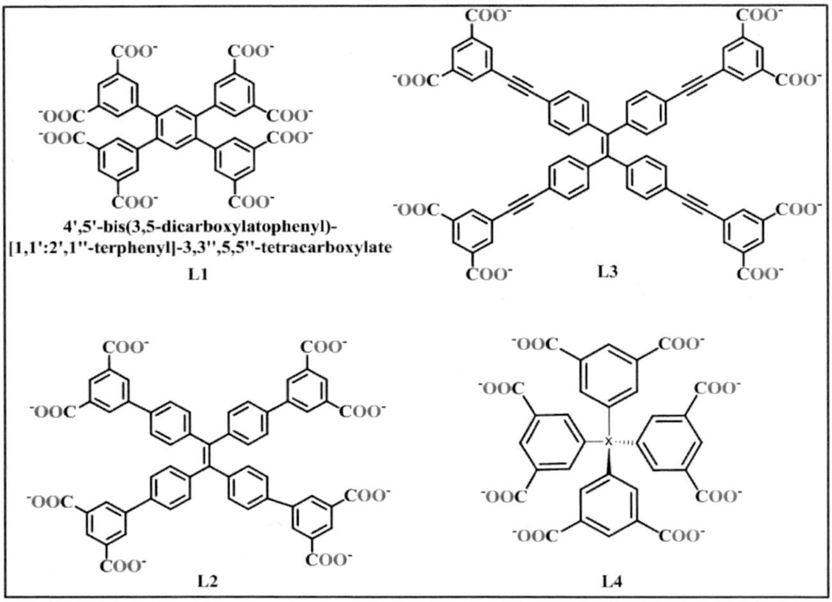

FIGURE 3.9 Schematic representation of **L1, L2, L3**, and **L4** linker.

using zinc SBU and 4′,5′-bis(3,5-dicarboxylatophenyl)-[1,1′:2′,1″-terphenyl]-3,3″,5,5″-tetracarboxylate octatopic linker [43]. Each metal and crosslinks unit had four and eight connected two, respectively, forming a quasi-4,8-connected network with ***flu*** configuration. The moderate pores are utilized to absorb the acetylene, ethylene, and ethane gas with high selectivity. Another USTA-34b was fabricated by using the same linker with dicopper paddle wheel which has been shown high separation capacity with ***ybh*** topology [44]. Microporous MOF PCN-921 was synthesized by using ligand **L2** and a dizinc paddle wheel with a new ***scu*** topology [45]. The ethylene group, neighboring phenyl ring and vicinal phenyl ring lies in different plane due to steric effect with non-planer dihedral angle ranging from 40 degree to 50 degree (Fig. 3.10). Ethylene extended structure of linker **L3** exhibits cylindrical symmetry. The same geometry adopted by the linker **L2** and formed isoreticular PCN-921 [45]. These factors such as minimization of torsional angle, dihedral angle during design of linker, affect the certain physical properties. The tetrahedral geometry containing linkers are always used on the top of MOF fabrication because of preinstalled 3D geometry. The solvothermal produced MOF formed three new nanopores MOF with interesting topographic characteristics and composition by leveraging linker 5,5′,5″,5″-silanetetrayltetraisophthalate (**L4**) and different SBUs [46].

UNLPF-1 and PCN-80 were also built using an 8,4-connected dicopper surfboard node with ***scu*** topology. The lower bulkiness of benzyloxyl, ethoxy,

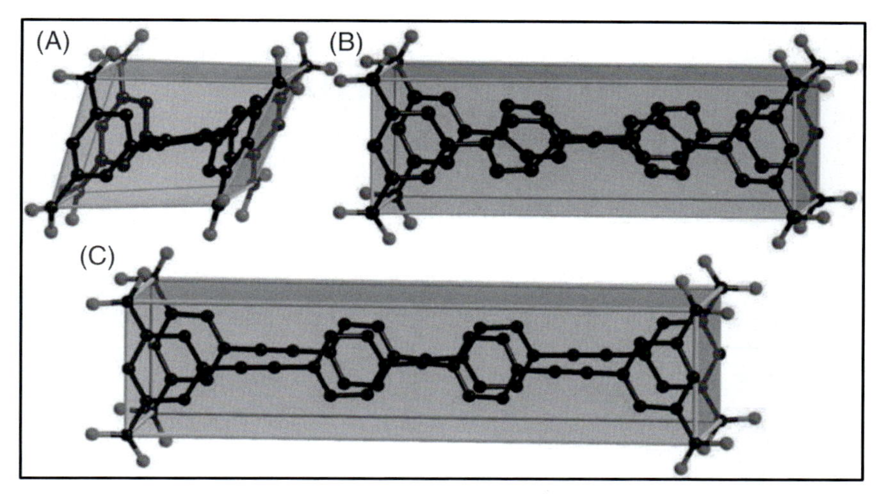

FIGURE 3.10 Geometrical representation of octadentate linker [47].

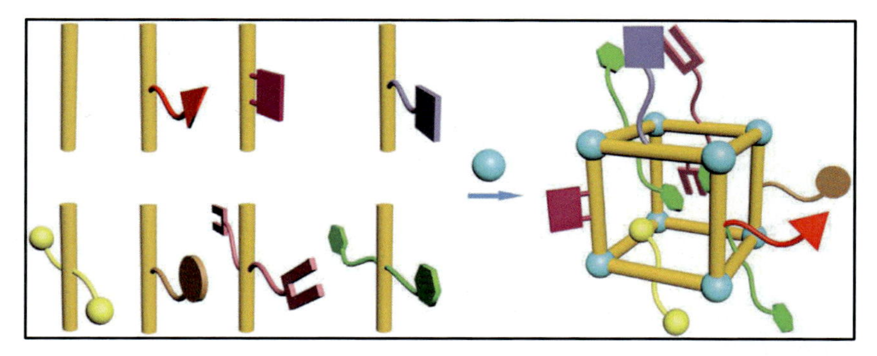

FIGURE 3.11 MTV-MOF-5 framework representation to various coordination functionality [48].

and hydroxyl at the position of the circle of the linkers revealed the porous structure of MOF, which could be governed by the different aspect arms all through the spacer design.

3.2.1.5 Mixed dicarboxylate-based linker

The mixed-linker coordination copolymers could be favored with better response to enhance the properties compared to homopolymer. MTVMOF-5 was built in a regular order with the frameworks' zinc oxide and phenyl, and produce approximately units. The substituent was dispersed in a disturbed manner in three main forms based on spacer conversations [48] (Fig. 3.11). A mixed-linker strategy was used to summarize the biphenyl-based IRMOFs. For example, 9,10-bis(triisopropylsilyloxy)phenanthrene-2,7-dicarboxylate (tpdc) and

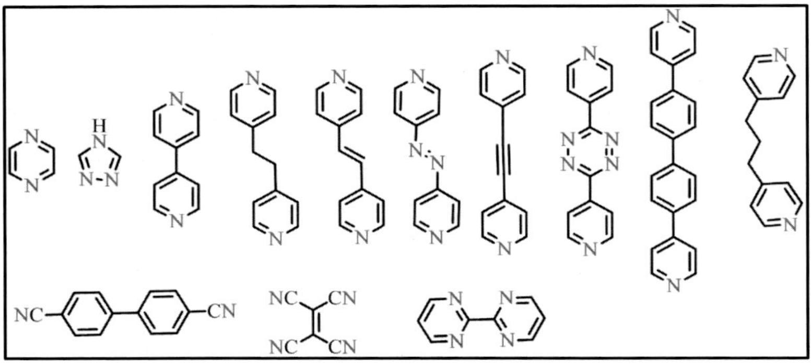

FIGURE 3.12 Schematic representation of neutral ligands.

3,3′,5,5′-tetramethyl-4,4′-biphenyldicarboxylate (Me_4bpdc) have been used to produce MOF with $Zn(NO_3)_2$ in N,N'-diethylformamide (DEF) solution. Here, a less bulky linker (Me_4bpdc) is used to increase the percentage of adsorbent surface, which has a larger surface area (up to 3000 m^2/g) than the aromatic ring $Zn_4O(tpdc)_3$ and interpenetrated $Zn_4O(Me_4bpdc)_3$ structures.

3.3 N-heterocyclic-based linkers or neutral organic linkers

In transition metal complexes, nitrogen-coordinated bonds are stronger than that of oxygen coordinated in solution [50]. Designed organic linkers with nitrogen donating sites (i.e., neutral organic ligand) (Fig. 3.12) have been widely used to synthesize highly stable metal–organic frameworks.

3.3.1 Ditopic N-heterocyclic linkers

The most common binding affinities of neutral linkers are showed by pyrazine (pyz) and 4,4′-pyridile (bpy), which are used as pillars to convert two-dimensional (2D) layer upon layer to three-dimensional (3D) layers [51]. Bipyridyl organic linkers with different lengths, such as 4,4′-dipyridylacetylene and pyrazine, have been used to create reticular chemistry which governed the pore diameter. ZIF-95 ($Zn(cbIm)_2$; cbIm = 5-chlorobenzimidazolate) and ZIF-100 ($Zn_{20}(cbIm)_{39}(OH)$) have fixed porous structure with new framework and adsorption surface area of 1240 m^2/g and 780 m^2/g, as well as selective CO_2 gas adsorption isotherms [52]. The stability of ZIFs in organic and aqueous media makes it possible to perform organic responses at the crystal level with repaired configuration. The aldehyde group is presented in structure of the ZIF-90 and 92, which reduced by the $NaBH_4$ and react with methanolic solution of ethanolamine, respectively (Fig. 3.13), which indicated that the crystals can be utilized as molecule in organic reaction due to chemical inertness [53].

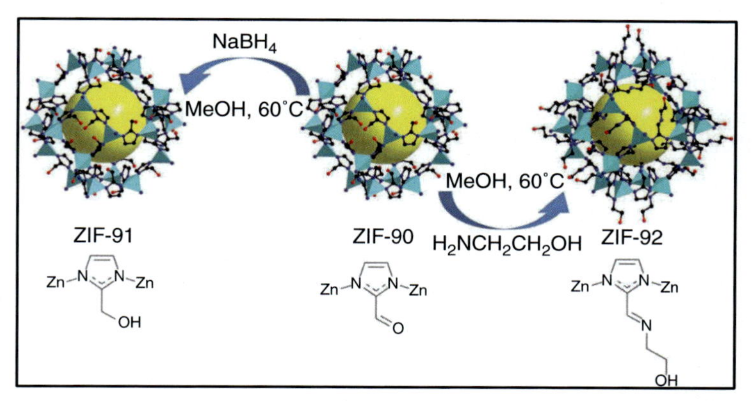

FIGURE 3.13 Depiction of ZIF-90 transformed to ZIF-91 by alkyne decrease with $NaBH_4$ and to ZIF-92 by phosphatidyl response [53].

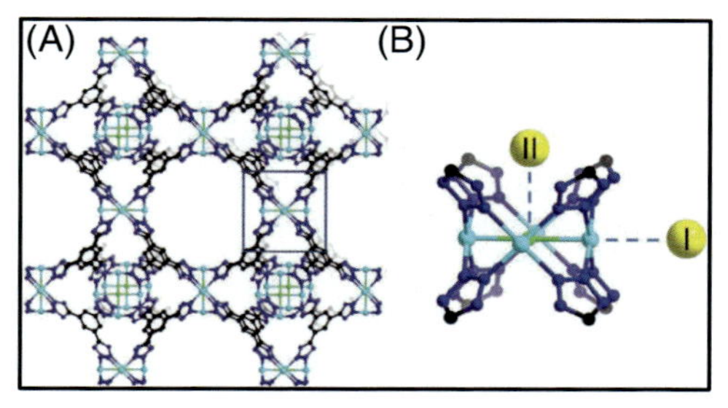

FIGURE 3.14 (A) Mn, Fe, Cu type, and (B) a representation of two prominent D2 binding sites as observed by diffraction patterns [55].

Using various metals ions (M = Ni, Cu, Zn, and Co) and H3btp = 1,3,5-tris(1H-pyrazol-4-yl) benzene linker, four pyrazolate-bridged MOFs $M_3(btp)_2{\cdot}X$ solvent were produced. The stimulation of these MOFs disclosed micro porous structure with BET surface of 1650, 1860, 930, and 1027 m^2/g, including both, as well as good thermal stability. $Zn_3(btp)_2$ proved high thermal stability up to 500°C, so although $Ni_3(btp)_2$ proved superior chemical stability in boiled aqueous solution at various pH values (from 2 to 14) [54].

Several MOFs have been formed using the tetra dentate tetrazolate spacer btt (btt = 1,3,5-benzenetristetrazolate) with the combination of multiple metal ions. This linker's MOF sequence disclosed a nitrogen-bonded coordination sphere. Mn-btt proved a 3,8-connected micro-porous anionic material with fixed high porosity for H_2 gas absorption (6.9 wt.% at 77 K and 90 bar). The high gas uptake capacity revealed the framework's solidity and unsaturated metal centers. The diffraction pattern showed two major binding sites, one of which has a strong affinity to absorb H_2, and another dish cavity on the chlorine ion of the cluster, which is populated site (Fig. 3.14). Other experimental method also prepared

M-btt systems for the synthesis of possible targets such as Zn-btt, that could be useful [55].

Chloride-centered tetranuclear SBU-based MOF material was produced using (tetrakis(4-tetrazolylphenyl) methane) and $CuCl_2$, resulting in a 4,8-connected network with ***flu*** topology. This MOF had an anionic nature and transformed into a neutral MOF. This MOF was activated using a solvent evaporation method, generating an elevated BET surface area of 2506 m^2/g with substantial H_2 gas take-up (2.8 wt.% at 77 K and 1.2 bar). The chlorine ion is removed from the framework through solvent exchange, which is affirmed by slightly shorter Cu–N bonds and *trans* Cu…Cu represents the distance, and offers high binding sites for molecules [56].

3.4 Conclusion

The emergence of new MOF structures in this era has widely been largely inadvertent. Now a day, MOF are used in real applications. The increasing interest of underlying net of metal coordination environments, organic linker orientation and topology helps us to understand the synthetic efforts specifically. The different methods are used to develop the metal organic framework. Different method of synthesis is giving different network for MOF with different underlying net. The non-interpenetrated structure resists the integrity of topological framework. Neutral linker could be more useful to fabricate the MOF than that of carboxylate-based linker due to their stability. The higher number of connecting arms of linker are formed high coordination sphere. In this chapter, we have discussed the connecting mode of the linker with metal ion and their underlying net. Overall, higher functionality of the linker could be provided a high connecting network with enhanced porosity and surface area.

References

[1] J.R. Long, O.M. Yaghi, The pervasive chemistry of metal–organic frameworks, Chem. Soc. Rev. 38 (2009) 1213–1214.

[2] N. Batten, X.-M. Chen, J. Garcia-Martinez, S. Kitagawa, L. Öhrström, M.O. Keeffe, et al., Terminology of metal–organic frameworks and coordination polymers (IUPAC recommendations 2013), Pure Appl. Chem. 85 (2013) 1715–1924.

[3] H. Furukawa, K.E. Cordova, M.O. Keeffe, O.M. Yaghi, The chemistry and applications of metal-organic frameworks, Science 341 (2013) 1230444.

[4] M. Eddaoudi, D.B. Moler, H. Li, B. Chen, T.M. Reineke, M.O. Keeffe, O.M. Yaghi, Modular chemistry: secondary building units as a basis for the design of highly porous and robust metal−organic carboxylate frameworks, Acc. Chem. Res. 34 (2001) 319–330.

[5] S. Kitagawa, R. Kitaura, S.-I. Noro, Functional porous coordination polymers, Angew. Chem. Int. Ed. 43 (2004) 2334–2375.

[6] G. Férey, Hybrid porous solids: past, present, future, Chem. Soc. Rev. 37 (2008) 191–214.

[7] O.M. Yaghi, Reticular chemistry in all dimensions, Mol. Front. J. 3 (2019) 66–83.

[8] H. Li, M. Eddaoudi, M.O. Keeffe, O.M. Yaghi, Design and synthesis of an exceptionally stable and highly porous metal-organic framework, Nature 402 (1999) 276–279.

[9] S.S.-Y. Chui, S.M.-F. Lo, J.P.H. Charmant, A.G. Orpen, I.D. Williams, A chemically functionalizable nanoporous material $[Cu_3(TMA)_2(H_2O)_3]n$, Science 283 (1999) 1148–1150.
[10] G. Férey, C. Mellot-Draznieks, C. Serre, F. Millange, J. Dutour, S. Surblé, et al., A chromium terephthalate-based solid with unusually large pore volumes and surface area, Science 309 (2005) 2040–2042.
[11] N.L. Rosi, J. Kim, M. Eddaoudi, B. Chen, M.O. Keeffe, O.M. Yaghi, Rod packings and metal−organic frameworks constructed from rod-shaped secondary building units, J. Am. Chem. Soc. 127 (2005) 1504–1518.
[12] S.R. Caskey, A.G. Wong-Foy, A.J. Matzger, Dramatic tuning of carbon dioxide uptake via metal substitution in a coordination polymer with cylindrical pores, J. Am. Chem. Soc. 130 (2008) 10870–10871.
[13] S. Lee, E.A. Kapustin, O.M. Yaghi, Coordinative alignment of molecules in chiral metal-organic frameworks. Science 353 (2016) 808–811.
[14] T. Islamoglu, S. Goswami, Z. Li, A.J. Howarth, O.K. Farha, J.T. Hupp, Postsynthetic tuning of metal-organic frameworks for targeted applications, Acc. Chem. Res. 50 (2017) 805–813.
[15] J.-P. Zhao, B.-W. Hu, Q. Yang, T.-L. Hu, X.-H. Bu, On the nature of the reversibility of hydration–dehydration on the crystal structure and magnetism of the ferrimagnet $[Mn^{II}(enH)(H_2O)][Cr^{III}(CN)_6]\cdot H_2O$, Inorg. Chem. 48 (2009) 7111–7116.
[16] S. Yuan, Y.-P. Chen, J.-S. Qin, W. Lu, L. Zou, Q. Zhang, et al., Linker installation: engineering pore environment with precisely placed functionalities in zirconium MOFs, J. Am. Chem. Soc. 138 (2016) 8912–8919.
[17] C.-X. Chen, Z.-W. Wei, J.-J. Jiang, S.-P. Zheng, H.-P. Wang, Q.-F. Qiu, et al., Dynamic spacer installation for multirole metal–organic frameworks: a new direction toward multifunctional MOFs achieving ultrahigh methane storage working capacity, J. Am. Chem. Soc. 139 (2017) 6034–6037.
[18] J. Pang, S. Yuan, J.-S. Qin, C.T. Lollar, N. Huang, J. Li, et al., Tuning the ionicity of stable metal–organic frameworks through ionic linker installation, J. Am. Chem. Soc. 141 (2019) 3129–3136.
[19] O.M. Yaghi, M.O. Keeffe, N.W. Ockwig, H.K. Chae, M. Eddaoudi, J. Kim, Reticular synthesis and the design of new materials, Nature 423 (2003) 705–714.
[20] B.F. Hoskins, R. Robson, Design and construction of a new class of scaffolding-like materials comprising infinite polymeric frameworks of 3D-linked molecular rods. A reappraisal of the zinc cyanide and cadmium cyanide structures and the synthesis and structure of the diamond-related frameworks $[N(CH_3)_4][Cu^{I}Zn^{II}(CN)_4]$ and CuI[4,4′,4″,4‴-tetracyanotetraphenylmethane]$BF_4.xC_6H_5NO_2$, J. Am. Chem. Soc. 112 (1990) 1546–1554.
[21] S.R. Venna, M.A. Carreon, Metal organic framework membranes for carbon dioxide separation, Chem. Eng. Sci. 124 (2015) 3–19.
[22] Q. Pang, B. Tu, Q. Li, New linker installation in metal–organic frameworks, Dalton Trans. 48 (2019) 12000–12008.
[23] Y. Bai, Y. Dou, L.-H. Xie, W. Rutledge, J.-R. Li, H.-C. Zhou, Zr-based metal–organic frameworks: design, synthesis, structure, and applications, Chem. Soc. Rev. 45 (2016) 2327–2367.
[24] Y. Zhang, X. Zhang, J. Lyu, K.-I. Otake, X. Wang, L.R. Redfern, et al., A flexible metal–organic framework with 4-connected Zr_6 nodes, J. Am. Chem. Soc. 140 (2018) 11179–11183.
[25] Z. Chen, S.L. Hanna, L.R. Redfern, D. Alezi, T. Islamoglu, O.K. Farha, Reticular chemistry in the rational synthesis of functional zirconium cluster-based MOFs, Coord. Chem. Rev. 386 (2019) 32–49.

[26] H. Furukawa, J. Kim, N.W. Ockwig, M.O. Keeffe, O.M. Yaghi, Control of vertex geometry, structure dimensionality, functionality, and pore metrics in the reticular synthesis of crystalline metal–organic frameworks and polyhedra, J. Am. Chem. Soc. 130 (2008) 11650–11661.

[27] M.-H. Choi, H.J. Park, D.H. Hong, M.P. Suh, Comparison of gas sorption properties of neutral and anionic metal–organic frameworks prepared from the same building blocks but in different solvent systems, Chem. Eur. J. 19 (2013) 17432–17438.

[28] H.K. Chae, D.Y. Siberio-Perez, J. Kim, Y. Go, M. Eddaoudi, A.J. Matzger, et al., A route to high surface area, porosity and inclusion of large molecules in crystals, Nature 427 (2004) 523–527.

[29] G. Férey, C. Serre, C. Mellot-Draznieks, F. Millange, S. Surblé, J. Dutour, et al., A hybrid solid with giant pores prepared by a combination of targeted chemistry, simulation, and powder diffraction, Angew. Chem. Int. Ed. 43 (2004) 6296–6301.

[30] P. Horcajada, S. Surble, C. Serre, D.-Y. Hong, Y.-K. Seo, J.-S. Chang, et al., Synthesis and catalytic properties of MIL-100(Fe), an iron(iii) carboxylate with large pores, Chem. Commun. 27 (2007) 2820–2822.

[31] C. Volkringer, D. Popov, T. Loiseau, G.R. Férey, M. Burghammer, C. Riekel, et al., Stability of metal-organic frameworks under gamma irradiation, Chem. Mater. 21 (2009) 5695–5697.

[32] A. Lieb, H. Leclerc, T. Devic, C. Serre, I. Margiolaki, F. Mahjoubi, et al., A mesoporous vanadium metal organic framework with accessible metal sites, Micropor. Mesopor. Mater. 157 (2012) 18–23.

[33] M. Li, D. Li, M.O. Keeffe, O.M. Yaghi, Topological analysis of metal–organic frameworks with polytopic linkers and/or multiple building units and the minimal transitivity principle, Chem. Rev. 114 (2014) 1343–1370.

[34] A.G. Wong-Foy, O. Lebel, A.J. Matzger, Porous crystal derived from a tricarboxylate linker with two distinct binding motifs, J. Am. Chem. Soc. 129 (2007) 15740–15741.

[35] J. Duan, Z. Yang, J. Bai, B. Zheng, Y. Li, S. Li, Highly selective CO_2 capture of an agw-type metal–organic framework with inserted amides: experimental and theoretical studies, Chem. Commun. 48 (2012) 3058–3060.

[36] F. Nouar, J.F. Eubank, T. Bousquet, L. Wojtas, M.J. Zaworotko, M. Eddaoudi, Supermolecular building blocks (SBBs) for the design and synthesis of highly porous metal-organic frameworks, J. Am. Chem. Soc. 130 (2008) 1833–1835.

[37] Y. Yan, X. Lin, S. Yang, A.J. Blake, A. Dailly, N.R. Champness, et al., Exceptionally high H_2 storage by a metal–organic polyhedral framework, Chem. Commun. 9 (2009) 1025–1027.

[38] X. Song, S. Jeong, D. Kim, M.S. Lah, Transmetalations in two metal–organic frameworks with different framework flexibilities: kinetics and core–shell heterostructure, CrystEngComm 14 (2012) 5753–5756.

[39] Z. Guo, H. Wu, G. Srinivas, Y. Zhou, S. Xiang, Z. Chen, et al., A metal–organic framework with optimized open metal sites and pore spaces for high methane storage at room temperature, Angew. Chem. Int. Ed. 50 (2011) 3178–3181.

[40] X. Zhao, X. Wang, S. Wang, J. Dou, P. Cui, Z. Chen, et al., Novel metal–organic framework based on cubic and trisoctahedral supermolecular building blocks: topological analysis and photoluminescent property, Cryst. Growth Des. 12 (2012) 2736–2739.

[41] L. Li, S. Tang, X. Lv, M. Jiang, C. Wang, X. Zhao, An rht type metal–organic framework based on small cubicuboctahedron supermolecular building blocks and its gas adsorption properties, New J. Chem. 37 (2013) 3662–3670.

[42] Z. Chen, S. Xiang, T. Liao, Y. Yang, Y.-S. Chen, Y. Zhou, et al., A new multidentate hexacarboxylic acid for the construction of porous metal–organic frameworks of diverse structures and porosities, Cryst. Growth Des. 10 (2010) 2775–2779.

[43] Y. He, Z. Zhang, S. Xiang, F.R. Fronczek, R. Krishna, B. Chen, A microporous metal–organic framework for highly selective separation of acetylene, ethylene, and ethane from methane at room temperature, Chem Eur. J. 18 (2012) 613–619.

[44] Y. He, Z. Zhang, S. Xiang, H. Wu, F.R. Fronczek, W. Zhou, et al., High separation capacity and selectivity of C_2 hydrocarbons over methane within a microporous metal–organic framework at room temperature, Chem Eur. J. 18 (2012) 1901–1904.

[45] Z. Wei, W. Lu, H.-L. Jiang, H.-C. Zhou, a route to metal–organic frameworks through framework templating, Inorg. Chem. 52 (2013) 1164–1166.

[46] Y.-S. Xue, F.-Y. Jin, L. Zhou, M.-P. Liu, Y. Xu, H.-B. Du, et al., Structural diversity and properties of coordination polymers built from a rigid octadentenate carboxylic acid, Cryst. Growth Des. 12 (2012) 6158–6164.

[47] N.B. Shustova, A.F. Cozzolino, M. Dinca, Conformational locking by design: relating strain energy with luminescence and stability in rigid metal–organic frameworks, J. Am. Chem. Soc. 134 (2012) 19596–19599.

[48] H. Deng, C.J. Doonan, H. Furukawa, R.B. Ferreira, J. Towne, C.B. Knobler, et al., Multiple functional groups of varying ratios in metal-organic frameworks, Science 327 (2010) 846–850.

[49] T.-H. Park, K. Koh, A.G. Wong-Foy, A.J. Matzger, Nonlinear properties in coordination copolymers derived from randomly mixed ligands, Cryst. Growth Des. 11 (2011) 2059–2063.

[50] R.D. Hancock, A.E. Martell, Ligand design for selective complexation of metal ions in aqueous solution, Chem. Rev. 89 (1989) 1875–1914.

[51] J. Seo, R. Matsuda, H. Sakamoto, C. Bonneau, S. Kitagawa, A pillared-layer coordination polymer with a rotatable pillar acting as a molecular gate for guest molecules, J. Am. Chem. Soc. 131 (2009) 12792–12800.

[52] B. Wang, A.P. Cote, H. Furukawa, M.O. Keeffe, O.M. Yaghi, Colossal cages in zeolitic imidazolate frameworks as selective carbon dioxide reservoirs, Nature 453 (2008) 207–211.

[53] W. Morris, C.J. Doonan, H. Furukawa, R. Banerjee, O.M. Yaghi, Crystals as molecules: postsynthesis covalent functionalization of zeolitic imidazolate frameworks, J. Am. Chem. Soc. 130 (2008) 12626–12627.

[54] V. Colombo, S. Galli, H.J. Choi, G.D. Han, A. Maspero, G. Palmisano, et al., High thermal and chemical stability in pyrazolate-bridged metal–organic frameworks with exposed metal sites, Chem. Sci. 2 (2011) 1311–1319.

[55] K. Sumida, D. Stück, L. Mino, J.-D. Chai, E.D. Bloch, O. Zavorotynska, et al., Impact of metal and anion substitutions on the hydrogen storage properties of m-btt metal–organic frameworks, J. Am. Chem. Soc. 135 (2012) 1083–1091.

[56] M. Dincă, A. Dailly, J.R. Long, Structure and charge control in metal–organic frameworks based on the tetrahedral ligand tetrakis(4-tetrazolylphenyl)methane, Chem. Eur. J. 14 (2008) 10280–10285.

Chapter 4

Microwave-assisted synthesis of metal–organic frameworks

SK Khalid Rahaman [a], Taposi Chatterjee [a,b] and Seikh Mafiz Alam [a]

[a] *Department of Chemistry, Aliah University, Kolkata, West Bengal, India,* [b] *Department of Basic Science & Humanities, Techno International New Town, Kolkata, West Bengal, India*

4.1 Introduction

Porous coordination networks (PCNs) or porous coordination polymers (PCPs) are the new domain of porous crystalline solid materials, in the field of crystal engineering, comprising of metal ions or clusters and organic linkers, popularly known as metal–organic frameworks (MOFs) [1–3]. On the other hand, zeolites happen to be microporous, aluminosilicate network compounds commonly used as commercial adsorbents and catalysts. There exist stark similarities between MOF as well as zeolites in terms of the pore size, pore structure, and coordination space. For the last two decades, inorganic synthetic chemists have been focusing day in and day out on the synthesis of MOFs because of their remarkable topological designs and diverse applications. In terms of shape, functionality, flexibility, and symmetry, the structural features of the synthesized compounds can be modified by an incisive choice of metal-containing secondary building units and bridging organic linkers [4–6]. MOFs have appealed extensive attention over traditional zeolites due to their chemical and thermal toughness, tunable structure, ultra-high porosity, and functionality. It has diverse applications in many fields, including gas storage [7–9], gas separation [10], energy storage and conversion [11,12], drug delivery [13], chemical sensors [14,15], catalysis [16–21], optical luminescence [22,23], light harvesting [24], and so on.

The most widely explored synthetic routes for MOF synthesis are conventional solvothermal, hydrothermal methods, and slow diffusion methods [25,27,26]. Thermal stability remains the most important criterion for MOF synthesis. Existing hydrothermal or solvothermal reaction techniques require prolonged reaction time up to several days. To reduce the reaction time without compromising the quality of MOFs, alternative techniques are highly desirable. Microwave irradiation finds itself to provide rapid, inexpensive, and commercially viable routes toward the construction of these compounds to enable their use in various applications. The reaction time of MOFs synthesis is reduced

Synthesis of Metal–Organic Frameworks via Water-Based Routes: A Green and Sustainable Approach.
DOI: https://doi.org/10.1016/B978-0-323-95939-1.00009-5

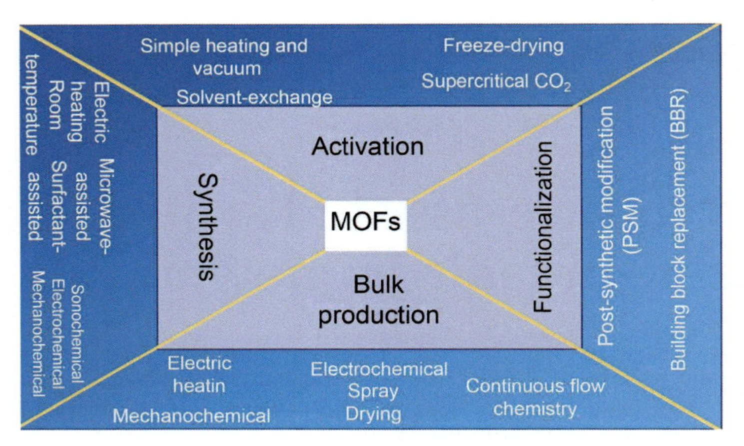

FIGURE 4.1 Schematic illustrations of fundamental chemistry of metal–organic frameworks.

to a few minutes by microwave irradiation hydrothermal method because of fast kinetics, rapid and homogeneous heating of the inserted reaction mixture (Fig. 4.1) [28–32].

4.2 Historical developments of MOFs

With the advancements of technologies and industrial applications in several parts of day-to-day life, the development of porous materials is an obvious choice. Due to the high surface area and a huge fraction of void space with low density, researchers can add much functionality based on the anticipated applications. The synthesis of MOF can be recognized as a young field of modern science and technology. It was derived from the area of coordination and solid-state or zeolite chemistry. There has been a long journey from coordination chemistry to coordination polymers which were designed by joining metal ions as connectors and organic ligands as linkers. Synthetic zeolites have been extensively studied since the 1940s; consist of exclusively inorganic components such as silicates and aluminates.

In the late 1980s and early 1990s, the first crystalline porous materials with pore sizes greater than 1 nm and 2 nm were synthesized and reported respectively [33,34]. In 1989 and 1990, Hoskins and Robson, in their work, set the future foundation of MOF [35,36].

It was Yaghi and Li in 1995 [37], who introduced MOFs, a unique class of crystalline porous materials with large surface areas and enduring porosity due to the strong bonds between metal ions and charged organic ligands. The newly invented inorganic–organic hybrid materials with desirable properties unlocked a new path in the design and application of porous materials. In 1997, Kitagawa et al. [26] reported the first three-dimensional MOF that exhibited gas sorption properties at room temperature under high pressure.

In 1998, Yaghi reported the first MOF showing permanent porosity with nitrogen adsorption/desorption isotherms at 77 K and low pressure: a procedure that has been employed before for probing permanent porosity in other porous materials such as zeolites, porous silica, and porous carbon—and derived the first values of apparent surface area and pore volume in a MOF. In their early work, Yaghi and Li were succeed only to exchange or removal of guest molecules which supports the presence of porosity in MOFs. But in the later work, they got huge success by eventually measuring the gas adsorption by MOFs in the desolvated state. This work is a landmark in the field of producing porous MOFs and laid the basis for achieving permanent porosity in MOFs.

Two most studied MOFs, HKUST-1 and MOF-5, were synthesized in the year 1999. Non-flexible as well as flexible porous MOFs, i.e., MIL-47 and MIL-53 or MIL-88 respectively, were reported by Ferey et al. in 2002 [38,39]. The concept of reticular chemistry, (from the Latin "reticulum" meaning "having the form of a net" or "netlike"), established by Yaghi (2002), is the study of linking distinct constructing units (molecules and clusters) by strong bonds to make large and extended crystalline structures. Under the monarchy of reticular chemistry, definitely, MOFs are the most prominent class of materials. Inorganic polynuclear clusters [termed secondary building units (SBUs)] and organic linkers are connected by strong bonds which extend the crystalline structures. Reticular chemistry was popularly established after the production of Zn dicarboxylates and the idea was extended to other materials too. Specifically, the mixed-linker compounds $[M_2(\text{dicarboxylate})_2(\text{diamine})]$ (M = Zn, Cu) have shown to be a flexible class as the choice of two organic components have no limits. The first successful working model using mixed linkers was reported in 2001. The kingdom of MOFs was extended to imidazolate-based compounds which are currently recognized as zeolitic imidazole frameworks (ZIFs) after 2001. Over 100,000 structures have been reported in the "MOF subset" of the Cambridge Structural Database (CSD) in the past few decades and the research publications have been continuously increasing worldwide in this field (Fig. 4.2) [40,41].

4.3 Conventional synthesis of MOFs

Methodologies adopted for the synthesis of MOF gradually developed in three phases. Slow evaporation method emerged as the foremost stage. It takes almost weeks to months to prepare large single crystal in minute quantities. In the second phase, solvothermal method was implemented. It reduces the reaction time to few days but produce microcrystalline powders, from where separation of single crystal was a challenging task. It was obvious that researcher hunted alternate approach to reduce the reaction time further to few hours, prepare multifunctional materials in larger amounts and also wanted to increase reaction yields. Microwave assisted synthesis [42,43], sonochemical synthesis [44,45], electrochemical synthesis [46,47], mechanochemical synthesis [48], etc. stepped in the third place as effective synthetic routes for MOF synthesis. The techniques

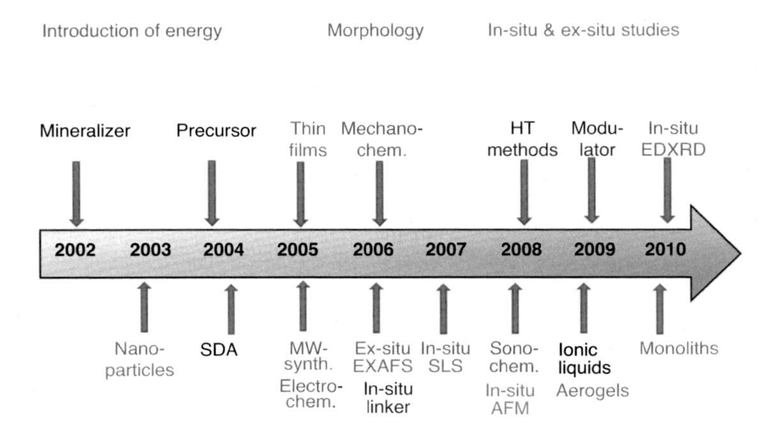

FIGURE 4.2 The developments of different synthetic routes are summarized on a time line.

used in the first two phases can be categorized under the domain of conventional synthesis, where the last one comes under the spectrum of non-conventional synthesis.

The reactions executed by conventional electric heating without any parallelization of reactions are usually termed as conventional synthesis. The reaction temperature is one of the key factors in the synthesis of MOFs. Solvothermal and non-solvothermal processes use different ranges of temperatures. The general definition of solvothermal reactions has yet not been established but Rabenau proposed one definition-reactions taking place above the boiling point of the solvent in closed vessels under autogenous pressure [49]. Contrarily, non-solvothermal reactions are performed below or at the boiling point under ambient pressure, which provides these techniques simple and viable. These can be classified as (1) reactions performed under room temperature and (2) reactions performed under high temperature. Hoskins and Robson's initial work mainly follow the low-temperature routes. Thus, slow evaporation of the solvent or precipitation reaction followed by recrystallization was reported to create simple molecular or ionic crystals by changing the reaction conditions, i.e., the rate of nucleation and crystal growth. By evaporating the solvent or by changing the temperature, the concentration of the reactants is controlled properly so that the critical nucleation concentration can be overcome. It is the basis to grow crystals from clear solutions. When particles surpassing the critical radius are produced, crystal growth takes place. The concentration gradient plays an important role in MOFs formation. Various methods and techniques like slow diffusion of reactants into each other or layering of solutions, and evaporation of solvent of a solution of reactants are adopted to control concentration gradients. By applying a temperature gradient or slow cooling of the reaction mixture, i.e., using temperature as a variable, one can also adjust the concentration gradients

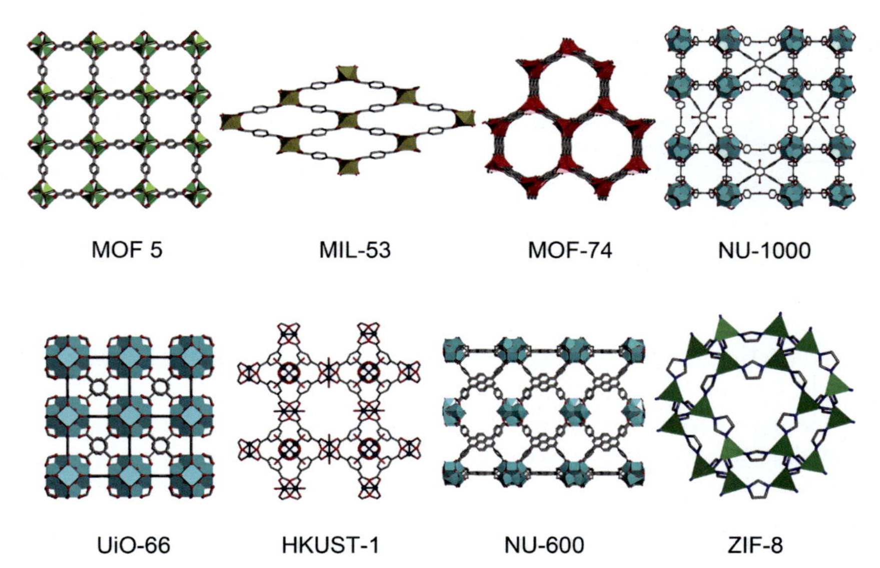

FIGURE 4.3 Representative metal–organic framework structures.

which form big crystals fit for structural elucidation. MOF-5, HKUST-1, MOF-74, or ZIF-8, etc. were the renowned and extensively studied MOFs that were synthesized at room temperature by just facile mixing of the starting materials [50,51]. Kinetically inert ions may require high reaction temperature in order to attain suitable crystallinity and reaction rates. Unfortunately, solvothermal synthesis for the synthesis of MOF comes up with many shortcomings: (1) it requires a longer reaction time (week to months), (2) higher energy consumption, (3) possible degradation of the MOF due to prolonged reaction times, and (4) it exploits heavy instrumentation. Several alternative techniques like electrochemical essays [52] and solvothermal synthesis with immiscible solvents [53] have come into the segment to overcome such drawbacks. Microwave-assisted synthesis [54] finds itself as a new, more promising technique as compared to its counterparts (Fig. 4.3).

4.4 Microwave-assisted synthesis of MOFs

The microwave-assisted synthesis techniques have been extensively applied in organic chemistry. In the last few decades, it has been widely used for the synthesis of inorganic nanoporous materials-zeolites, and MOFs, under hydrothermal conditions [55–57]. It can be recognized as a highly energy-efficient heating method that uses microwave electromagnetic radiation (EMR).

In conventional solvothermal methods, heat is transmitted from the source to the solution through the reaction container. Contrarily, in MW synthesis, the electromagnetic irradiation intermingles directly with the reactants present

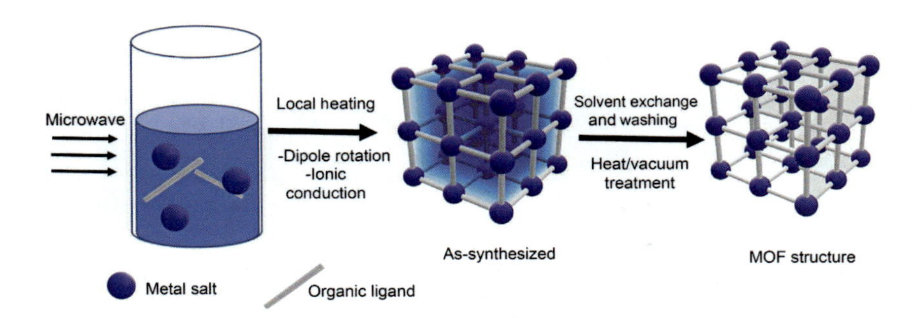

FIGURE 4.4 Microwave-assisted solvothermal synthesis of metal–organic framework structures.

inside the container thereby producing more efficient and faster heating. The underlying theory of this method is the interaction of electromagnetic waves with any material containing mobile electric charges, such as polar or ionic molecules present in a solvent or conducting ions in a solid. In the solid phase, an electric current is produced by electromagnetic radiation, and thermal energy is formed due to the electric resistance of the solid. The heat production mechanism is somewhat different for the solution phase. Polar molecules (e.g., water, ethanol, etc.) try to align themselves by the influence of an electromagnetic field and an oscillating field so that the molecules change their alignments permanently. By applying the suitable frequency, the oscillating molecules will collide with each other, which produces high kinetic energy, i.e., the temperature of the system will increase. Direct interaction of the radiation with the solution or reactants in MW-assisted heating makes itself a low energy consumption method of heating. In order to achieve homogeneous heating throughout the sample and high heating rates, this technique is highly acceptable. This synthesis method also allows monitoring of pressure and temperature during the reaction which confirms more precise control of reaction conditions. Fast crystallization, significant phase selectivity, precise control over the crystal morphology and size ranging from microcrystals to nanoparticles can be attained with high yield by MW-assisted synthesis (Fig. 4.4).

We will now discuss in detail the supremacy of MW-assisted synthesis over other conventional techniques used so far in the field of MOFs synthesis. Our discussion mainly focuses on: (1) the quantitative degree of acceleration in the rapid synthesis of MOFs, (2) the phase-selective synthesis of MOFs, and (3) production of nano-sized or smaller crystals with MW with larger quantities.

4.4.1 Faster Synthesis

Cr-MIL-100 (chromium-benzenetricarboxylate) was the first reported MOF which was synthesized by microwave radiation [58]. The synthesis time was 4 hours and the reaction temperature was 220°C which produce a 44% yield. In conventional hydrothermal synthesis, it took 4 days at 220°C to synthesize the MOF. Microwave accelerated the reaction by almost 20 times compared with the conventional electric (CE) heating at the same temperature. The rapid synthesis of Cr-MIL-100 via MW was elucidated by the combination of the (1) fast dissolution of the precursors, especially metallic chromium, and (2) acceleration of the condensation of the oxygen–metal networks. Faster dissolution of less soluble metal and organic precursors through the careful heating of dipolar species by MW irradiation, reduced the nucleation period.

After this successful attempt, this method was further implemented to produce Cr-MIL-101(chromium benzenedicarboxylate) which took less than 60 minutes at 210°C. The crystallization starts in the first minute of the reaction. The reported material has comparable physicochemical and textural properties compared with the standard material synthesized using the conventional electrical heating method [42]. It indicated high and rapid adsorption of benzene. MOF-5 (zinc benzenedicarboxylate) was reported using MW irradiation and the effect of irradiation time on the crystal growth was investigated by Ni and Masel [59]. The reaction only took 25 seconds with MW which confirms very fast kinetics. Though, conventional hydrothermal synthesis by conventional electric heating took 7–24 hours, as reported [60]. Excitingly, no crystal of MOF-5 was detected within first 20 seconds, though, the crystal growth occurred in just 5 seconds, i.e., the synthesis was completed in 25 seconds. The very fast crystal growth via MW can be explained as the construction of hot spots by the superheating of the solvent which was the basis of rapid crystallization throughout the bulk solution. On the other hand, crystal growth via the conventional solvothermal method took a long time as the creation of crystals took place near the walls or on dust particles, which were very few seeds. Synthesis of Fe-MIL-53 (iron benzenedicarboxylate) was studied by microwave-assisted synthesis and conventional electric heating at various reaction temperatures, and it was clearly observed that the rates of nucleation and crystal growth were higher in MW-assisted synthesis than conventional electric (CE) heating [61]. The MW-assisted enhancement was discussed in terms of the kinetic parameters by Haque et al. for Fe-MIL-53 [62] (Table 4.1).

To calculate activation energy (E_a) and pre-exponential factor (A), the Arrhenius equation (Eq. 4.1) was used.

Arrhenius equation:

$$k = A.e^{-(E_a/RT)} \tag{4.1}$$

$$\ln k = \ln A - E_a/RT \tag{4.2}$$

where k = kinetic constant, A = pre-exponential factor, E_a = activation energy, R = gas constant, T = absolute temperature.

The E_a and A were derived from the slope and intercept, respectively, of the logarithm of the Arrhenius equation (Eq. 4.2). From the equation, it is obvious that the reaction rates should be increased in the order of MW $<$ CE if only the E_a value is taken into consideration and taking A values as constant in all

TABLE 4.1 Nucleation and crystal growth rates with two synthesis methods at various temperatures and calculated pre-exponential factors (*A*) and activation energies (E_a) for the synthesis of Fe-MIL-53.

Synthesis method	Temperature (°C)	Nucleation time (min)	Nucleation rate (/min)[b]	*A* for nucleation (/min)[a]	E_a for nucleation (kJ/mol)[a]	Crystal growth rate (/min)[c]	*A* for crystal growth (/min)[a]	E_a for crystal growth (kJ/mol)[a]
MW	60	44	2.27×10^{-2}	1.19×10^{10}	74.8	8.01×10^{-3}	1.23×10^{12}	90.6
	65	32	3.13×10^{-2}			1.11×10^{-2}		
	70	20	5.00×10^{-2}			2.08×10^{-2}		
CE	70	310	3.23×10^{-3}	3.05×10^{3}	39.2	3.78×10^{-4}	4.78×10^{6}	66.4
	75	250	4.00×10^{-3}			5.16×10^{-4}		
	80	210	4.76×10^{-3}			7.31×10^{-4}		

[a] *Pre-exponential factors (A) and activation energies (E_a) are calculated from rates at three temperatures.*
[b] *Calculated from the 1/(nucleation time).*
[c] *Calculated from the slope of a crystallization curve (between 20% and 80% crystallinity).*

cases. Though, the overall rates decreased in the opposite order (MW ≫ CE). Consequently, it can be concluded that the increased reaction rate of MW over CE was due to an increased A rather than a decreased E_a.

HKUST-1, one of the highly investigated MOF, was also synthesized very fast by Microwave-assisted heating in the range of 10–20 μm in high yields (~90%) within 1 hour [63]. Fe-MIL-101-NH_2, IRMOF-3 (H2BDC-NH_2) [64], and ZIF-8 (HMeIm) [65] were the well-known MOFs which were produced using microwave-assisted synthesis methods.

4.4.2 Phase-selective synthesis

By changing external reaction parameters such as reaction time, temperature, etc., materials with different structures, physical properties, and chemical compositions from the same reaction component has been reported. Phase selective synthesis can be defined as the selective synthesis of a particular phase of products with precise and characteristic crystal structures from the same reaction components. Specific phases of crystals were synthesized and reported by microwave assisted synthesis, on contrary, conventional heating would produce different or other phases under the same reaction conditions [66–70]. Less stable phase (kinetically favorable phase) crystals can be prepared by MW irradiation due to the extremely fast crystallization process. But in conventional electric heating, the less stable phase is tough to synthesize because of the interconversion of the less stable phase into the more stable or dense phases. Microwave irradiation has vast applications in the phase-selective synthesis of porous materials like MOFs. Prolonged microwave irradiation creates comparatively more stable structures due to conversion of a less stable phase into more stable phase [71,72]. To understand the phase selectivity and phase-transitions, two highly porous MOFs Cr-MIL-53 and Cr-MIL-101 were taken into consideration and crystallization has been explored. Kinetically favorable Cr-MIL-101 (lower-density phase) was obtained during the initial stage of the reaction, while the thermodynamically favorable Cr-MIL-53 phase (higher density phase) was obtained at the expenditure of Cr-MIL-101 with longer reaction times. In opposition, preventing the conversion into the thermodynamically favorable phase Cr-MIL-53, under parallel conditions, MW heating provided the selective synthesis of Cr-MIL-101 with comparatively higher yield. Very fast kinetics of nucleation and crystallization under MW heating, is the main reason behind the selective formation of the kinetically favored Cr-MIL-101.A coordination network, which could only be separated by MW irradiation from $Cu(HCO_2)_2 \cdot xH_2O$ and 2,2-dipyridyldisulfide, was reported by Delgado and Zamora [73].

2,2-dipyridyldisulfide partially oxidized to sulfate over MW heating which produces three different coordination complexes having Cu(I), Cu(I/II), and Cu(II) as the central metal ions. Maniam and co-workers reported that the selective synthesis of $Ni_2(BDC)_2(DABCO) \cdot (DMF)_4(H_2O)_4$, one of the polymorphs of the pillared-layered $Ni_2(BDC)_2(DABCO)$ framework, was achieved due to the fast nucleation achievable with MW irradiation [74]. Three zirconium oxide-based MOFs (MIL-140A, MIL-140B, and MIL-140A-NH_2 with 1,4-benzenedicarboxylate, 2,6-naphthalenedicarboxylate and 2-amino-1,4-benzenedicarboxylate linkers, respectively) was carefully studied by Liang et al. using MW and CE heating. In this comparative study, they observed that more crystalline MIL-140A was produced by MW heating in an efficient manner. Contrarily, along with MIL-140A, the construction of another MOF (UiO-66 or zirconium-benzene dicarboxylate) as an impurity, was obtained by CE heating. The improved phase purity and crystallinity of MIL-140A by MW-assisted synthesis were because of the very fast kinetics of MW irradiation. Additionally, MIL-140A-NH_2 (analogous to MIL-140A) was selectively isolated with irradiation which was impossible by CE heating.

It was interestingly observed that by changing the reaction parameters (like reactant concentration, reactant composition, reaction time, etc.) in CE heating, only UiO-66-NH_2 could be obtained rather than MIL-140A-NH_2. Jhung et al. [67] synthesized hydrothermally two porous MOFs (nickel glutarates) with analogous physicochemical properties under various conditions by microwave irradiation and CE heating. CE heating took several hours or days for the synthesis of Cubic nickel glutarate, $[Ni_{20}(C_5H_6O_4)_{20}(H_2O)_8]\cdot 40H_2O$ where similar phase (cubic) was observed by MW heating in the initial stage of the reaction, and finally recrystallized to a more stable new phase of nickel glutarate $[Ni_{22}(C_5H_6O_4)_{20}(OH)_4(H_2O)_{10}]\cdot 38H_2O$ with a tetragonal structure. Hence, we can conclude that MW heating can serve as a better phase-selective synthetic route for less stable or kinetically controlled phases in the production of MOFs from the same reactant mixtures at the same reaction temperatures.

4.4.3 Crystal size reduction

Due to rapid and homogenous heating of the reaction mixture by microwave irradiation, small crystals can be synthesized. The size of a produced material can easily be adjusted as per the requirements by took control over nucleation or crystal growth rate and nanocrystals are obtained when the nucleation rate is high [75–77]. The key strength of MW is uniform superheating which creates hot spots, thereby improving the dissolution of reactants within a very small timescale and leading to the creation of nuclei in high concentration. Ultimately, small particles are usually found and particle sizes are homogeneous. On the other hand, large crystals in large amounts of industrial scale can be prepared if a sufficient amount of time is allowed for crystallization.

The size and shape are the two important parameters of porous materials like MOFs in different applications such as drug delivery, separations, catalysis, imaging contrast agents, and membranes [78–82]. It is observed that nanoparticles of MOFs are more efficient in the field of drug delivery and membrane. Thus, the synthesis of nanosized MOFs are very important. The assurance of a single nucleation event is very important for the synthesis of uniform nanocrystals but extra nucleation during the successive crystal growth process should be avoided [83]. Nucleation or crystal growth steps are mainly controlled by reaction temperature and heating method and therefore, govern the size and shape of the final product.

The synthesis of small MOF-5 crystals from zinc and 1,4-benzene-dicarboxylic acid using MW irradiation was reported by Ni et al. MOF-5 crystals in the range of sub-micrometer scale were detected with MW heating, whereas, normal CE heating yields micrometer scale crystals [84]. CE synthesis produced 50–200 μm-sized crystals, whereas, MW-assisted solvothermal synthesis gave rise to the formation of 5–20 μm-sized crystals with a narrow size distribution and relatively uniform shape. Similarly, the synthesis of Cu-BTC crystals via MW irradiation was reported where crystals with much smaller dimensions were created compared with conventional heating [85].

4.5 Factors affecting MOF synthesis by MW irradiation

MOF-5, IRMOF-1, IRMOF-2 and IRMOF-3, $[Cu_3(BTC)_2(H_2O)_3]$ and $[Cu_2(OH)(BTC)(H_2O)]\cdot 2nH_2O$ [63] are the most studied porous materials and suitable for the specific inspection of morphology, crystallinity, particle size and the design of new applications. Modifications can be incorporated as the synthetic routes are already known for the above-mentioned material which can lead to open new possibilities for functionalization. Now, we will discuss in detail about the factors which affect the reaction conditions in MW-assisted synthesis of MOFs.

4.5.1 Temperature and reaction time

Reaction conditions in microwave irradiation differ considerably with respect to (1) the apparatus used, (2) the MOF crystal under investigation, and (3) the desires of the particular research group involved in the synthesis. Variations can be made in the specific parameters like the composition of the reactive mixtures, temperatures, irradiation power, and time of the reaction but the other parameters are precise to the system under examination. Classically, the synthesis is mainly executed above 100°C and the period of reaction is methodically varied in order to get the best conditions for the isolation of the desired product. The power of irradiation is the central instrumental parameter in microwave heating but major focus is given on how it is created, delivered, and controlled throughout the reaction. It mainly depends on the device engaged. The experimental set-up is the core reason for the different conclusions made as reported recently by Schlesinger et al. [63] and Seo et al. [64] in the formation of HKUST. The best-reported reaction conditions are tabulated in Table 4.2. Overall, it is noticed that the reactions occur between a few seconds to minutes, rarely exceeding 1 hour.

The crystallinity and morphology of MOF-5 was examined by Choi et al. by changing the temperature, power level, irradiation time, solvent concentration, and substrate composition by comparing it with the conventional heating. The reaction time varied between 10 minutes and 60 minutes, the temperature of

TABLE 4.2 Reaction conditions for selected microwave-assisted syntheses of metal–organic frameworks.

MOFs	Reaction conditions	References
IRMOF1, IRMOF2, IRMOF3	150 W, 25 s	[59]
$[Cu_3(BTC)_2(H_2O)_3][Cu_2(OH)(BTC)(H_2O)]\cdot 2nH_2O$	140°C, 60 min	[64]
$[Cu(H_2BTC)_2(H_2O)_2]\cdot 3H_2O$	170°C, 10 min	[64]
$[Ni_{20}(C_5H_6O_4)_{20}(H_2O)_8]\cdot 40H_2O$ $[Ni_{22}(C_5H_6O_4)_{20}(OH)_4(H_2O)_{10}]\cdot 38H_2O$	150–220°C within 1 min	[67]
$[Cd(HIDC)(bbi)_{0.5}]$	2.0 Mpa, 20 min	[86]
Cr-MIL-101	210°C, 1–40 min	[42]
$[Mn_3(BTC)_2(H_2O)]_6$	120°C, 10 min	[87]
$[Zn_2(NDC)_2(DPNI)]$	120°C, 1 h	[88]
$[Cu_2(pyz)_2(SO_4)(H_2O)_2]_n$	180°C, 20 min, 6 h	[89]
$[Co_3(NDC)_3(DMF)_4]$ and $[Mn_3(NDC)_3(DMF)_4]$	110°C, 30 min	[90]
MOF-5	95–135°C, 10–60 min	[91]

microwave heating between 95°C and 135°C, and the irradiation power was 600 W, 800 W, and 1000 W where conventional heating required 105°C, 24 hours. Some of the very interesting facts come out after the synthesis of this porous structure: (1) crystals are smaller than those obtained using conventional heating (20–25 mm vs 500 mm) (Fig. 4.5A and B) (2) the surface area and CO_2 sorption capacity are remained same in comparison with the conventional heating, (3) crystals forms after only 15 minutes while 12 hours are essential with the conventional method, (4) alteration of the composition of the reaction mixture (e.g., Zn^{2+} content) can adjust the size and quality of the crystals, (5) surface defects and crystal degradation happen when irradiation continues over 30 minutes (Fig. 4.5C). Lu et al. [92] also examined the properties of MOF-5 and matched the adsorption ability of crystals prepared under microwave irradiation and under reflux. As expected, the decision was that the former method yields enhanced developed crystals with a higher capacity to adsorb CO_2.

While forming Cr-MIL-101 as nano-sized crystals, Jhung et al. found that reaction time was of vital importance: crystal size and distribution enhanced with enhanced irradiation time, eventually allowing the isolation of high surface area particles (Fig. 4.6).

4.5.2 The pH of the reactive mixture

The pH of the reactive mixture also plays an important and crucial role in the MW-assisted synthesis of MOFs.

FIGURE 4.5 Scanning electron microscope images of MOF-5 prepared under (A) conventional solvothermal synthesis and (B) microwave heating. (C) MOF-5 crystal degradation with exposure to microwave irradiation.

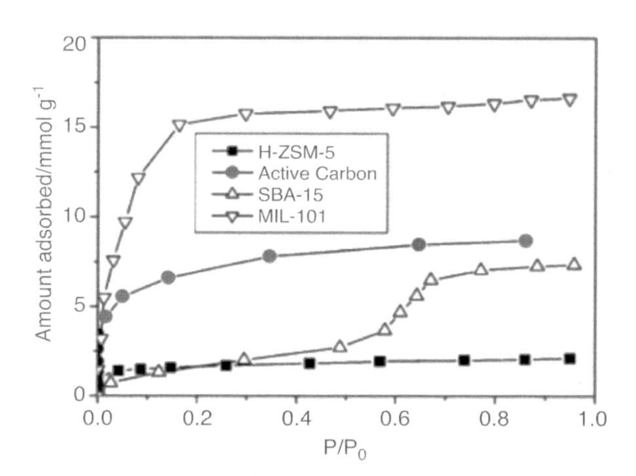

FIGURE 4.6 Comparison of the sorption isotherms of benzene of MIL-101 prepared using microwave heating, active carbon, zeolite HZSM-5 and mesoporous SBA-15.

Jhung et al. investigated the phase-selectivity of the microwave heating for the preparation of the nickel(II) glutarates $[Ni_{20}(C_5H_6O_4)_{20}(H_2O)_8]\cdot 40H_2O$ (MIL-77, cubic) and $[Ni_{22}(C_5H_6O_4)_{20}(OH)_4(H_2O)_{10}]\cdot 38H_2O$ (tetragonal). At the lower value of pH and at low temperature, the cubic phase is observed and isolated whereas with the advancement of the reaction tetragonal phase was observed and isolated at higher pH and at increased temperature. Recently, a one-dimensional polymer, reported by Delgado and Zamora, can only be isolated using microwave irradiation. Liu et al. reported an intense effect of the pH of the reactive mixtures [93]. Different MOFs were prepared by controlling the pH of the reaction mixtures in just 20 minutes. The synthesized MOFs were:

- $[Mg_2(BTEC)(H_2O)_8]$ (one-dimensional chain, pH 4.8).
- $[Mg_2(BTEC)(H_2O)_4]\cdot 2H_2O$ (framework-type structure, pH 4.6).
- $[Mg_2(BTEC)(H_2O)_6]$ (lamellar, pH 4.2).

4.6 Advantages of microwave-assisted synthesis over the conventional method

Microwave-assisted MOF synthesis has been preferentially adopted as one of the most efficient MOF syntheses. The followings are the advantages of microwave-assisted MOF synthesis over other conventional synthetic methods:

- Short reaction time ranges from a few seconds to 1 hour.
- The required temperature is above 100°C.
- Small crystallization time.
- Efficient Phase selectivity observed.
- Precise control over morphology.
- Nano-sized and homogeneous particle size distribution.
- Highly pure crystals.
- Specific pore volume.
- Quantitative yield.
- Industrial scale production.
- High surface area.
- Efficient sorption property.
- Low energy consumption.
- Greener approach.

4.7 Microwave-assisted synthesis: a green and sustainable approach

Green chemistry can be defined as the design, production, and application of chemical products that reduce or eliminate the use or generation of substances hazardous to human health or the environment. U.S. Environmental Protection Agency first introduced the term “green” and characterized it as the efficient employment of a set of principles to reduce or eliminate the use or generation of hazardous substances in the design, manufacture, and application of chemical products. This green protocol can be addressed by the use of 12 principles of green chemistry which are as follows:

(1) It is better to prevent waste than to treat or clean up waste after it has been created.
(2) Synthetic methods should be designed to maximize the incorporation of all materials used in the process, into the final product.
(3) Synthetic methods should be designed to use and generate less hazardous/toxic chemicals.
(4) Chemical products should be designed to affect their desired function while minimizing their toxicity.
(5) The use of solvents and auxiliary substances should be made unnecessary wherever possible and innocuous when used.
(6) Energy requirements of chemical processes should be minimized, and synthetic methods should be conducted at ambient temperature and pressure if possible.
(7) A raw material should be renewable rather than depleting whenever practicable.
(8) Unnecessary derivatization should be minimized or avoided if possible.
(9) Catalytic reagents are superior to stoichiometric reagents.
(10) Chemical products should be designed so that at the end of their function they break down into innocuous degradation products that do not persist in the environment.
(11) Analytical methodologies need to be further developed to allow for real-time, in-process monitoring and control prior to the formation of hazardous substances.
(12) Substances and the form of a substance used in a chemical process should be chosen to minimize the potential for chemical accidents [94–96].

Organic synthesis on a large scale uses the chemicals from petrochemical sector and after completion of the process, separation, purification, storage, packaging, distribution, etc. are the basic steps to be performed. Conventional methods of organic synthesis require tedious apparatus setup, larger heating time, and extreme use of solvents/reagents making itself low-cost efficient methods. Many problems arise related to health and safety for workers in addition to the environmental problems caused by their use and disposition as waste. Green chemistry stepped into the field of synthesis to assure less or no harm to human health and environment by using fewer toxic solvents, reduce the stages of the synthetic routes and curtail waste as far as virtually possible [97]. Microwave-assisted synthesis can be treated as green because of eco-friendliness. Its efficient energy utilization by directly couple with the reacting molecule reduces the power consumption. To synthesize a single compound by spending several days, microwave-assisted synthesis saves the time of researcher as the reaction can be performed in few minutes. The major problem of waste disposal of solvents is minimized by performing reactions without a solvent under microwave irradiation. Solvent-free conditions deliver clean chemical processes with the additional advantage of enhanced reaction rates, greater selectivity, higher yields, and greater ease of employment. Therefore, microwave-assisted synthesis acts as a potential tool for green chemistry.

4.8 Critical analysis

After the detailed discussion about microwave-assisted MOF synthesis, we are now highly envisioned to critically analyze the said method.

- The microwave-assisted synthesis of MOFs, as reported, mainly use the transition metal centers. Very few MOFs are reported that use other types of metallic centers like alkaline earth metals and lanthanides.
- Based on consistent and intensive microwave heating, the design and fabrication of MOFs by microwave-assisted synthesis take prominent control over other conventional synthetic methodologies.
- The microwave-assisted synthesis has several advantages such as high yield with significant phase selectivity and purity, the real opportunity of precise control over the crystal morphology, varied sizes ranging from microcrystals to nanoparticles, and short reaction time. Additionally, this method produces crystals with high yields with few or no side products with low energy consumption.
- This synthetic method has definitely some drawbacks. Reproducibility of results is the major concern with the particular instrumental apparatus. The different instrumental set up are incapable to provide the same parametric conditions such as reaction time, irradiation power, and temperature.

4.9 Conclusion and future prospects

The last two decades observed countless progress in the field of MOFs both in the synthesis of new structures and the development of new functional properties of these remarkable materials. The continually expanding research on the design and fabrication of MOFs can be efficiently synthesized by microwave heating. The detailed discussion is enough to conclude that this technique has several advantages over other conventional synthetic routes. Microwave-assisted synthesis has an unprecedented level of control over phase selectivity and purity, rapid reaction, high yields, the possibility to get control over crystal size, and low energy consumption with small or no chemical waste. The industrialist is fortunate enough to implement this efficient technique to synthesize MOF on a larger scale. In the last few years, the scientific community highly engaged to fabricate novel synthetic methods and discover the scale-up of MOFs synthesis by applying economically feasible strategies which use low-cost raw materials with less environmental issues. The twelve protocols of green chemistry are also closely achieved by this synthetic technique which makes it a good candidate as a green and sustainable approach towards MOFs synthesis.

Microwave heating also have some major drawbacks, starting with the instrumental apparatus itself. Different instrumental set up are incapable to deliver the same situations when reaction conditions are monitored by changing the temperature, irradiation power, and time of reaction. It affects highly the reproducibility property. Another efficient synthetic approach (ultrasound technique) has been reported and investigated. Ultrasound approach towards MOFs synthesis can compete with microwave assisted technique in terms of fast reaction time, crystal growth, and high surface area effective for sorption. At last, it is worthwhile to remind that MOFs are always constructed by self-assembly. It is thus a challenge doped on synthetic chemists to find best routes for MOFs synthesis abiding green protocols. We can conclude that, at present, microwave heating seems the most promising way to attain short reaction times and small crystalline sizes, essential for applications in devices.

References

[1] W. Lu, Z. Wei, Z.Y. Gu, T.F. Liu, J. Park, J. Park, et al., Tuning the structure and function of metal–organic frameworks via linker design, Chem. Soc. Rev. 43 (16) (2014) 5561–5593.

[2] M. Eddaoudi, J. Kim, N. Rosi, D. Vodak, J. Wachter, M. O'Keeffe, et al., Systematic design of pore size and functionality in isoreticular MOFs and their application in methane storage, Science 295 (5554) (2002) 469–472.

[3] J.R. Long, O.M. Yaghi, The pervasive chemistry of metal–organic frameworks, Chem. Soc. Rev. 38 (5) (2009) 1213–1214.

[4] B. Dutta, A. Dey, C. Sinha, P.P. Ray, M.H. Mir, Photochemical structural transformation of a linear 1D coordination polymer impacts the electrical conductivity, Inorg. Chem. 57 (14) (2018) 8029–8032.

[5] X.H. Bu, W. Chen, S.L. Lu, R.H. Zhang, D.Z. Liao, W.M. Bu, et al., Flexible meso-bis (sulfinyl) ligands as building blocks for copper (ii) coordination polymers: cavity control by varying the chain length of ligands, Angew. Chem. 113 (17) (2001) 3301–3303.

[6] D. Braga, F. Grepioni, G.R. Desiraju, Crystal engineering and organometallic architecture, Chem. Rev. 98 (4) (1998) 1375–1405.

[7] K. Sumida, D.L. Rogow, J.A. Mason, T.M. McDonald, E.D. Bloch, Z.R. Herm, et al., Carbon dioxide capture in metal–organic frameworks, Chem. Rev. 112 (2) (2012) 724–781.

[8] B. Chen, S. Xiang, G. Qian, Metal−organic frameworks with functional pores for recognition of small molecules, Acc. Chem. Res. 43 (8) (2010) 1115–1124.

[9] D.M. D'Alessandro, B. Smit, J.R. Long, Carbon dioxide capture: prospects for new materials, Angew. Chem. Int. Ed. 49 (35) (2010) 6058–6082.

[10] J.R. Li, R.J. Kuppler, H.C. Zhou, Selective gas adsorption and separation in metal–organic frameworks, Chem. Soc. Rev. 38 (5) (2009) 1477–1504.

[11] D.Y. Lee, S.J. Yoon, N.K. Shrestha, S.H. Lee, H. Ahn, S.H. Han, Unusual energy storage and charge retention in Co-based metal–organic-frameworks, Microporous Mesoporous Mater. 153 (2012) 163–165.

[12] T. Qiu, Z. Liang, W. Guo, H. Tabassum, S. Gao, R. Zou, Metal–organic framework-based materials for energy conversion and storage, ACS Energy Lett. 5 (2) (2020) 520–532.

[13] J. Della Rocca, D. Liu, W. Lin, Nanoscale metal–organic frameworks for biomedical imaging and drug delivery, Acc. Chem. Res. 44 (10) (2011) 957–968.

[14] L.E. Kreno, K. Leong, O.K. Farha, M. Allendorf, R.P. Van Duyne, J.T. Hupp, Metal–organic framework materials as chemical sensors, Chem. Rev. 112 (2) (2012) 1105–1125.

[15] Z. Hu, B.J. Deibert, J. Li, Luminescent metal–organic frameworks for chemical sensing and explosive detection, Chem. Soc. Rev. 43 (16) (2014) 5815–5840.

[16] J. Lee, O.K. Farha, J. Roberts, K.A. Scheidt, S.T. Nguyen, J.T. Hupp, Metal–organic framework materials as catalysts, Chem. Soc. Rev. 38 (5) (2009) 1450–1459.

[17] J.S. Seo, D. Whang, H. Lee, S.I. Jun, J. Oh, Y.J. Jeon, et al., A homochiral metal–organic porous material for enantioselective separation and catalysis, Nature 404 (6781) (2000) 982–986.

[18] L. Ma, C. Abney, W. Lin, Enantioselective catalysis with homochiral metal–organic frameworks, Chem. Soc. Rev. 38 (5) (2009) 1248–1256.

[19] J. Liu, L. Chen, H. Cui, J. Zhang, L. Zhang, C.Y. Su, Applications of metal–organic frameworks in heterogeneous supramolecular catalysis, Chem. Soc. Rev. 43 (16) (2014) 6011–6061.

[20] T. Zhang, W. Lin, Metal–organic frameworks for artificial photosynthesis and photocatalysis, Chem. Soc. Rev. 43 (16) (2014) 5982–5993.

[21] M. Yoon, R. Srirambalaji, K. Kim, Homochiral metal–organic frameworks for asymmetric heterogeneous catalysis, Chem. Rev. 112 (2) (2012) 1196–1231.

[22] Y. Cui, Y. Yue, G. Qian, B. Chen, Luminescent functional metal–organic frameworks, Chem. Rev. 112 (2) (2012) 1126–1162.

[23] J. Heine, K. Müller-Buschbaum, Engineering metal-based luminescence in coordination polymers and metal–organic frameworks, Chem. Soc. Rev. 42 (24) (2013) 9232–9242.

[24] X. Zhang, W. Wang, Z. Hu, G. Wang, K. Uvdal, Coordination polymers for energy transfer: preparations, properties, sensing applications, and perspectives, Coord. Chem. Rev. 284 (2015) 206–235.

[25] H. Li, M. Eddaoudi, T.L. Groy, O.M. Yaghi, Establishing microporosity in open metal−organic frameworks: gas sorption isotherms for Zn(BDC)(BDC= 1, 4-benzenedicarboxylate), J. Am. Chem. Soc. 120 (33) (1998) 8571–8572.

[26] M. Kondo, T. Yoshitomi, H. Matsuzaka, S. Kitagawa, K. Seki, Three-dimensional framework with channeling cavities for small molecules:{$[M_2(4,4'\text{-bpy})_3\ (NO_3)_4]\cdot xH_2O$}n (M=Co, Ni, Zn), Angew. Chem. Int. Ed. Engl. 36 (16) (1997) 1725–1727.

[27] C.M. Serre, C. Mellot-Draznieks, S. Surblé, N. Audebrand, Y. Filinchuk, G. Férey, Role of solvent-host interactions that lead to very large swelling of hybrid frameworks, Science 315 (5820) (2007) 1828–1831.

[28] O.M. Yaghi, M. O'Keeffe, N.W. Ockwig, H.K. Chae, M. Eddaoudi, J. Kim, Reticular synthesis and the design of new materials, Nature 423 (6941) (2003) 705–714.

[29] D.N. Dybtsev, H. Chun, K. Kim, Rigid and flexible: a highly porous metal–organic framework with unusual guest-dependent dynamic behavior, Angew. Chem. 116 (38) (2004) 5143–5146.

[30] K. Seki, W. Mori, Syntheses and characterization of microporous coordination polymers with open frameworks, J. Phys. Chem. B 106 (6) (2002) 1380–1385.

[31] K.S. Park, Z. Ni, A.P. Côté, J.Y. Choi, R. Huang, F.J. Uribe-Romo, et al., Exceptional chemical and thermal stability of zeolitic imidazolate frameworks, Proc. Natl. Acad. Sci. 103 (27) (2006) 10186–10191.

[32] Y.Q. Tian, C.X. Cai, Y. Ji, [Co5(im)10 x 2MB]infinity: a metal-organic open-framework with zeolite-like topology, Angew. Chem. Int. Edi. 41 (2002) 1384–1386.

[33] A.C. Kresge, M.E. Leonowicz, W.J. Roth, J.C. Vartuli, J.S. Beck, Ordered mesoporous molecular sieves synthesized by a liquid-crystal template mechanism, Nature 359 (6397) (1992 Oct) 710–712.

[34] M.E. Davis, C. Saldarriaga, C. Montes, J. Garces, C. Crowdert, A molecular sieve with eighteen-membered rings, Nature 331 (6158) (1988) 698–699.
[35] B.F. Hoskins, R. Robson, Infinite polymeric frameworks consisting of three dimensionally linked rod-like segments, J. Am. Chem. Soc. 111 (15) (1989) 5962–5964.
[36] B.F. Hoskins, R. Robson, Design and construction of a new class of scaffolding-like materials comprising infinite polymeric frameworks of 3D-linked molecular rods. A reappraisal of the zinc cyanide and cadmium cyanide structures and the synthesis and structure of the diamond-related frameworks $[N(CH_3)_4][CuIZnII\ (CN)_4]$ and $CuI[4,4',4'',4'''$-tetracyanotetraphenylmethane$]\ BF_4.xC_6H_5NO_2$, J. Am. Chem. Soc. 112 (4) (1990) 1546–1554.
[37] O.M. Yaghi, H. Li, Hydrothermal synthesis of a metal-organic framework containing large rectangular channels, J. Am. Chem. Soc. 117 (41) (1995 Oct) 10401–10402.
[38] K. Barthelet, J. Marrot, D. Riou, G. Férey, A breathing hybrid organic–inorganic solid with very large pores and high magnetic characteristics, Angew. Chem. Int. Ed. 41 (2) (2002) 281–284.
[39] C. Serre, F. Millange, C. Thouvenot, M. Nogues, G. Marsolier, D. Louër, G. Férey, Very large breathing effect in the first nanoporous chromium (III)-based solids: MIL-53 or CrIII (OH)⊙{O2C−C6H4−CO2}⊙{HO2C−C6H4−CO2H} x⊙ H2O y., J. Am. Chem. Soc. 124 (45) (2002) 13519–13526.
[40] P.Z. Moghadam, A. Li, S.B. Wiggin, A. Tao, A.G. Maloney, P.A. Wood, et al., Development of a Cambridge Structural Database subset: a collection of metal–organic frameworks for past, present, and future, Chem. Mater. 29 (7) (2017) 2618–2625.
[41] A. Li, R. Bueno-Perez, S. Wiggin, D. Fairen-Jimenez, Enabling efficient exploration of metal–organic frameworks in the Cambridge Structural Database, CrystEngComm 22 (43) (2020) 7152–7161.
[42] S.H. Jhung, J.H. Lee, J.W. Yoon, C. Serre, G. Férey, J.S. Chang, Microwave synthesis of chromium terephthalate MIL-101 and its benzene sorption ability, Adv. Mater. 19 (1) (2007) 121–124.
[43] H.Y. Cho, D.A. Yang, J. Kim, S.Y. Jeong, W.S. Ahn, CO_2 adsorption and catalytic application of Co-MOF-74 synthesized by microwave heating, Catal. Today 185 (1) (2012) 35–40.
[44] D.W. Jung, D.A. Yang, J. Kim, J. Kim, W.S. Ahn, Facile synthesis of MOF-177 by a sonochemical method using 1-methyl-2-pyrrolidinone as a solvent, Dalton Trans. 39 (11) (2010) 2883–2887.
[45] J. Kim, S.T. Yang, S.B. Choi, J. Sim, J. Kim, W.S. Ahn, Control of catenation in CuTATB-n metal–organic frameworks by sonochemical synthesis and its effect on CO_2 adsorption, J. Mater. Chem. 21 (9) (2011) 3070–3076.
[46] M. Hartmann, S. Kunz, D. Himsl, O. Tangermann, S. Ernst, A. Wagener, Adsorptive separation of isobutene and isobutane on $Cu_3(BTC)_2$, Langmuir 24 (16) (2008) 8634–8642.
[47] T.R. Van Assche, G. Desmet, R. Ameloot, D.E. De Vos, H. Terryn, J.F. Denayer, Electrochemical synthesis of thin HKUST-1 layers on copper mesh, Microporous Mesoporous Mater. 158 (2012) 209–213.
[48] T. Friščić, D.G. Reid, I. Halasz, R.S. Stein, R.E. Dinnebier, M.J. Duer, Ion-and liquid-assisted grinding: improved mechanochemical synthesis of metal–organic frameworks reveals salt inclusion and anion templating, Angew. Chem. 122 (4) (2010) 724–727.
[49] A. Rabenau, The role of hydrothermal synthesis in preparative chemistry, Angew. Chem. Int. Ed. Engl. 24 (12) (1985) 1026–1040.

[50] D.J. Tranchemontagne, J.R. Hunt, O.M. Yaghi, Room temperature synthesis of metal-organic frameworks: MOF-5, MOF-74, MOF-177, MOF-199, and IRMOF-0, Tetrahedron 64 (36) (2008) 8553–8557.
[51] J. Cravillon, S. Münzer, S.J. Lohmeier, A. Feldhoff, K. Huber, M. Wiebcke, Rapid room-temperature synthesis and characterization of nanocrystals of a prototypical zeolitic imidazolate framework, Chem. Mater. 21 (8) (2009) 1410–1412.
[52] G. Férey, Hybrid porous solids: past, present, future, Chem. Soc. Rev. 37 (1) (2008) 191–214.
[53] P.M. Forster, P.M. Thomas, A.K. Cheetham, Biphasic solvothermal synthesis: a new approach for hybrid inorganic–organic materials, Chem. Mater. 14 (1) (2002) 17–20.
[54] D.E. Clark, W.H. Sutton, Microwave processing of materials, Annu. Rev. Mater. Sci. 26 (1) (1996) 299–331.
[55] I. Bilecka, M. Niederberger, Microwave chemistry for inorganic nanomaterials synthesis, Nanoscale 2 (8) (2010) 1358–1374.
[56] M. Gharibeh, G.A. Tompsett, K.S. Yngvesson, W.C. Conner, Microwave synthesis of zeolites: effect of power delivery, J. Phys. Chem. B 113 (26) (2009) 8930–8940.
[57] Z. Hu, T. Kundu, Y. Wang, Y. Sun, K. Zeng, D. Zhao, Modulated hydrothermal synthesis of highly stable MOF-808 (Hf) for methane storage, ACS Sustain. Chem. Eng. 8 (46) (2020) 17042–17053.
[58] G. Férey, C. Serre, C. Mellot-Draznieks, F. Millange, S. Surblé, J. Dutour, et al., A hybrid solid with giant pores prepared by a combination of targeted chemistry, simulation, and powder diffraction, Angew. Chem. 116 (46) (2004) 6456–6461.
[59] Z. Ni, R.I. Masel, Rapid production of metal–organic frameworks via microwave-assisted solvothermal synthesis, J. Am. Chem. Soc. 128 (38) (2006) 12394–12395.
[60] S.H. Jhung, J.H. Lee, J.S. Chang, Microwave synthesis of a nanoporous hybrid material, chromium trimesate, Bull. Korean Chem. Soc. 26 (6) (2005) 880–881.
[61] J. Gordon, H. Kazemian, S. Rohani, Rapid and efficient crystallization of MIL-53 (Fe) by ultrasound and microwave irradiation, Microporous Mesoporous Mater. 162 (2012) 36–43.
[62] E. Haque, N.A. Khan, J.H. Park, S.H. Jhung, Synthesis of a metal–organic framework material, iron terephthalate, by ultrasound, microwave, and conventional electric heating: a kinetic study, Chem. A Eur. J. 16 (3) (2010) 1046–1052.
[63] M. Schlesinger, S. Schulze, M. Hietschold, M. Mehring, Evaluation of synthetic methods for microporous metal–organic frameworks exemplified by the competitive formation of [$Cu_2(btc)_3(H_2O)_3$] and [$Cu_2(btc)(OH)(H_2O)$], Microporous Mesoporous Mater. 132 (1–2) (2010) 121–127.
[64] Y.K. Seo, G. Hundal, I.T. Jang, Y.K. Hwang, C.H. Jun, J.S. Chang, Microwave synthesis of hybrid inorganic–organic materials including porous $Cu_3(BTC)_2$ from Cu(II)-trimesate mixture, Microporous Mesoporous Mater. 119 (1–3) (2009) 331–337.
[65] J.H. Park, S.H. Park, S.H. Jhung, Microwave-syntheses of zeolitic imidazolate framework material, ZIF-8, J. Korean Chem. Soc. 53 (5) (2009) 553–559.
[66] N.A. Khan, S.H. Jhung, Phase-transition and phase-selective synthesis of porous chromium-benzene dicarboxylates, Cryst. Growth Des. 10 (4) (2010) 1860–1865.
[67] S.H. Jhung, J.H. Lee, P.M. Forster, G. Férey, A.K. Cheetham, J.S. Chang, Microwave synthesis of hybrid inorganic–organic porous materials: phase-selective and rapid crystallization, Chem. A Eur. J. 12 (30) (2006) 7899–7905.
[68] S.H. Jhung, J.S. Chang, J.S. Hwang, S.E. Park, Selective formation of SAPO-5 and SAPO-34 molecular sieves with microwave irradiation and hydrothermal heating, Microporous Mesoporous Mater. 64 (1–3) (2003) 33–39.

[69] S.H. Jhung, J.H. Lee, J.W. Yoon, J.S. Hwang, S.E. Park, J.S. Chang, Selective crystallization of CoAPO-34 and VAPO-5 molecular sieves under microwave irradiation in an alkaline or neutral condition, Microporous Mesoporous Mater. 80 (1–3) (2005) 147–152.
[70] N.A. Khan, J.H. Park, S.H. Jhung, Phase-selective synthesis of a silicoaluminophosphate molecular sieve from dry gels, Mater. Res. Bull. 45 (4) (2010) 377–381.
[71] P.M. Forster, A.R. Burbank, C. Livage, G. Férey, A.K. Cheetham, The role of temperature in the synthesis of hybrid inorganic–organic materials: the example of cobalt succinates, Chem. Commun. (4) (2004) 368–369.
[72] W. Liang, D.M. D'Alessandro, Microwave-assisted solvothermal synthesis of zirconium oxide based metal–organic frameworks, Chem. Commun. 49 (35) (2013) 3706–3708.
[73] S. Delgado, A. Santana, O. Castillo, F. Zamora, Dynamic combinatorial chemistry in a solvothermal process of Cu(I, II) and organosulfur ligands, Dalton Trans. 39 (9) (2010) 2280–2287.
[74] P. Maniam, N. Stock, Investigation of porous Ni-based metal–organic frameworks containing paddle-wheel type inorganic building units via high-throughput methods, Inorg. Chem. 50 (11) (2011) 5085–5097.
[75] Z.A. Lethbridge, J.J. Williams, R.I. Walton, K.E. Evans, C.W. Smith, Methods for the synthesis of large crystals of silicate zeolites, Microporous Mesoporous Mater. 79 (1-3) (2005) 339–352.
[76] S. Qiu, J. Yu, G. Zhu, O. Terasaki, Y. Nozue, W. Pang, et al., Strategies for the synthesis of large zeolite single crystals, Microporous Mesoporous Mater. 21 (4-6) (1998) 245–251.
[77] T.O. Drews, M. Tsapatsis, Progress in manipulating zeolite morphology and related applications, Curr. Opin. Colloid Interface Sci. 10 (5–6) (2005) 233–238.
[78] I. Imaz, M. Rubio-Martínez, L. García-Fernández, F. García, D. Ruiz-Molina, J. Hernando, et al., Coordination polymer particles as potential drug delivery systems, Chem. Commun. 46 (26) (2010) 4737–4739.
[79] W.J. Rieter, K.M. Taylor, H. An, W. Lin, W. Lin, Nanoscale metal−organic frameworks as potential multimodal contrast enhancing agents, J. Am. Chem. Soc. 128 (28) (2006) 9024–9025.
[80] A. Carne, C. Carbonell, I. Imaz, D. Maspoch, Nanoscale metal–organic materials, Chem. Soc. Rev. 40 (1) (2011) 291–305.
[81] R.C. Huxford, J. Della Rocca, W. Lin, Metal–organic frameworks as potential drug carriers, Curr. Opin. Chem. Biol. 14 (2) (2010) 262–268.
[82] P. Horcajada, R. Gref, T. Baati, P.K. Allan, G. Maurin, P. Couvreur, et al., Metal–organic frameworks in biomedicine, Chem. Rev. 112 (2) (2012) 1232–1268.
[83] J. Park, J. Joo, S.G. Kwon, Y. Jang, T. Hyeon, Synthesis of monodisperse spherical nanocrystals, Angew. Chem. Int. Ed. 46 (25) (2007) 4630–4660.
[84] B. Panella, M. Hirscher, H. Putter, U. Muller, Hydrogen adsorption in metal–organic frameworks: Cu-MOFs and Zn-MOFs compared, Adv. Funct. Mater. (2006) 520–524.
[85] Z. Xiang, D. Cao, X. Shao, W. Wang, J. Zhang, W. Wu, Facile preparation of high-capacity hydrogen storage metal-organic frameworks: a combination of microwave-assisted solvothermal synthesis and supercritical activation, Chem. Eng. Sci. 65 (10) (2010) 3140–3146.
[86] W. Liu, L. Ye, X. Liu, L. Yuan, X. Lu, J. Jiang, Rapid synthesis of a novel cadmium imidazole-4,5-dicarboxylate metal–organic framework under microwave-assisted solvothermal condition, Inorg. Chem. Commun. 11 (10) (2008) 1250–1252.
[87] K.M. Taylor, W.J. Rieter, W. Lin, Manganese-based nanoscale metal−organic frameworks for magnetic resonance imaging, J. Am. Chem. Soc. 130 (44) (2008) 14358–14359.

[88] Y.S. Bae, K.L. Mulfort, H. Frost, P. Ryan, S. Punnathanam, L.J. Broadbelt, et al., Separation of CO_2 from CH_4 using mixed-ligand metal−organic frameworks, Langmuir 24 (16) (2008) 8592–8598.

[89] P. Amo-Ochoa, G. Givaja, P.J. Miguel, O. Castillo, F. Zamora, Microwave assisted hydrothermal synthesis of a novel CuI-sulfate-pyrazine MOF, Inorg. Chem. Commun. 10 (8) (2007) 921–924.

[90] B. Liu, R.Q. Zou, R.Q. Zhong, S. Han, H. Shioyama, T. Yamada, et al., Microporous coordination polymers of cobalt (II) and manganese (II) 2, 6-naphthalenedicarboxylate: preparations, structures and gas sorptive and magnetic properties, Microporous Mesoporous Mater. 111 (1–3) (2008) 470–477.

[91] J.S. Choi, W.J. Son, J. Kim, W.S. Ahn, Metal–organic framework MOF-5 prepared by microwave heating: factors to be considered, Microporous Mesoporous Mater. 116 (1–3) (2008) 727–731.

[92] C.M. Lu, J. Liu, K. Xiao, A.T. Harris, Microwave enhanced synthesis of MOF-5 and its CO_2 capture ability at moderate temperatures across multiple capture and release cycles, Chem. Eng. J. 156 (2) (2010) 465–470.

[93] H.K. Liu, T.H. Tsao, Y.T. Zhang, C.H. Lin, Microwave synthesis and single-crystal-to-single-crystal transformation of magnesium coordination polymers exhibiting selective gas adsorption and luminescence properties, CrystEngComm 11 (7) (2009) 1462–1468.

[94] M. Lancaster, Green Chemistry: An Introductory Text, Royal Society of Chemistry, 2020.

[95] U.J. Joshi, K.M. Gokhale, A.P. Kanitkar, Green chemistry: need of the hour, Indian J. Pharm. Edu. Res. 45 (2) (2011) 168–174.

[96] P.T. Anastas, J.C. Warner, Green Chemistry: Theory and Practice, Oxford University Press, Oxford, 2000, pp. 2–5.

[97] J.H. Clark, D.J. Macquarrie (Eds.), Handbook of Green Chemistry and Technology, John Wiley & Sons, 2008.

Chapter 5

Hydrothermal synthesis of metal–organic frameworks

Mohd Muslim, Arif Ali and Musheer Ahmad
Department of Applied Chemistry, ZHCET, Faculty of Engineering and Technology, Aligarh Muslim University, Aligarh, Uttar Pradesh, India

5.1 Introduction

Metal–organic frameworks (MOFs) have piqued the interest of many researchers because of their tunable porosity, composition, and functionality, which are suitable for a diverse range of applications [1]. Increased emphasis has been directed toward the discovery of stable frameworks with high nuclearity building blocks and structural features of the framework [2]. The targeted synthesis of MOFs can be further enhanced by a better understanding of the relationship between reaction parameters and a better synthesis approach. The last decade has seen significant progress in the understanding of MOF synthesis and its environmental application via chemical, physical, and electronic means [3]. Hydrothermal synthesis is the most popular technique for crystallizing substances at high temperatures and ambient pressure in aqueous solutions [4]. German geologist Karl Emil von Schafhäutl published the first report on the growth of hydrothermally produced crystals in 1845. The use of laboratory simulations to study natural systems such as rocks, minerals, and ore deposits can be traced back to this era [5]. Hydrothermal experiments were conducted in glass tubes until 1881. Recent research has shown how the nucleation of MOF and growth occurs, as well as reaction parameters and their impact on the crystallization process [6]. The process of acquiring, developing, and enhancing the MOF integration response remains very complex and sets significant barriers to its implementation [7]. Several studies have been conducted regarding the synthesis of MOF using organic molecules, with particular emphasis on the recent advances in synthetic technology [8]. This chapter carefully reviews and analyzes MOF nucleation findings, growth studies, and concludes with a summary of the several frameworks obtained from the hydrothermal synthesis of MOFs. Hydrothermological research began in Japan and spread to China and India after World War II. The Industrial Revolution aided the spread of hydrothermal technology from

Synthesis of Metal–Organic Frameworks via Water-Based Routes: A Green and Sustainable Approach.
DOI: https://doi.org/10.1016/B978-0-323-95939-1.00011-3

Europe to North America. In 1926, Dr. Tominosuke Katsurai conducted the first hydrothermological research in Japan [9].

The history of thermal and solvothermal synthesis methods goes back more than a century, from the synthesis of minerals and the extraction of nutrients from minerals. Generally, hydrothermal and solvothermal methods are used in the zeolite and quartz industries. These methods have been widely used in the integration of conventional and advanced materials, waste treatment, and simulation of geothermal and bio-hydrothermal processes [10]. High-tech materials and biomolecules are increasingly based on liquid processes, such as hydrothermal or solvothermal processing, which differ from the solid processing process. If crystallization is involved in the synthesis process, the correlation of the reaction with the crystal growth should be considered. The solid-state reactions via hydrothermal and solvothermal reactions can lead to different properties of the final product, even when the same reactants are used. Chemistry is involved in preparing and assembling special compounds or materials treated with a different solutions. The hydrothermal reactions can provide an alternative synthetic route from the solid-state reaction [11]. More importantly, they can produce high-quality MOFs with unique functional properties and framework structures that cannot be obtained in the usual way of other synthesis methods. Hydrothermal synthesis methods can be used to combine new materials and crystals used in many technological fields. The chemical and physical properties of the resulting structures are different and better as compared to the solvothermal synthesis. We can combine a variety of active substances that are not possible in reacting to harsh conditions due to the evaporation of the reactants at high temperatures. MOF synthesis methods may be classified as conventional and non-standard methods. Conventional methods of assembling MOF have many flaws, the main problem is the requirement for a large number of solvents, harmful chemicals, and high energy costs. The methods listed in Fig. 5.1 are updated methods for future MOF synthesis with reduced energy requirements [12].

Hydrothermal synthesis can produce MOFs with special valence, metastable structure, condensed, and aggregation states with a low melting point, vapor pressure, and thermal stability [14]. Hydrothermal and solvothermal synthetic techniques are superior to other synthetic methods. However, the efficiency and durability of these methods are other factors in the development of synthetic frameworks and the physical properties of MOF materials [15]. A hydrothermal reaction recrystallizes substances in sealed vessels at high pressure and temperature that are not soluble to dissolve under normal circumstances. It has been used to synthesize MOFs with no upper pressure or temperature limit [16]. The process was initially developed to take advantage of the tunable chemical and physical properties of MOFs but has now broadened into a range of organic functional materials [17]. The method of single-crystal synthesis is based on the precursor solubility in hot water under high pressure and temperature. The crystal growth takes place on slow evaporation at room temperature and sometimes crystal growth are appeared in an autoclave directly (Fig. 5.2). Hydrothermal reactions are now considered a subset of chemical transport reactions and this

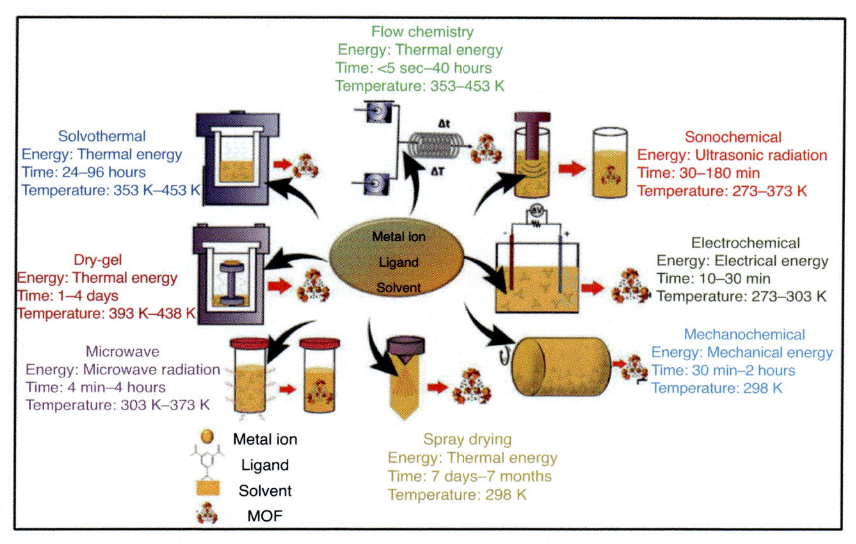

FIGURE 5.1 Conventional and non-conventional synthesis methods and conditions used for the synthesis of metal–organic frameworks. *(Reproduced with permission from ref. [13])*.

method can also use for the synthesis of microporous (ionic conductors, zeolites, fluorides, complex oxides, inorganic–organic hybrid materials, and nanomaterials) materials with high yield and porous behavior [18]. MOF researchers have used a wide range of experimental and computational techniques to investigate how crystals are formed, including powder X-ray diffraction (PXRD), dynamic light scattering (DLS)/static light scattering (SLS), small angle X-ray scattering (SAXS)/wide angle X-ray scattering (WAXS), and other scattering methods. Therefore, scattering is a spatially averaged technique and provides quantitative insight into crystallization, while microscopy is usually spatially localized and used for microscopic analysis [19].

Crystallization is the formation of monolastic MOFs or MOF families and it has long been a serious problem in chemistry and architectural engineering. The nucleation and crystallization of MOF materials derived from zeolites, as well as zeotype theory, are given special attention [21]. To avoid ambiguity, the nucleation and growth of MOFs framework hypotheses are discussed about mechanisms based on decoupling nucleation and crystal growth processes. Metal precursors are introduced into a solution and react with ligands to form small MOF clusters in the formation of MOF (nucleation stage). The formation of large crystals from these clusters is hampered by an unfavorable environment for crystal formation. The clusters then continue to grow into MOF crystals (the growth stage), with the amount of metal added and other factors influencing their growth [22]. The MOF crystals form directly from small clusters (stationary stage), the size of the resulting MOF crystal is determined by the binding mode of the organic linker and the concentration of added metal precursors. The

FIGURE 5.2 Hydrothermal reaction setup with different constituent assemblies for the synthesis of metal–organic frameworks and other functional materials. *(Reproduced with permission from ref. [20])*.

nucleation, crystal growth, stationary stage, and physical downsizing all steps are responsible for the synthesis of MOFs depicted in Fig. 5.3.

5.2 Hydrothermal synthesis of MOFs

The growth of hydrothermal synthesis of MOFs allows the formation of single crystals that are large, clean, and immovable at room temperature. The accumulation and growth of large numbers of individual crystals are one of the most important processes in hydrothermal synthesis [24]. The viscosity of hydrothermal medium is much lower than the proximal solvothermal method, and the separation of these methods have little effect on crystal growth [25]. The hydrothermal synthesis technique entails immersing precursor in hot water at pressures greater than 1 atm. There are several comprehensive reviews

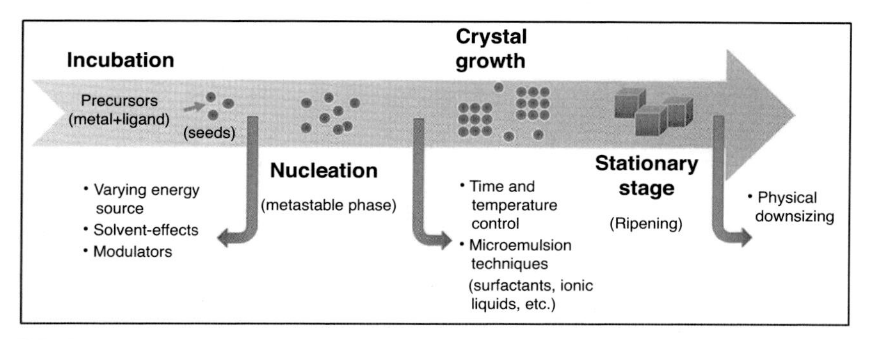

FIGURE 5.3 Flow diagram for crystallization stages and physical downsizing of the metal–organic framework synthesis and the influencing factor up to crystallization of crystal. *(Reproduced with permission from ref. [23]).*

available to help you understand the hydrothermal synthesis of MOFs and their biomedical applications [26]. Exogenous ligands such as imidazolate, terephthalic, and trimesic are the most commonly reported linkers that are used in the synthesis of MOFs. Reactions to this liquid phase compaction are carried out at high temperatures (above 100°C) and pressed for several hours or days [27]. The main advantage of this method is that it allows for high precursor solubility and the formation of high-quality MOF crystals. According to Zhou et al. [28], high-dimensional frameworks were produced by the hydrothermal reaction of multidentate ligand 2,2′-bi benzimidazole (H_2bbim) with a cobalt nitrate $Co(NO_3)_2{\cdot}6H_2O$ metal salt. The crystal growth process is carried out in an autoclave, which is a steel pressure vessel (Fig. 5.4A). Hydrothermal reaction with various metal salts, cadmium nitrate, zinc chloride, and cobalt nitrate with deprotonated terephthalic acid has led to the development of a new MOF at the University of Bristol laboratory; $\{[Zn_2(L1)(tp)(formate)_2]{\cdot}H_2O\}_n$ (**1**), $\{[Cd_2(L1)(ip)_2]{\cdot}2H_2O\}_n$ (**2**), $\{[Co_2(L1)(obba)_2]\}_n$ (**3**) where, (L1=1,2-bis {2,6-bis [(1H-imidazol-1-yl) methyl]-4-methylphenoxy} ethane} propane)]. These frameworks are made of the 3D frame with a two-fold interpenetrated form and an L28 net-connected framework (Fig. 5.4B) [29]. The three-dimensional framework of MOFs has been used in biomedical systems after dealing with serious problems, such as size, area charge, composition, stability, and toxicity [30]. In this chapter, some hydrothermal syntheses are reported in selected metals such as Fe, Zr, Cu, and Zn-based MOFs as well as hydrothermal compounds of three nano-bioMOFs, including copper serinate, copper prolinate, and copper threonine.

5.2.1 Synthesis of Cu-MOFs

5.2.1.1 Synthesis of copper serinate

The initial reaction between serine and copper chloride in the presence of water leads to the formation of a new substance known as copper serinate. To mix

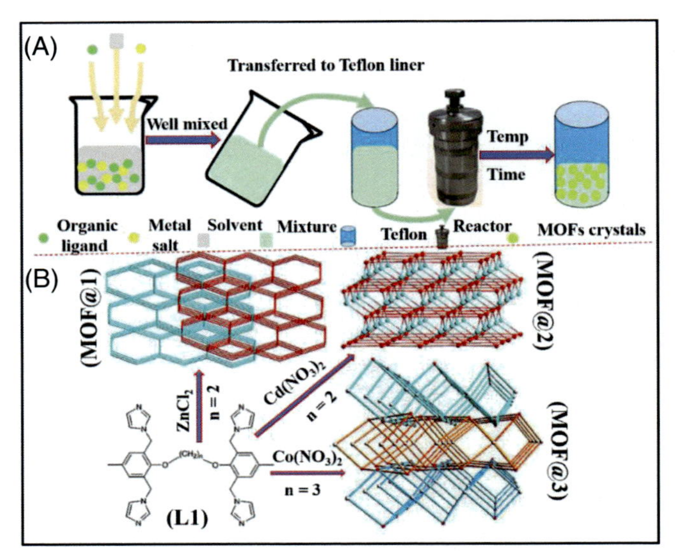

FIGURE 5.4 (A) Hydrothermal reaction procedure in autoclave up to crystallization of metal–organic frameworks crystal; and (B) three MOFs (MOFs@1-3) with their topological network, which were synthesized by hydrothermal synthesis method. (*Reproduced with permission from ref. [31]*).

his mixture, serin linker (0.15 g, 0.2 mmol) and copper chloride, $CuCl_2 \cdot H_2O$ (0.35 g, 0.2 mmol), were dissolved in 10 mL of water in a beaker. This solution is vigorously mixed for 30 minutes and a basic pH solution was maintained using a solution of sodium carbonate. This solution was transferred to 25 mL of teflon-lined autoclave and the autoclave was placed in the oven at 150°C for one day. It produced a colloidal blue solution after one day of slow evaporation. The solid blue particles were separated by sonication after centrifugation and dried at room temperature. These dry powder-like substances were then stored for later analysis.

5.2.1.2 Synthesis of copper prolinate and copper threoninate

The process of synthesis of copper prolinate and copper threoninate was similar to the synthesis of copper serinate differing from using prolinate and threoninate as a precursor. Three new nanosized materials are assembled hydrothermally and obtained by scanning electron microscopy for morphology (Fig. 5.4A and B). Due to its very small size, structural testing of these innovative crystalline crystals is not possible with a single-crystal XRD analysis. Mass spectrometric studies were used to determine the molecular weight of these three nano-sized bioMOF substances. As these newly synthesized nano bioMOFs were intended for in vitro absorption of the drug in additives such as essential amino acids (serine, proline, and threonine). Therefore, it is clear that these drug carriers will not have any side effects at all if they are used in the absorption of drugs because

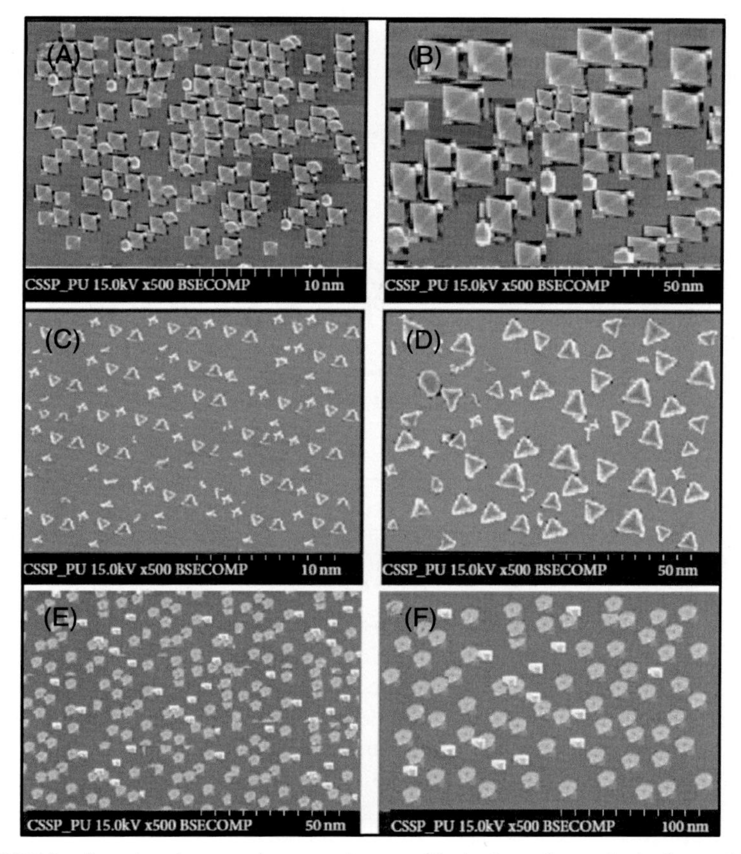

FIGURE 5.5 Scanning electron microscopy images of hydrothermally synthesized materials, (A and B) copper serinate, (C and D) copper prolinate, and (E and F) copper threoninate, showing the formation of nanoparticles of the nano-bio metal–organic frameworks. (*Reproduced with permission from ref. [33]*).

the human body is already needed by the components where these substances are synthesized (Fig. 5.5A–F) [32].

Researchers at the University of Bristol in the United Kingdom have developed the first copper-based metal–organic framework (Cu-MOF) to treat wastewater and benzene-1,3,5-tricarboxylic acid dissolved in ethanol. Cu-MOFs are synthesized hydrothermally with assembled to form a MOF nanosheet and then inserted using a layer-by-layer LTL-coated glass-coated coating layer with a thin layer of TiO_2 (Fig. 5.6A). Extracting the Cu-MOF nanosheet containing analyte sorbent enhances MOF film performance and is required to reduce the transmission resistance of the TiO/MOF/electrolyte optical connector (Fig. 5.6B) [34].

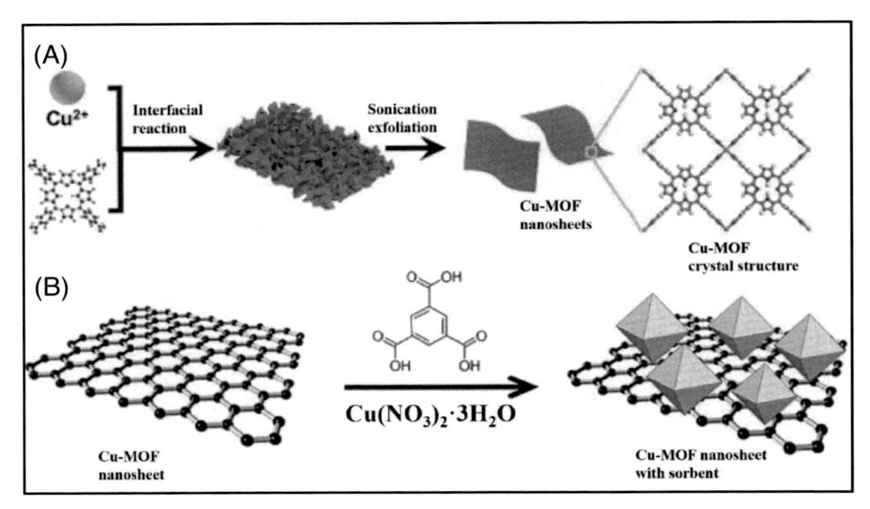

FIGURE 5.6 (A) Synthetic procedure and crystal structure of copper-based metal–organic frameworks (Cu-MOFs) nanosheet; and (B) analyte sorbent onto Cu-MOF nanosheet. *(Reproduced with permission from ref. [35])*.

A new electrode of nitrite for electro-oxidation has been developed by scientists at the University of Bristol in the United Kingdom. Cu-MOF modified carbon paste electrode (CPE) lowers nitrite oxidation, and introduces a new nitrite oxidation platform. A Cu-MOF was created to investigate the electro-oxidation of nitrite and the raw materials used to make nitric oxide. Researchers at the University of Aberystwyth used electro-reduction and absorption to strengthen the microstructures of gold (Au) on the surface of Cu-MOF. Gold microspheres (Au) accelerate the electron transfer rate of electrochemical reactions and reduce nitrite oxidation. The modified Au/Cu-MOF/CPE showed a low detection limit, wide line width, and good stability when determining nitrite (Fig. 5.7).

5.2.2 Synthesis of Fe-MOFs

This chapter reveals a detailed look at the hydrothermal synthesis of MOFs, which shows structural features towards the removal of organic and inorganic pollutants from wastewater [37]. A review of Fe-MOF-based water treatment methods, with emphasis on advertising, heterogeneous catalysis, and photocatalysis, provides a clearer understanding of the processes involved [38]. The challenges of using Fe-MOFs to fix water due to structural instability are discussed in a review by the US Environmental Protection Agency (EPA) and the US Army Corps of Engineers (ApoE) [39]. In regards to the above problem, Institut Lavoisier has created a series of iron-based organic compounds (Fe-MOFs) such as MIL-14, MIL-88A, and MIL-100 materials using the hydrothermal synthesis method (Fig. 5.8A). In addition to iron-based nano, MOF materials showed

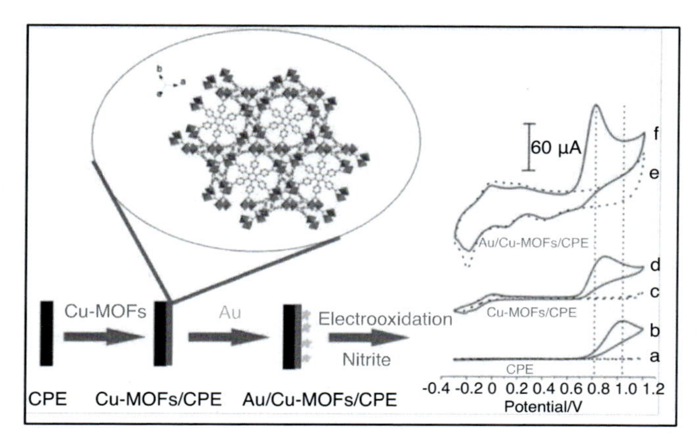

FIGURE 5.7 A copper-based metal–organic framework (Cu-MOF) was electrocatalytically applied as a novel sensing platform for enhanced nitrite electro-oxidation. (*Reproduced with permission from ref. [36]*).

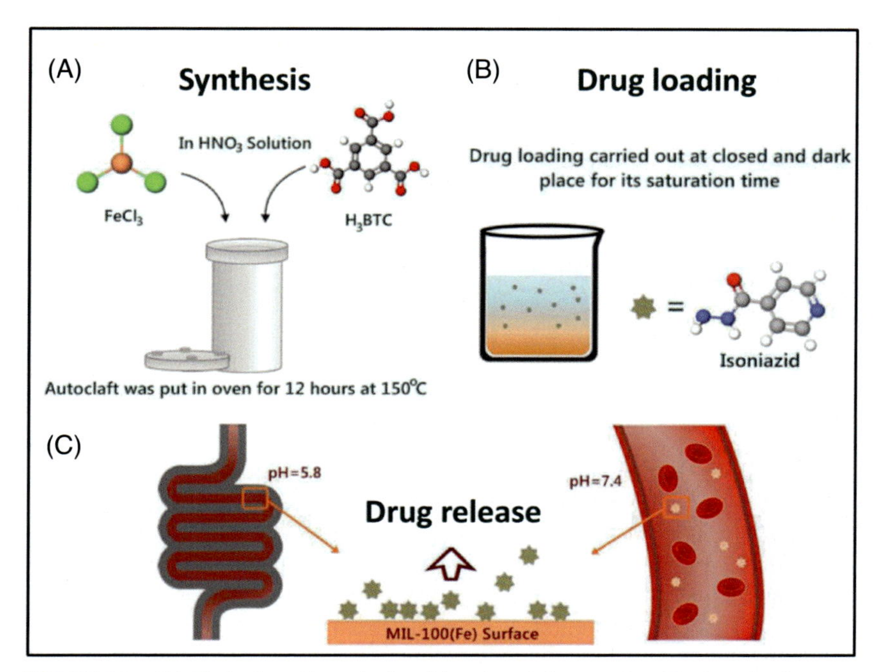

FIGURE 5.8 (A) Hydrothermal synthesis of iron-based frameworks; (B) drug loading onto MIL-100(Fe) carried out at closed and dark place for its saturation time, and (C) drug release from MIL-100(Fe) framework at different pH level. (*Reproduced with permission from ref. [41]*).

significant in vitro degradation after seven days of incubation at 37°C, under underbody conditions. In vitro nano, MOF injection showed that nano MOFs were non-toxic due to a lack of immune and inflammatory responses. The drug loading efficiency of nano MOFs (Doxo) was first tested and the adhesion to smaller MOFs was lower than MIL-100 but much higher than existing MIL-14 and MIL-88A (Fig. 5.8B). Nano MOFs can be used separately due to the presence of metal atoms and MIL-100 (Fe) can also be used to deliver docetaxel (DTX) to breast cancer cells as a nanocarrier novel [40]. Graft Fast reaction was used to prepare the surface area of hydrothermal mesoporous iron trimesate iron trimesate nano MOF MIL-100 (Fe). Modified nano MOFs showed higher 5-FU drug uptake and lower drug release in vitro, and MIL-53 showed a pH dependence during drug release at pH 5.8, compared to first-order kinetics at pH 7.4 (Fig. 5.8C). The iron-based nano MOFs of various sizes and shapes appear to have proton affinity and brightness enhancement in MRI images and the use of these MOFs may enhance the effectiveness of brimonidine drugs.

The components of iron-based metal–organic frameworks (Fe-MOFs) have been extensively studied to purify water among the many available MOFs. The formation of Fe-MOFs is relatively easy due to the low toxicity and high saturation of the metal in the earth's crust. Fe-MOFs also have large metal structures with high catalytic activity, making them suitable for use in advanced oxidation processes (AOPs), especially in the decomposition of Fenton and Fenton-like pollutants. Qualitative Fe-MOF catalysts have excellent electron transfer properties and the Fe (III)/Fe (II) redox pairs. Low-toxic and stable Fe-MOFs have excellent potential for almost all water treatment methods, especially photocatalysis under bright light, Fenton, and Fenton-like heterogeneous catalysis (Fig. 5.9A and B) [42]. Fe-MOFs have become an important water treatment tool due to their high catalytic activity, abundance of active sites, and advertising related to pollution. The deterioration was done in the presence of foreign chemicals, photolytic, mechanical, and thermal systems of the structure.

5.2.3 Synthesis of Zr-MOFs

Clicking on a chemical reaction can change the surface of a nano MOF to change its chemical composition [44]. The beneficial effect is indicated by the contrast with the melting point of the fragrant dicarboxylic acid. Zirconium-based MOF, UiO-66 showed a half-maximal inhibitory concentration (IC_{50}) of 1.50, 0.15 mg/mL against the HeLa cell line (Fig. 5.10A). With the use of trimethylamine as a module, the hole size of the MOF can be increased. Caveolin-pathway infiltration into Zr-MOFs was facilitated by naphthalene-2,6-dicarboxylic acid, which inhibits lysosomal degradation and increased therapeutic activity [45]. Hydrothermal conditions and modulators such as p-azidomethyl benzoic acid are used to create high-performance MOF hcp UiO-66 (Fig. 5.10B) [4]. PEGylated UiO-66 nano MOFs have the potential for use in the delivery of stimulant drugs in cancer patients. PEGylation improved nanoparticle stability

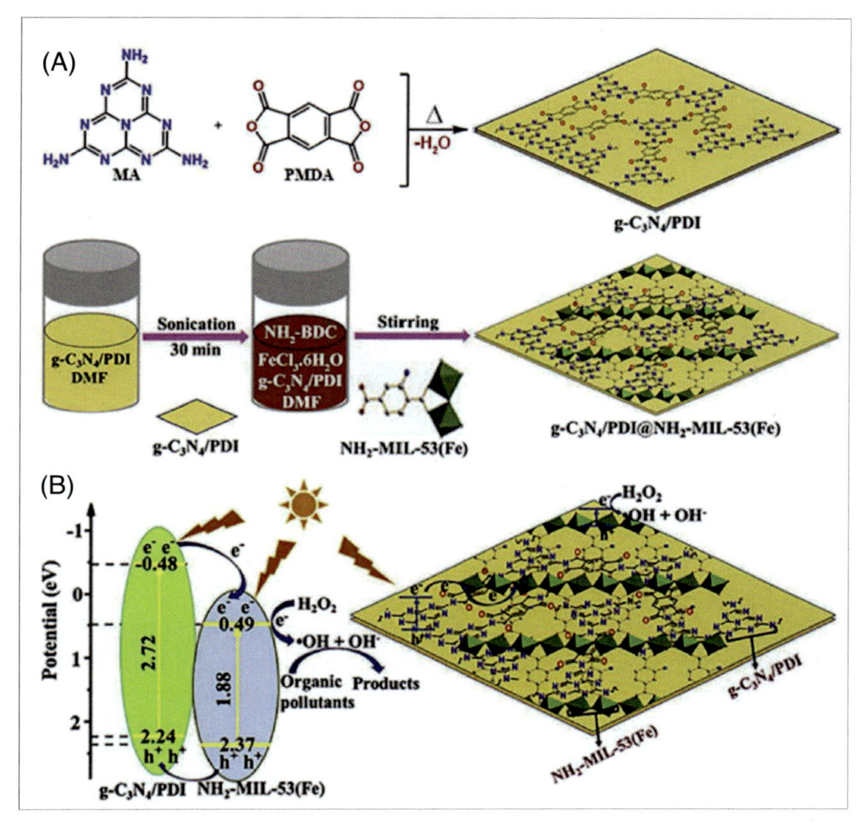

FIGURE 5.9 (A) Synthesis of g-C_3N_4/PDI and g-C_3N_4/PDI@NH_2-MIL-53 (Fe); Preparation of g-C_3N_4/PDI: (i) a mixture of pyromellitic dianhydride (PMDA) and melem (MA) by grinding in agate mud, (ii) transferring this mixture to porcelain crucible and heating at 325°C for 4 hours, (iii) concentrating and washing the obtained yellow product, and (iv) filtering and drying the yellow powder of g-C_3N_4/PDI. Preparation of g-C_3N_4/PDI@NH_2-MIL-53 (Fe): (1) Dispersion of g-C_3N_4/PDI powder into dimethylformamide by ultrasonication for 30 minutes, (2) addition of ferric chloride hexahydrate ($FeCl_3 \cdot 6H_2O$) and 2-aminoterephthalic acid (NH_2-BDC) in this suspension by stirring for 60 minutes, (3) then cool in the surrounding area and temperature, and IV) the formation of g-C_3N_4/PDI@NH_2-MIL-53 (Fe) after centrifugation, washing, drying, and setting. (B) the potential for photocatalytic degradation of pollutants using g-C_3N_4/PDI @ NH_2-MIL-53 (Fe) under visible light rays. (*Reproduced with permission from ref. [43]*).

in biological solutions and mimicked the transition from blood circulation to the intrusion of cancer cells [46]. The integration of MOFs demonstrates how to control cluster nuclear units in secondary building units (SBUs) to guide phases based on Zr_6-BDC and Zr_{12}-BDC (Fig. 5.10C) [47].

Zirconium-based connectors produce strong bonds to MOFs that have been reported to have excellent water and structural stability [48]. Zr-based MOFs are designed for a diverse range of applications, including photovoltaics,

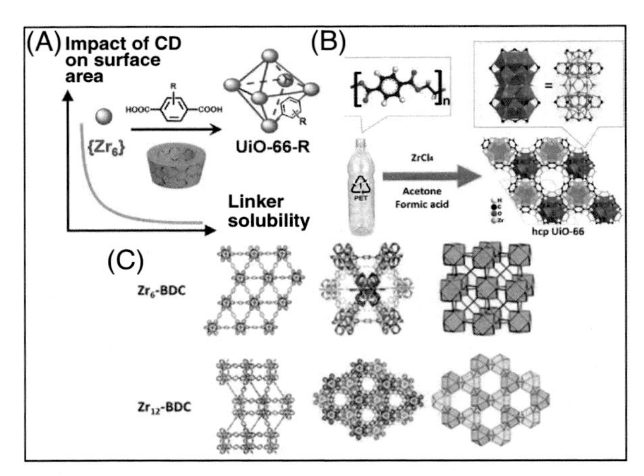

FIGURE 5.10 (A) Impact of surface area and linker solubility onto zirconium-based metal–organic-framework (Uio-66-R); (B) reducing waste by adsorption of PET onto Zr-based MOF (hcp-Uio-66) framework, and (C) various types of a topological framework for Zr-based MOFs (Zr_6-BDC and Zr_{12}-BDC). *(Reproduced with permission from ref. [33]).*

solar cells, and medicine. In aqueous solutions of acidic, neutral, and one of the standard MOFs (UiO-66) has excellent thermal and chemical resistance (Fig. 5.11A) [49]. Electrochemical oxygen depletion is essential for converting fuel cells into electrochemical energy. Because oxygen can be reduced by employing two electrons to produce hydrogen peroxide, the number of transferred electrons is significant for these electrocatalysts (Fig. 5.11B) [50]. The high kinetic barrier and the high potential for oxygen exposure compared to hydrogen emissions, so, better catalysts are needed. Even if the primary purpose of water separation is to produce hydrogen gas instead of oxygen, the development of these stimuli is still important [51]. The IrO_2 and RuO_2 are two common electrocatalysts for heterogeneous oxygen evolution. At relatively low power, they show remarkable activity in both acid and alkaline environments [52]. Photoelectrochemical water separation has received considerable attention as a renewable energy source (Fig. 5.12C). Electrocatalysts are used to speed up a half-reaction that converts water into hydrogen gas and oxygen into carbon dioxide. They also reduce the force beyond your control of both the partial reaction and the photosensitizers that make the separation of the charge possible (Fig. 5.11D) [53].

5.2.4 Synthesis of Zn-MOFs

Zinc-based metal–organic frameworks (Zn-MOFs) are a novel type of live cell molecule and a sensory nerve candidate, which show unusual photoluminescence behavior [55]. The magnitude of the change in fluorescence pressure is determined by the various interactions with the solvents. They can detect

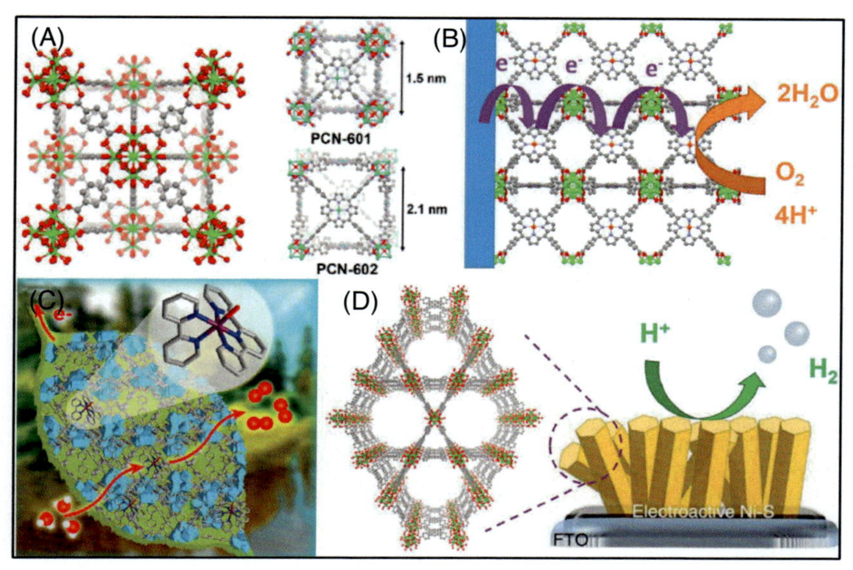

FIGURE 5.11 (A) Topological networks of zirconium-based metal–organic frameworks (Zr-MOFs); (B) (Zr-MOFs) for electrocatalytic oxygen reduction process; (C) the use of a Zr-MOF incorporated with ruthenium-based complex for electrocatalytic oxygen evolution; and (D) NU-1000 thin film is electrodeposited with a nickel layer to provide an optical view of the structure. (*Reproduced with permission from ref. [54]*).

a variety of fragrant compounds, with 1,4-DNB compounds exhibiting potent extinguished behavior [56]. A few benefits have been made to the design and principles of a targeted drug delivery system for therapeutic compounds and enhanced cell images. Moreover, Zn-MOFs are a major addition to the highly effective in medicinal practice. Some zinc-based MOFs (ZIF-7, ZIF-8, ZIF-90, MOF-5, MOF-1, and MOF-2) are hydrothermally synthesized by the combination of Zn^{2+} as a central metal ion [57]. Zn-MOF with natural molecules has different topological structures to form a small molecule (Fig. 5.12A and B). This opens the door to the construction of electrons rich in electrons with luminescence features. Luminescent MOFs can be promising candidates for use in small molecules for biological applications. The Zn-MOF structure is thought to contribute to the fluorescence extinguishing effect of nitro compounds and the nitro-retardant group of electrons, allowing for the more sensitive detection of nitroaromatics (Fig. 5.12C). The materials have a different solvent-based photoluminescence output that makes them available with X-ray diffraction microscopes [58].

MOF is the most widely used component of drug delivery systems, cell imaging, and diagnostic analyses, especially cancer biomarkers. Because of their excellent biocompatibility, easy operation, high storage capacity, and excellent biodegradability decay, these MOFs are considered to be a completely new class

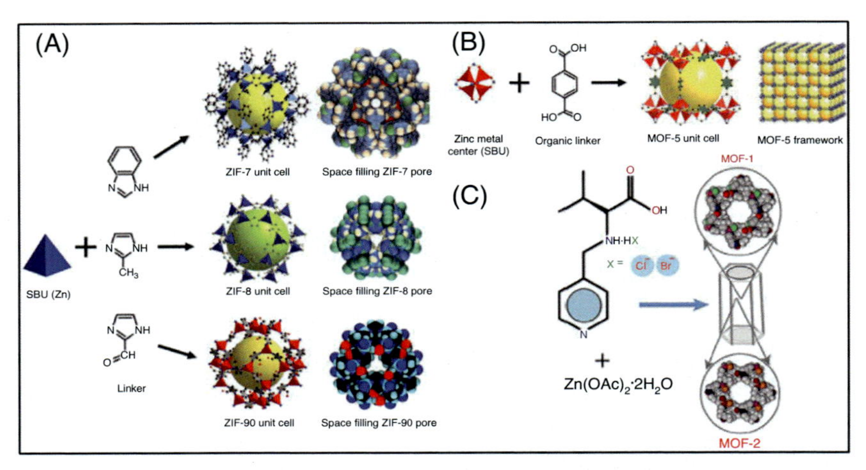

FIGURE 5.12 The hydrothermal synthesis scheme of zinc-based metal–organic frameworks (Zn-MOFs) with a different type of organic linker (A) ZIF-7, ZIF-8, ZIF-90 Reproduced with permission from ref. [59]; (B) MOF-5; and (C) MOF-1 and MOF-2. *(Reproduced with permission from ref. [60])*.

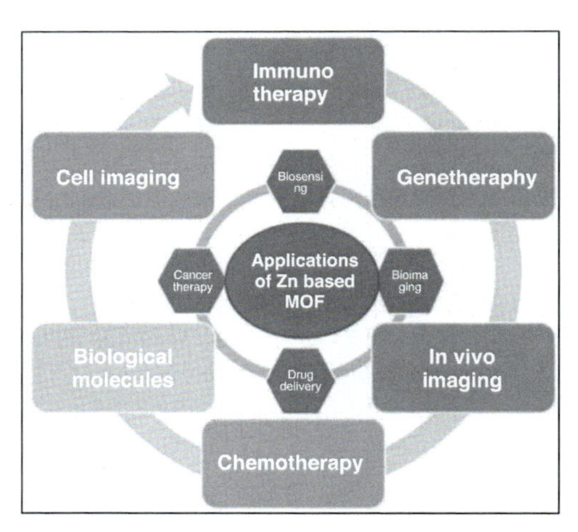

FIGURE 5.13 Biological application of zinc-based metal–organic frameworks (Zn-based MOFs). *(Reproduced with permission from ref. [62])*.

of material [61]. The MOF-based heterogeneous catalyst is robust and can be reused several times without loss of function. Zn-metal centers in MOFs were acquired through improved performance and improved drug delivery. The heterogeneous catalyst is a potent stimulant that can be used for drug development. Zn-MOFs have natural cells with unique characteristics that make them suitable carriers for drug delivery, cell imaging, and chemosensory sensations (Fig. 5.13). This chapter provides a systematic approach to understanding the development of Zn-MOFs as active drug carriers. Zinc has emerged as a promising candidate

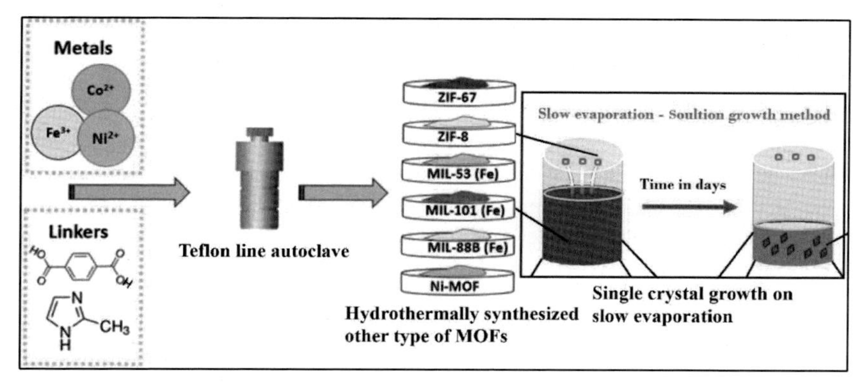

FIGURE 5.14 Hydrothermal synthesis of other types of metal–organic frameworks (ZIF-67, ZIF-8, MIL-53 (Fe), MIL-101 (Fe), MIL-888 (Fe), and Ni-MOF) and their crystal growth on slow evaporation method at room temperature. *(Reproduced with permission from ref. [66])*.

for the preparation of zinc-based MOF for biological use, particularly as a drug carrier. In addition, zinc has been widely used in dermatology as a means of moisturizing the skin with astringent, anti-bacterial, and anti-inflammatory properties [57].

5.2.5 Synthesis of other MOFs

The metal ion-based MOFs have been synthesized using the hydrothermal method for a variety of applications. Herein, we displayed six interpenetrating frameworks MOFs that were synthesized using different central metal ions and long heterocyclic aromatic ligands [25]. A study has shown that different types of MOFs can be synthesized using different carboxyl groups in coordination modes with transition metals and lanthanides contractions [63]. The adaptability of carboxylic groups is thought to have a significant impact on how the MOFs are prepared for use in the production of new functional materials [64]. All demonstrated six MOFs have three-dimensional (3D) network structures and are synthesized using a hydrothermal technique, i.e., crystallized on slow evaporation at room temperatures (Fig. 5.14) [65].

5.3 Conclusion and prospects

The primary goal of MOFs synthesis is to produce definite inorganic building blocks that do not decompose organic linkers. The MOFs and their synthetic chemistry are rapidly evolving and resulting in the discovery of advanced functional MOFs with unusual properties through the manipulation of physical parameters such as composition and size. The hierarchical assemblies are critical for regulating MOF behaviors like separation, transportation, and controlling the formation of MOFs with various morphologies. Despite the development of numerous MOF synthetic techniques, a one-pot hydrothermal batch synthesis strategy remains the most commonly used. Hydrothermal synthesis conditions such as synthetic temperature, additional additives, and the amount of solvent influence the formation of metal–organic architectures. Under conditions of low temperature, the synthesis temperature has little effect on morphology but has a significant impact on the size of the framework. The hydrothermal synthesis method has significant advantages over others MOFs conventional and non-conventional synthesis methods. Interestingly, functional materials that are highly stable at high temperatures can be synthesized through hydrothermal synthesis. Moreover, hydrothermal method can construct MOFs with high vapour pressures and minimal reagents loss, which are becoming popular in a diverse range of industrial applications such as drug delivery, gas storage, separation, and heterogeneous catalysis, and so on.

Acknowledgment

The authors are especially grateful to the Department of Applied Chemistry, ZHCET, Faculty of Engineering and Technology, Aligarh Muslim University, Aligarh, Uttar Pradesh, India for providing extensive assistance during the completion of this project.

References

[1] Y. Sakamaki, M. Tsuji, Z. Heidrick, O. Watson, J. Durchman, C. Salmon, et al., Preparation and applications of metal-organic frameworks (MOFs): a laboratory activity and demonstration for high school and/or undergraduate students, J. Chem. Educ. 97 (2020) 1109–1116.

[2] Y.R. Lee, J. Kim, W.S. Ahn, Synthesis of metal-organic frameworks: a mini review, Korean J. Chem. Eng. 30 (2013) 1667–1680.

[3] X. Lan, N. Huang, J. Wang, T. Wang, A general and facile strategy for precisely controlling the crystal size of monodispersed metal-organic frameworks: via separating the nucleation and growth, Chem. Commun. 54 (2018) 584–587.

[4] L. Zhou, S. Wang, Y. Chen, C. Serre, Direct synthesis of robust hcp UiO-66(Zr) MOF using poly(ethylene terephthalate) waste as ligand source, Microporous Mesoporous Mater. 290 (2019) 109674.

[5] F. Containing, L. Rectangular, Hydrothermal synthesis of a metal-organic framework containing large rectangular channels, J. Am. Chem. Soc. 117 (1995) 10401–10402.

[6] M. Chang, Y. Wei, D. Liu, J.X. Wang, J.F. Chen, A general strategy for instantaneous and continuous synthesis of ultrasmall metal–organic framework nanoparticles, Angew Chemie Int. Ed. 60 (2021) 26390–26396.

[7] S.R. Jambovane, S.K. Nune, R.T. Kelly, B.P. McGrail, Z. Wang, M.I. Nandasiri, et al., Continuous, one-pot synthesis and post-synthetic modification of nanoMOFs using droplet nanoreactors, Sci. Rep. 6 (2016) 1–9.

[8] V. Pascanu, G. González Miera, A.K. Inge, B. Martín-Matute, Metal-organic frameworks as catalysts for organic synthesis: a critical perspective, J. Am. Chem. Soc. 141 (2019) 7223–7234.

[9] P.S. Sharanyakanth, M. Radhakrishnan, Synthesis of metal-organic frameworks (MOFs) and its application in food packaging: a critical review, Trends Food Sci. Technol. 104 (2020) 102–116.
[10] S. Dai, A. Tissot, C. Serre, Metal-organic frameworks: from ambient green synthesis to applications, Bull. Chem. Soc. Jpn. 94 (2021) 2623–2636.
[11] T.A. Tabish, S. Zhang, Graphene quantum dots: syntheses, properties, and biological applications, in: Reference Module in Materials Science and Materials Engineering, Elsevier Ltd., 2019, pp. 1–5.
[12] S.B. Peh, Y. Cheng, J. Zhang, Y. Wang, G.H. Chan, J. Wang, et al., Cluster nuclearity control and modulated hydrothermal synthesis of functionalized Zr_{12} metal-organic frameworks, Dalt. Trans. 48 (2019) 7069–7073.
[13] J. Joseph, S. Iftekhar, V. Srivastava, Z. Fallah, E.N. Zare, M. Sillanpää, Iron-based metal-organic framework: synthesis, structure and current technologies for water reclamation with deep insight into framework integrity, Chemosphere 284 (2021) 131171.
[14] L. Pessanha de Carvalho, J. Held, E.J.T. de Melo, Essential and nonessential metal effects on extracellular Leishmania amazonensis in vitro, Exp. Parasitol. 209 (2020) 107826.
[15] J.D. Evans, B. Garai, H. Reinsch, W. Li, S. Dissegna, V. Bon, et al., Metal–organic frameworks in Germany: from synthesis to function, Coord. Chem. Rev. 380 (2019) 378–418.
[16] Q. Ren, F. Wei, H. Chen, D. Chen, B. Ding, Preparation of Zn-MOFs by microwave-assisted ball milling for removal of tetracycline hydrochloride and Congo red from wastewater, Green Process. Synth. 10 (2021) 125–133.
[17] M.J. Van Vleet, T. Weng, X. Li, J.R. Schmidt, In situ, time-resolved, and mechanistic studies of metal-organic framework nucleation and growth, Chem. Rev. 118 (2018) 3681–3721.
[18] Y. Jin, C. Zhao, Z. Sun, Y. Lin, L. Chen, D. Wang, et al., Facile synthesis of Fe-MOF/RGO and its application as a high performance anode in lithium-ion batteries, RSC Adv. 6 (2016) 30763–30768.
[19] J.L. Crane, K.E. Anderson, S.G. Conway, Hydrothermal synthesis and characterization of a metal-organic framework by thermogravimetric analysis, powder X-ray diffraction, and infrared spectroscopy: an integrative inorganic chemistry experiment, J. Chem. Educ. 92 (2015) 373–377.
[20] S. Liang, Q. Shi, H. Zhu, B. Peng, W. Huang, One-step hydrothermal synthesis of W-Doped VO_2 (M) nanorods with a tunable phase-transition temperature for infrared smart windows, ACS Omega 1 (2016) 1139–1148.
[21] D. Osypiuk, B. Cristóvão, A. Bartyzel, New coordination compounds of Cu(II) with schiff base ligands—crystal structure, thermal, and spectral investigations, Crystals 10 (2020) 1–20.
[22] M.T. Conato, A.J. Jacobson, Control of nucleation and crystal growth kinetics of MOF-5 on functionalized gold surfaces, Microporous Mesoporous Mater. 175 (2013) 107–115.
[23] K.A.S. Usman, J.W. Maina, S. Seyedin, M.T. Conato, L.M. Payawan, L.F. Dumée, et al., Downsizing metal–organic frameworks by bottom-up and top-down methods, NPG Asia Mater. 12 (2020).
[24] C.T. Pereira da Silva, B.N. Safadi, M.P. Moisés, J.G. Meneguin, P.A. Arroyo, S.L. Fávaro, et al., Synthesis of Zn-BTC metal organic framework assisted by a home microwave oven and their unusual morphologies, Mater. Lett. 182 (2016) 231–234.
[25] Y. Qi, F. Luo, Y. Che, J. Zheng, Hydrothermal synthesis of metal-organic frameworks based on aromatic polycarboxylate and flexible bis(imidazole) ligands, Cryst. Growth Des. 8 (2008) 606–611.

[26] S. Bhattacharjee, J.S. Choi, S.T. Yang, S.B. Choi, J. Kim, W.S. Ann, Solvothermal synthesis of Fe-MOF-74 and its catalytic properties in phenol hydroxylation, J. Nanosci. Nanotechnol. 10 (2010) 135–141.

[27] R. Ye, M. Ni, Y. Xu, H. Chen, S. Li, Synthesis of Zn-based metal-organic frameworks in ionic liquid microemulsions at room temperature, RSC Adv. 8 (2018) 26237–26242.

[28] Y.L. Zhou, F.Y. Meng, J. Zhang, M.H. Zeng, H. Liang, Mononuclear, tetra-,penta- 3d molecular clusters based on the variability of ss-1,2-bis(1H-benzimidazole-2-yl)-1,2-ethanediol ligand arising from hydroponic and hydrothermal conditions: structure, crystal growth, and magnetic properties, Cryst. Growth Des. 9 (2009) 1402–1410.

[29] M. Kariem, M. Yawer, H.N. Sheikh, Solvent induced synthesis, structure and properties of coordination polymers based on 5-hydroxyisophthalic acid as linker and 1,10-phenanthroline as auxiliary ligand, J. Solid State Chem. 231 (2015) 239–247.

[30] M. Sajid, Toxicity of nanoscale metal organic frameworks: a perspective, Environ. Sci. Pollut. Res. 23 (2016) 14805–14807.

[31] Y. Liang, W.G. Yuan, S.F. Zhang, Z. He, J. Xue, X. Zhang, et al., Hydrothermal synthesis and structural characterization of metal-organic frameworks based on new tetradentate ligands, Dalt. Trans. 45 (2016) 1382–1390.

[32] H. Wei, E. Wang, Nanomaterials with enzyme-like characteristics (nanozymes): next-generation artificial enzymes, Chem. Soc. Rev. 42 (2013) 6060–6093.

[33] T. Sattar, M. Athar, M.N. Haq, Hydrothermal synthesis, characterization, and in vitro drug adsorption studies of some nano-bioMOFs, J. Nanomater. 2016 (2016) 1–10.

[34] H.V. Doan, Y. Fang, B. Yao, Z. Dong, T.J. White, A. Sartbaeva, et al., Controlled formation of hierarchical metal-organic frameworks using CO_2-expanded solvent systems, ACS Sustain. Chem. Eng. 5 (2017) 7887–7893.

[35] S. Wu, L. Qin, K. Zhang, Z. Xin, S. Zhao, Ultrathin 2D metal-organic framework nanosheets prepared via sonication exfoliation of membranes from interfacial growth and exhibition of enhanced catalytic activity by their gold nanocomposites, RSC Adv. 9 (2019) 9386–9391.

[36] B. Yuan, J. Zhang, R. Zhang, H. Shi, N. Wang, J. Li, et al., Cu-based metal–organic framework as a novel sensing platform for the enhanced electro-oxidation of nitrite, Sens. Actuators B Chem. 222 (2016) 632–637.

[37] J. Tang, J. Wang, Fe-based metal organic framework/graphene oxide composite as an efficient catalyst for Fenton-like degradation of methyl orange, RSC Adv. 7 (2017) 50829–50837.

[38] C. Zhang, L. Ai, J. Jiang, Solvothermal synthesis of MIL-53(Fe) hybrid magnetic composites for photoelectrochemical water oxidation and organic pollutant photodegradation under visible light, J. Mater. Chem. A 3 (2015) 3074–3081.

[39] D. Fatta-Kassinos, S. Meric, A. Nikolaou, Pharmaceutical residues in environmental waters and wastewater: current state of knowledge and future research, Anal. Bioanal. Chem. 399 (2011) 251–275.

[40] M. Rezaei, A. Abbasi, R. Varshochian, R. Dinarvand, M. Jeddi-Tehrani, NanoMIL-100(Fe) containing docetaxel for breast cancer therapy, Artif. Cells Nanomed. Biotechnol. 46 (2018) 1390–1401.

[41] S. Lu, L. Liu, H. Demissie, G. An, D. Wang, Design and application of metal-organic frameworks and derivatives as heterogeneous Fenton-like catalysts for organic wastewater treatment: a review, Environ. Int. 146 (2021) 106273.

[42] Y. Yuan, L. Wang, L. Gao, Nano-sized iron sulfide: structure, synthesis, properties, and biomedical applications, Front. Chem. 8 (2020) 818.

[43] D. Zhao, W. Zhang, Z.H. Wu, H. Xu, Nanoscale metal−organic frameworks and their nanomedicine applications, Front. Chem. 9 (2022) 834171.

[44] Y. Sun, L. Sun, D. Feng, H.C. Zhou, An in situ one-pot synthetic approach towards multivariate zirconium MOFs, Angew. Chemie Int. Ed. 55 (2016) 6471–6475.

[45] C. Fu, H. Zhou, L. Tan, Z. Huang, Q. Wu, X. Ren, et al., Microwave-activated Mn-doped zirconium metal-organic framework nanocubes for highly effective combination of microwave dynamic and thermal therapies against cancer, ACS Nano. 12 (2018) 2201–2210.

[46] I. Abánades Lázaro, S. Haddad, S. Sacca, C. Orellana-Tavra, D. Fairen-Jimenez, R.S. Forgan, Selective surface PEGylation of UiO-66 nanoparticles for enhanced stability, cell uptake, and pH-responsive drug delivery, Chem 2 (2017) 561.

[47] P. Hirschle, T. Preiß, F. Auras, A. Pick, J. Völkner, D. Valdepérez, et al., Exploration of MOF nanoparticle sizes using various physical characterization methods-is what you measure what you get? Cryst. Eng. Comm. 18 (2016) 4359–4368.

[48] P. Deria, D.A. Gómez-Gualdrón, W. Bury, H.T. Schaef, T.C. Wang, P.K. Thallapally, et al., Ultraporous, water stable, and breathing zirconium-based metal-organic frameworks with FTW topology, J. Am. Chem. Soc. 137 (2015) 13183–13190.

[49] X.J. Kong, J.R. Li, An overview of metal–organic frameworks for green chemical engineering, Engineering 7 (2021) 1115–1139.

[50] L. Peng, Z. Wei, Catalyst engineering for electrochemical energy conversion from water to water: water electrolysis and the hydrogen fuel cell, Engineering 6 (2020) 653–679.

[51] S. Wang, A. Lu, C.J. Zhong, Hydrogen production from water electrolysis: role of catalysts, Nano Converg. 8 (2021) 1–23.

[52] M. Plevová, J. Hnát, K. Bouzek, Electrocatalysts for the oxygen evolution reaction in alkaline and neutral media: a comparative review, J. Power Sources 507 (2021) 230072.

[53] X. Li, L. Zhao, J. Yu, X. Liu, X. Zhang, H. Liu, et al., Water splitting: from electrode to green energy system, Nano Micro Lett. 12 (2020) 1–29.

[54] M. Cheng, C. Lai, Y. Liu, G. Zeng, D. Huang, C. Zhang, et al., Metal-organic frameworks for highly efficient heterogeneous Fenton-like catalysis, Coord. Chem. Rev. 368 (2018) 80–92.

[55] R. Diana, L. Sessa, S. Concilio, S. Piotto, B. Panunzi, Luminescent Zn (II)-based nanoprobes: a highly symmetric supramolecular platform for sensing of biological targets and living cell imaging, Front. Mater. 8 (2021) 450.

[56] A. Mishra, R. Dheepika, P.A. Parvathy, P.M. Imran, N.S.P. Bhuvanesh, S. Nagarajan, Fluorescence quenching based detection of nitroaromatics using luminescent triphenylamine carboxylic acids, Sci. Rep. 11 (2021) 19324.

[57] S. He, L. Wu, X. Li, H. Sun, T. Xiong, J. Liu, et al., Metal-organic frameworks for advanced drug delivery, Acta Pharm. Sin. B 11 (2021) 2362–2395.

[58] R.X. Yao, X. Cui, X.X. Jia, F.Q. Zhang, X.M. Zhang, A luminescent zinc(II) metal-organic framework (MOF) with conjugated π -electron ligand for high iodine capture and nitro-explosive detection, Inorg. Chem. 55 (2016) 9270–9275.

[59] E.V. Perez, C. Karunaweera, I.H. Musselman, K.J. Balkus, J.P. Ferraris, Origins and evolution of inorganic-based and MOF-based mixed-matrix membranes for gas separations, Processes 4 (2016) 32.

[60] Z. Xie, W. Xu, X. Cui, Y. Wang, Recent progress in metal–organic frameworks and their derived nanostructures for energy and environmental applications, Chem. Sus. Chem. 10 (2017) 1645–1663.

[61] C. Liu, X. Xu, J. Zhou, J. Yan, D. Wang, H. Zhang, Redox-responsive tumor targeted dual-drug loaded biocompatible metal–organic frameworks nanoparticles for enhancing anticancer effects, BMC Mater. 2 (2020) 1–11.

[62] R. Safdar Ali, H. Meng, Z. Li, Zinc-based metal-organic frameworks in drug delivery, cell imaging, and sensing, Molecules 27 (2021) 100.

[63] M. Karimi, Z. Mehrabadi, M. Farsadrooh, R. Bafkary, H. Derikvandi, P. Hayati, et al., Metal–organic framework, Interface Sci. Technol. 33 (2021) 279–387.
[64] J. Li, W. Ye, C. Chen, Removal of toxic/radioactive metal ions by metal-organic framework-based materials, Interface Sci. Technol. 29 (2019) 217–279.
[65] Y.B. Lu, J. Huang, X.R. Yuan, S.J. Liu, R. Li, H. Liu, et al., A three-dimensional porous Mn(II)-metal-organic framework based on a caged structure showing high room-temperature proton conductivity, Cryst. Growth Des. 22 (2022) 1045–1053.
[66] V.R. Remya, M. Kurian, Synthesis and catalytic applications of metal–organic frameworks: a review on recent literature, Int. Nano Lett. 9 (2019) 17–29.

Chapter 6

Mechanochemical synthesis of metal–organic frameworks

Bhaskar Nath
A. M. School of Education, Assam University, Silchar, Assam, India

Abbreviations

BET	Brunauer–Emmett–Teller
bipy	4,4′-bipyridine
bpe	trans-1,2-bis(4-pyridyl)-ethene
BTB	4,4′,4″-benzenetribenzoate
cnge	cyanoguanidine
DABCO	diazabicyclo[2.2.2]octane
Dace	trans-1, 4-diaminocyclohexane
DMF	N, N-dimethylformamide
DMSO	dimethylsulphoxide
FDC	9-fluorenone-2,7-dicarboxylate
GO	graphite oxide
g-C_3N_4	graphitic carbon nitride
H_3BTC	1,3,5-benzenetricarboxylic acids (trimesic acid)
HEtIm	2-ethylimidazole
HKUST-1	copper(II) trimesate
Hta	terephthalic acid
Him	imidazole
HMeIm	2-methylimidazole
H_2fum	fumeric acid
H_4dhta	2,5-dihydroxyterephthalic acid (or 2,5-dihydroxy-1,4-dicarboxylic acid)
ILAG	ion- and liquid-assisted grinding
INA	isonicotinic acid
IR	infrared
IR MOF	isorecticular MOF
LAG	liquid-assisted grinding
MIL	Materials of Institute Lavoiser Frameworks
MOF	metal–organic framework
NG	neat grinding
NMR	nuclear magnetic resonance
OMC	ordered mesoporous carbon

Synthesis of Metal–Organic Frameworks via Water-Based Routes: A Green and Sustainable Approach.
DOI: https://doi.org/10.1016/B978-0-323-95939-1.00014-9

PXRD	powder X-ray diffraction
SSA	specific surface area
UiO	zirconium 1,4-dicarboxybenzene MOF
VOCs	volatile organic compounds
ZIF	zeolitic imidazolate framework

6.1 Introduction

6.1.1 Metal–organic frameworks (MOFs)

Metal–organic frameworks are a class of highly ordered and crystalline materials composed of metal ions/clusters and polydentate organic ligands. The most common ligands used in MOF designing are aromatic di-, tri- or tetra-carboxylic acids, N-containing aromatics, etc. [1,2]. Because of their superior crystallinity, great porosity (as much as 90% free volume), and enormous internal surface areas (>6000 m^2/g), MOFs have increasingly gained recognition as smart materials [3,4]. Additionally, by changing the metal and ligands, and by regulating the reaction conditions during their production, the pore diameters of these compounds may be altered. These capabilities, together with the unusual degree of flexibility of different components of their structures, have made them appealing materials in the fields of energy storage [5–7], gas adsorption [8,9], hydrocarbon adsorption/separation [10], catalysis [11,12], sensor [13], magnetic [14], drug delivery [15], luminescence [16], biomedical imaging [17], etc. Furthermore, MOFs can also be functionalized further to cater to the requirements of individual applications. The pores of MOFs can also be modified post-synthetically to contain functional groups with various characteristics as needed [18].

Despite the fact that hundreds of novel MOF structures with intriguing uses have been reported during the last few decades, only a small number of them have been manufactured at a massive level for application scenarios. The primary causes of this are (1) the inability of the majority of MOFs to withstand humidity and elevated temperature; (2) raw ingredients are expensive, and (3) the challenge to scale up the synthesis. MOFs are traditionally manufactured in a one-pot reaction involving metal ions and multi-dentate ligands at high temperatures and pressure [19]. However, mild reaction conditions with a controlled rate of reaction are always favorable for obtaining single-crystal MOF materials. The most prevalent synthetic methodologies developed over the last two decades are represented in Fig. 6.1. The development of new synthetic methods has played a significant contribution to the evolution of the MOF research area. The most popular method of synthesis, known as solvothermal synthesis, involves combining ligands with metal ions or clusters in a variety of solvents, then allowing the mixture to react at a high temperature within a Teflon-lined bomb. The foremost benefit of this approach is that it produces MOFs with extremely high crystallinity, which enables the synthesis of a wide range of MOFs with diverse sizes, morphologies, and crystalline structures. The majority of

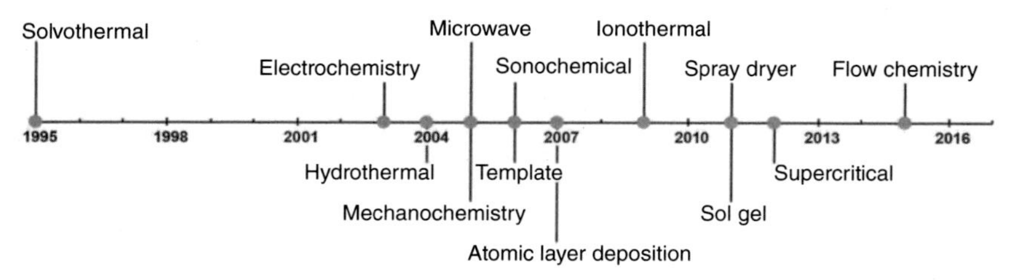

FIGURE 6.1 Chronology of different synthetic means for the construction of metal–organic frameworks [21].

well-known MOFs, for example, IRMOF, MILs, ZIFs, and UiOs are synthesized by this solvothermal method [20]. However, this method is not feasible for MOF production on a large scale, since the nucleation of MOFs occurs at the surface of reactor vessels. A considerable fall in the surface-to-volume ratio occurs as the reactor vessel is scaled up, which lowers the reaction's efficiency. According to Rubio-Martinez et al., there are several critical factors to be considered in order to achieve industrial manufacturing of MOFs: (1) to discover a flexible approach to fit most MOF structures using the same apparatus, (2) limiting extreme reaction circumstances, lowering operational costs and easing safety concerns, (3) it is desirable to switch from batch-wise synthesis to a continuous process, which would increase the rate of productivity, and (4) to attain a high space–time-yield factor, which determines the quantity of MOF generated per unit volume [21]. The aforementioned issues make scaling up MOF manufacturing difficult, which has driven numerous scientists to create new and economically feasible techniques for producing MOFs effectively and affordably.

6.1.2 Mechanochemical synthesis

The term mechanochemistry describes a reaction that occurs when mechanical energy is applied to reactants, such as when they are ground in a ball mill. Mechanical forces can spark a variety of physical phenomena and also chemical reactions. In this type of synthesis, applied mechanical force causes the rupture of intramolecular interactions followed by chemical reactions. Even though mechanochemical processes are known since 400 BC, the most advancement in this area has been growing speedily in the last few years only. Recent studies have demonstrated that the use of mechanical force, such as milling or grinding, maybe a great complement to the more traditional solution-based and solvothermal approaches for different types of organic as well as inorganic syntheses [22]. As a result, this technique has been rapidly adopted in many branches of chemistry, including catalysis, pharmaceutical synthesis, and inorganic chemistry, since it may stimulate quantitative reactions between solids with little or no additional solvent [23–26]. The mechanochemical approach offers the benefits of having quick product formation, huge preparation quantities, using a minimal amount of solvent or without solvent at ambient temperature, and encouragement of the industrialization of MOFs [27]. Furthermore, the solid-solid reaction offers immense benefits in terms of easy handling and large-scale manufacturing, and it produces powdered compounds with quantifiable yields. Mechanosynthesis is one of the most ecologically benign procedures, and it is anticipated as an economically viable process for the fabrication of a range of unique materials.

6.1.3 Mechanochemistry for MOF synthesis

Among the various routes known for MOF synthesis, the solvothermal route is the most often employed because it allows for exact control of particle size, shape, and crystallinity. Nevertheless, there are some shortcomings of this solvothermal method; namely, (1) it is a time-consuming route as the reaction time is too long, (2) it needs severe reaction conditions (elevated temperature and pressure), and (3) this method uses a large amount of organic solvents which may cause environmental pollutions. In addition, the MOFs prepared by the solvothermal method incorporate solvent molecules within their pores whose removal by heating may cause a structural collapse of the framework. Accordingly, the tedious post-synthetic process is also a difficult problem in the solvothermal method.

Looking at the emergent applications of MOFs, it is necessary to develop a new synthetic technique that enables their scalable synthesis in an environmentally friendly and commercial manner. In this context, we should utilize safer and less expensive metal precursors instead of thermally sensitive, poisonous, and expensive organic solvents. Therefore, it is beneficial to switch out hazardous metal salts like nitrates and chlorides with safer metal sources like metal sulfates or oxides. The usage of metal oxides as a source of metal precursor has a significant advantage and is environmentally beneficial because the only reaction byproduct in most cases is water. In addition, metal oxides are expected to be cheap and widely accessible. Unfortunately, due to the low solubility of metal oxides, they are rarely employed in solvent-based processes for MOF production. The aforementioned problems faced during the synthesis of MOFs by the solvothermal method can be effortlessly overcome by mechanochemical synthesis (Fig. 6.2). Indeed, mechanochemistry has drawn the attention of material chemists as one of the most competent synthetic approaches for achieving large-scale MOF manufacturing [27,28]. Mechanochemical synthesis either eliminates the need for organic solvent entirely or decreases its quantity to a catalytic or near-stoichiometric level, enabling the reactions to take place independently of the solubilities of starting components and products [29]. Thus, less soluble metal precursors such as metal carbonate and oxides can be employed in MOF synthesis using the mechanochemical method. The newly introduced mechanochemical methods, namely, liquid-assisted grinding (LAG) and ion- and liquid-assisted grinding (ILAG) methods are especially appealing for that purpose [22,29]. These two approaches are effective for

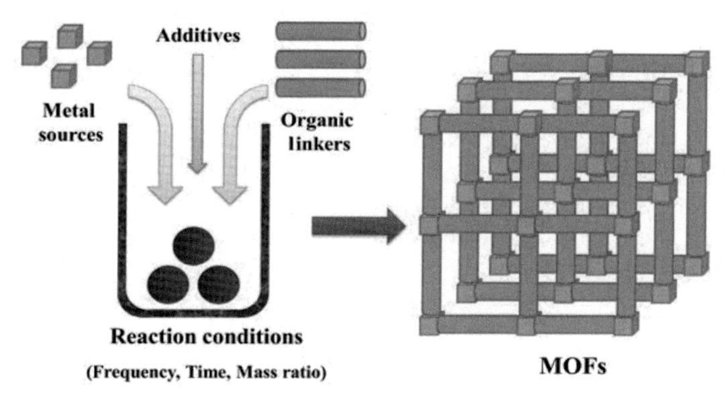

FIGURE 6.2 Schematic presentation of metal–organic framework synthesis by ball milling [28].

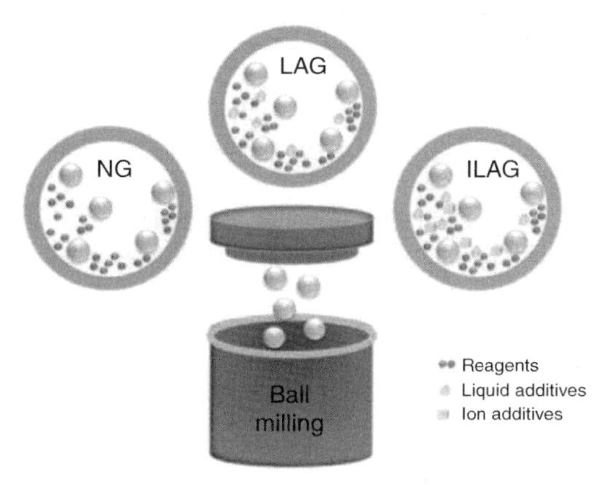

FIGURE 6.3 Presentation of different mechanochemical techniques in a schematic format, including neat grinding, liquid-assisted grinding, and ion- and liquid-assisted grinding method [27].

systems that are initially unreactive or have a poor yield. In fact, almost all the examples of MOFs provided in this chapter are obtained by using exact stoichiometric amounts of reagents (either ligand or metal precursor); whereas the conventional solution-based solvothermal method always uses excess reagents.

6.2 Methods of mechanochemical synthesis of MOFs

The mechanochemical reactions used to synthesize MOFs can be divided into 3 categories (Fig. 6.3): (1) Neat grinding (NG), that is, it does not require any solvent; (2) Liquid-assisted grinding method, that is, addition minimal amount of solvent (few drops) to activate the reactant at the molecular level, thereby accelerating the mechanochemical reaction. The added solvent could be quite

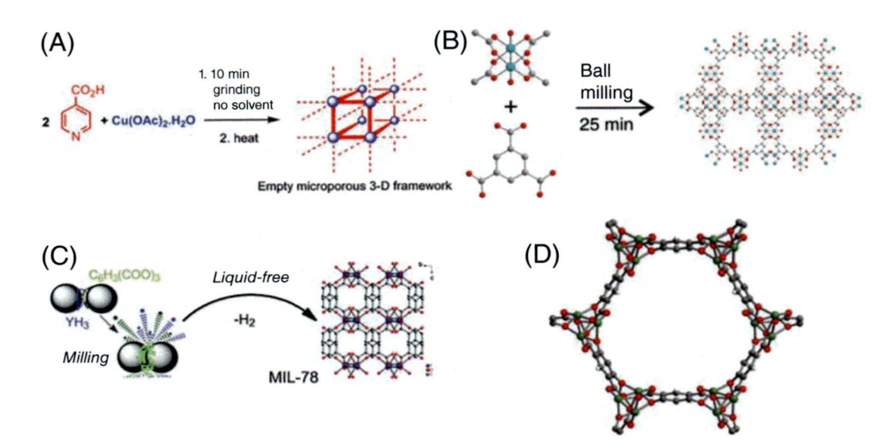

FIGURE 6.4 Synthesis of different metal–organic frameworks by neat grinding method: (A) porous and 3D framework of $[Cu(INA)_2]$ [30]; (B) copper(II) trimesate was created using a copper salt and H_3BTC; (C) MIL-78 [34]; (D) Mg_2(m-dhta) metal–organic framework [36]. *(Reproduced with permission from ref. [32]. Copyright (2010), American Chemical Society).*

important for shaping the MOF's ultimate structure; (3) The ion- and liquid-assisted grinding approach, accelerates the production of MOFs by simultaneously using a tiny quantity of solvent and salt ions [22]. This chapter will outline the contemporary mechanochemical synthesis of MOFs employing any of the aforesaid methodologies, as well as choose some exemplary cases for discussion and analysis.

6.2.1 Neat grinding

This is the most basic type of mechanochemical synthesis, and it entails the physical mixing of two or more reactants in solid form. The benefit of using this NG approach is that it is fully non-polluting and clean and does not call for any auxiliary liquid. Grinding can be performed physically with a mortar and pestle or using an electronically controlled programmable motorized ball mill. The findings of the earlier procedure, however, are hard to repeat due to a number of changeable elements including researcher strength, humidity, grinding frequency, and temperature of surroundings. Conversely, the later method has the ability to repeat the reaction in an identical way every time and thus gives reproducible products. The first solid-state synthesis of MOF $[Cu(INA)_2]$ having a microporous structure was reported by Pichon et al. in the year 2006 [30]. The co-grinding of Cu(II)-acetate and isonicotinic acid, together with heating to eliminate the byproducts water and acetic acid, results in an excellent production of $[Cu(INA)_2]$ that was hollow, crystalline, porous, and having a three-dimensional structure (Fig. 6.4A). Since then, this new method attracted much attention, and subsequently, several research groups had synthesized different MOFs by the mechanochemical method. James and Emmerling groups independently applied a similar reaction and constructed one exemplary MOF, for instance, HKUST-1 by milling a combination of trimesic acids (H_3BTC) and copper (II) acetate monohydrate (Fig. 6.4B) [31–33]. Singh et al. followed Pichon's lead and synthesized Yttrium-based MIL-74 by a mechanochemical reaction of Yttrium hydride with trimesic acid. Notably, this is the first instance where metal hydride has been employed as a metal source in the creation of MOFs (Fig. 6.4C) [34]. In 2018, Alammar et al. synthesized three different MIL-78 ($Ln_{0.5}Gd_{0.5}$, Ln=Eu, Tb, and Dy) MOFs following the solid state mechanochemical method, which show characteristic fluorescence of red, green, and yellow for Eu^{3+}, Tb^{3+}, and Dy^{3+}, respectively [35]. Thermogravimetric analysis shows that these MOFs decompose at 600°C, indicating that these materials have good thermal stability. Wang and colleagues investigated the mechanochemical production of MOF-74 mimics by merely grinding an excess of $Mg(NO_3)_2.6H_2O$ with H_4dhta for 5 minutes at ambient temperature when an exogenous base is present [36]. By using PXRD, it was demonstrated that adding 4.4 equivalents of Hünig's base to the reaction mixture resulted in the clear production of highly crystalline Mg_2(dhta) (Fig. 6.4D). They had decided to employ Hünig's base because it is less coordinating and volatile than Me_2NH ($pK_b = 3.3$) and Me_3NH ($pK_b = 3.2$), both of which have been used earlier to make analogs of MOF-74 [37,38]. This finding demonstrates that Mg_2(dhta) may be prepared in the solid state using just metal salt, an organic linker, and an amine base (Table 6.1).

6.2.2 Liquid-assisted grinding method

The addition of solvent in catalytic quantity of can accelerate the milling reaction. LAG is a solvent-catalyzed reaction, where small amount of solvent provide lubrication for molecular diffusion [29,39]. The commonly used solvents in the LAG method are water [40], ethanol [41], DMF [42], their mixture [43], etc. In 2006, Braga et al. used the LAG approach to synthesize a 1D coordination polymer with a porous structure, namely, $CuCl_2$(Dace). It was further demonstrated that adding modest quantities of water or DMSO may significantly increase the rate of $CuCl_2$(Dace) production as well as its crystallinity [44]. Klimakow et al. synthesized HKUST-1 and MOF-14 by LAG reaction using EtOH as grinding liquid and analyzed them using PXRD and other spectroscopic methods (Fig. 6.5A) [32]. The framework structures produced by this green chemistry method include readily removable guest molecules, resulting in reproducible open pores for use in various applications. The specific surface areas of these two compounds synthesized via ball milling were obtained by N_2-adsorption and confirmed that MOFs synthesized by the LAG method have higher specific surface area than their corresponding MOFs synthesized by the solvothermal method. Moreover, this method gives an easy procedure, quantitative yields, and high-quality materials. After that, Klimakow et al. performed simple washing with ethanol for cleaning and activation on HKUST-1 and MOF-14 prepared

TABLE 6.1 Selected examples of metal–organic framework prepared by mechanochemical reaction.

MOF	Method	Content	References
$[Cu(INA)_2]$	NG	Solvent-free grinding was used to create MOF.	[30]
HKUST-1		The morphology and reactivity of MOF produced via grinding are being investigated.	[31]
HKUST-1, MOF-14	LAG	MOF with a large surface area are produced using mechanochemical synthesis.	[32]
MOF-14	NG	Synthesis of $Cu_3(BTB)_2$ by ball milling.	[33]
Yttrium based MIL–78	NG	The synthesis of a MOF has been done for the first time using a metal hydride.	[34]
Ln-MIL-78	NG	Lanthanide based MIL-78 was created mechanochemically and its luminescent characteristics were studied.	[35]
MOF-74 analogs	NG	Exogenous organic base-induced rapid mechanochemical synthesis of Mg-MOFs.	[36]
Rare-earth(III) MOFs	LAG	Homo- as well as hetero-nuclear rare earth MOFs was synthesized by grinding for 20 minutes.	[40]
$CuCl_2$(Dace)	LAG	A one-dimensional coordination network for small molecule absorption.	[44]
MFeLAG	LAG	A MOF based on iron carboxylate and its application in diesel fuel desulfurization.	[45]
MOF-525 or PCN-223	LAG	Mechanochemistry is used to control polymorphism and topological modification in porphyrin based Zr containing MOFs.	[46]
UiO-66 and UiO-66-NH_2	NG, LAG	MOFs syntheses via milling or just by exposing solid mixtures of reactants to organic vapor.	[47]
Bimetallic MM′-MOF-74	LAG	Mechanochemical reactions of pre-assembled intermediate complexes with ZnO or Zn-acetate yield bimetallic MOFs.	[49]

(*continued on next page*)

TABLE 6.1 Selected examples of metal–organic framework prepared by mechanochemical reaction—cont'd

MOF	Method	Content	References
Fluorenone-based Zn-MOF	LAG	A fluorenone-based 2D pillared MOF with polarized fluorescence property was synthesized mechanochemically.	[50]
Zn-MOF-74	NG	In situ monitoring of MOF-74 formation by mechanochemical reaction.	[51]
Zn Fumarate MOF	LAG	Fumaric acid and ZnO's LAG reactivity.	[52]
$Zn_2(ta)_2(DABCO)$	ILAG	Pillared MOF is created utilizing zinc oxide, hta, and DABCO while also adding a DMF as additives.	[59]
ZIFs	ILAG	Rapid room-temperature synthesis of zeolitic imidazolate frameworks using the ILAG method.	[62]

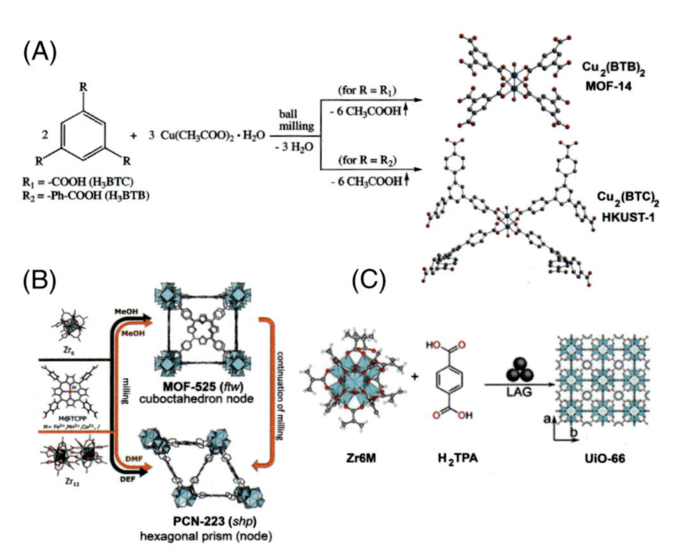

FIGURE 6.5 (A) Production of copper(II) trimesate and MOF-14 by mechanochemical means. (*Reproduced with permission from ref. [32], Copyright (2010) American Chemical Society;* (B) Structure changes in MOF-525 and PCN-223 during synthesis. *Reproduced with permission from ref. [46], Copyright (2019), American Chemical Society;* (C) Uio-66 synthesis through mechanochemical means. *Reproduced with permission from ref. [48]. This work is licensed under Creative Common licenses (CC-BY 4.0).*

by the LAG method and characterized them. It was confirmed that its specific surface area was greatly improved after activation (MOF-14 from 614 m^2/g to 1215 m^2/g; HKUST-1 from 758 m^2/g was increased to 1713 m^2/g), and a good microporous structure was obtained [33]. As well, Pilloni et al. successfully prepared an iron carboxylate-MOF structurally similar to MIL-100(Fe) [45] by liquid-assisted grinding of $Fe(NO_3)_3.9H_2O$ and trimesic acid (H_3BTC) at room temperature. This MOF possesses an enhanced specific surface area, larger pore volume, higher thermal stability, along with excellent crystallinity. The samples obtained by ball milling showed maximum adsorption capacity toward 4,6-dimethyldibenzothiophene from 4,6-DMDBT in n-heptane solutions compared to the commercially available one. In 2019, Karadeniz et al. revealed that various additives may regulate polymorphism in porphyrin-based zirconium MOFs, resulting in hexagonal PCN-223 and cubic MOF-525 structures after 20–60 minutes of grinding (Fig. 6.5B) [46]. Užarević et al. established a streamlined and unconventional mechanochemical technique for zirconium MOFs, which can produce UiO-66 and UiO-66-NH_2 in large amounts without using any strong acids or surplus of reactants at ambient temperatures [47]. In this case, the frameworks were constructed either by grinding or by merely giving exposure to organic vapor to the mixture of solid reactants.

Following that, Germann et al. employed the synchrotron powder XRD method to examine in situ real-time reactions that occur during the mechanochemical production of UiO-66 and UiO-66-NH_2 [48]. These real-time diffraction experiments allow for the evaluation of the growth of product crystallite size throughout the mechanochemical production of MOFs. Notably, the use of MeOH as the LAG additive resulted in considerable lattice shrinkage, whereas the use of DMF as the LAG additive led to a little expansion and contraction of the framework structures. Furthermore, it was revealed that the cell parameters determined from in situ studies were somewhat higher than the majority of those found in the literature.

Ayoub et al. employed the LAG approach to rapidly synthesize bimetallic MOFs with controlled stoichiometry [49]. In this method, they had shown how to produce bimetallic MM′-MOF-74 with various combinations of Zn^{2+}, Mg^{2+}, Co^{2+}, Ni^{2+}, Cu^{2+}, and Ca^{2+} through milling reactions of precursor compounds, namely, $Mg(H_2O)_5(H_2dhta).H_2O$, $[Ni\text{-}(H_2O)_4(H_2dhta).2H_2O]_n$ and $[Co(H_2O)_4(H_2dhta).2H_2O]_n$ with oxides or acetates of a different metal. They had synthesized a series of micro-porous mixed metal MOF-74 assemblies with different combinations of metal ions through liquid-assisted ball milling, as shown in Fig. 6.6A and B. This is the first mechanochemical approach for targeted oriented mixed metal MOF synthesis using specially created precursors.

By using an easy and sustainable method of liquid-assisted mechanochemical synthesis, Yan et al. successfully created 2D pillared Zn-based luminescent MOFs, namely, Zn-FDC-MOF (FDC=9-fluorenone-2,7-dicarboxylate) [50]. It was discovered from the PXRD experiment that dry milling of FDC and ZnO in a 1:1 ratio has no obvious diffraction peaks, indicating that they did not

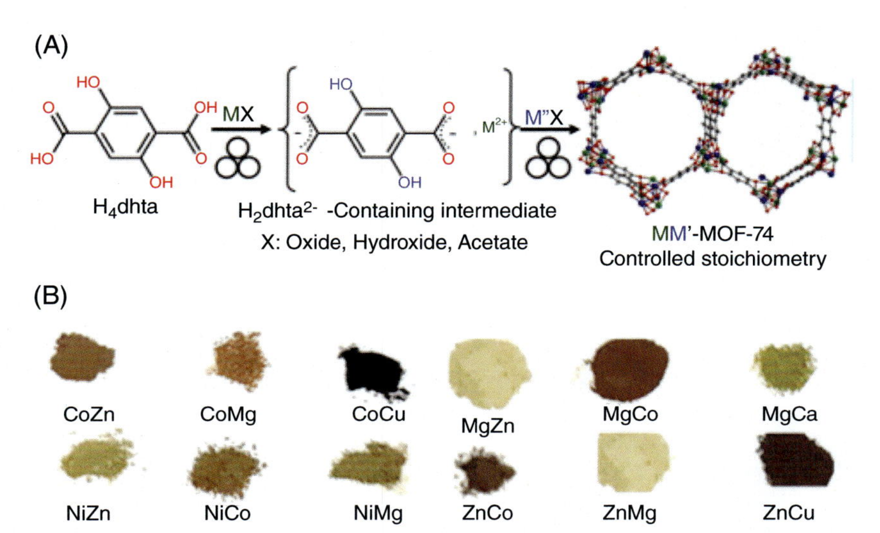

FIGURE 6.6 (A) Mechanochemical preparation of MM′-MOF-74; (B) Images of MM′- MOF-74 synthesized by mechanochemical method. *(Reproduced with permission from ref. [49]. Copyright (2019), American Chemical Society).*

react. However, after adding 30 μL of ethanol, chloroform, or acetone to the mixture and grinding it, the powder exhibits distinct PXRD patterns from its precursors, indicating the formation of new compositions, such as Zn-FDC MOF. The newly synthesized Zn-FDC MOF having pillared structure exhibits a blue-shifted photoluminescence with superior PLQY and fluorescence lifetime when compared to pure FDC sample, indicating a highly organized and uniform arrangement of the FDC molecules confined inside the MOF effectively diminishes the non-radiative relaxation process. Another Zn-based MOF, namely Zn-MOF-74 was synthesized by LAG reaction of ZnO with H_4dhta, the in situ real-time monitoring results show that water-assisted grinding can accelerate mechanochemical synthesis, while DMF-assisted grinding brought the BET surface area of MOF-74 to its maximum reported value of 1080–1145 m^2/g [51]. In this example, the milling reaction proceeds through an intermediate phase with a densely packed structure before it creates the porous MOF-74 framework. It was proposed that Zn coordinates with the carboxylate groups in the initial step to generate the densely packed intermediate before it coordinates with the phenol groups to form the ultimate porous structure.

Friščić et al. demonstrated an environmentally friendly synthesis of diverse metal–organic compounds using a simple technique and inexpensive metal oxide precursor [52]. They produced a variety of coordination polymers by grinding ZnO and fumaric acid together in the presence of diverse solvent additives (Fig. 6.7B). The milling of the mixture forms a 1D-linear polymer when water is present as an additive, whereas it forms a 3D closed-packed framework structure

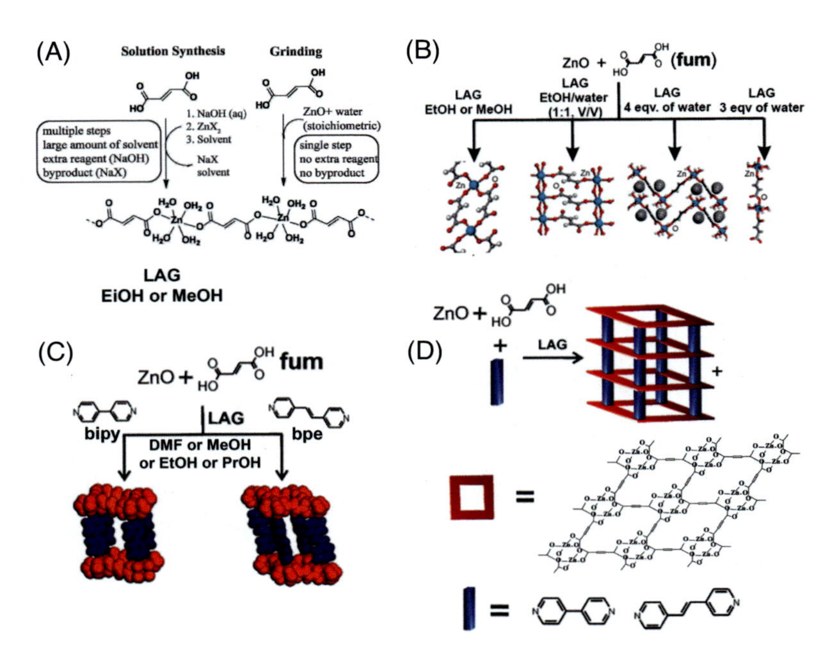

FIGURE 6.7 (A) Comparison between the solvent-based and the grinding methods; (B) Reactions of ZnO and fumaric acid by liquid-assisted grinding (LAG) method at different stoichiometric ratios (guest water molecules are depicted in *gray*); (C) Examples of Zn-fumarate based pillared MOFs prepared by LAG method and (D) Schematic presentation of the reaction [52].

in the presence of ethanol or methanol. The Addition of a second ligand bipy or bpe and subsequent grinding for 20 minutes in the presence of additives such as MeOH, EtOH, i-PrOH, or DMF at room temperature yields two pillared as shown in Fig. 6.7C. Thermal desolvation of these MOFs did not result in the collapse of their structures, as shown by PXRD studies, proving their good structural stability.

6.2.3 Ion- and liquid-assisted grinding method

The ILAG technique can be considered as an extension of the LAG method where both liquid as well as salts are added. The presence of these additives encourage the dissolution of the reactants making it a homogeneous reaction mixture and thus increases the substrate reactivity and boosts the milling process [27]. Additionally, the quality of the final product and reaction selectivity can both be improved by using the appropriate additives [53–58]. This approach is more promising when metal oxides are used as a precursor during the mechanochemical synthesis of MOF [29]. Friščić et al. failed in their attempt to create highly porous pillared MOF by simply grinding a combination of zinc oxide, hta, and DABCO in the presence of DMF additive. In contrast, the

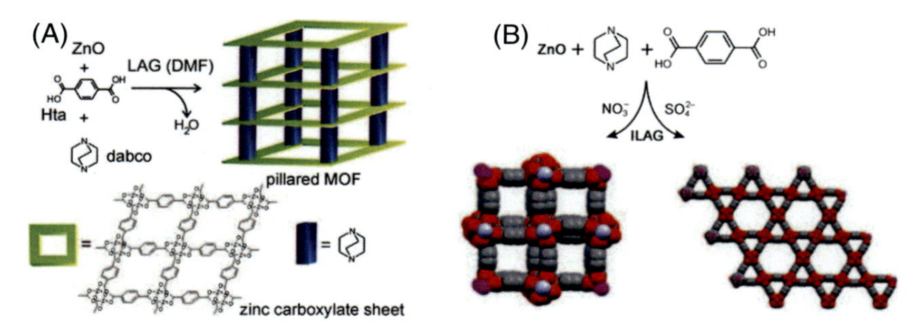

FIGURE 6.8 (A) Expected metal–organic framework assembly formed by ion- and liquid-assisted grinding reaction of ZnO, Hta, and diazabicyclo[2.2.2]octane; (B) two different types of isomer of Zn–MOF [59].

aforesaid reaction works nicely and yields pillared MOF, $Zn_2(ta)_2(DABCO)$, when a small quantity of ionic salt is added to it [59]. The use of sulfate salts [for example Na_2SO_4 or $(NH_4)_2SO_4$] produces a structure with hexagonal-shaped zinc terephthalate sheets, whereas the use of nitrate salts (for example KNO_3 or NH_4NO_3) produces a porous network based on a square grid of zinc terephthalate (Fig. 6.8A and B). These MOF structural alterations were brought on by the salt ion template effect, and it was validated by solid-state DP-MAS ^{15}N-NMR and vibrational spectroscopy.

The mechanochemical synthesis of ZIF from ZnO and imidazole derivative is likely the case study that has received the most attention. Initially, synthesis of metal imidazolates from its precursor compounds under solvent-free conditions was time-consuming and resulted in a mixture of residual material as well as weakly crystalline products or the reaction did not happen at all, regardless of reaction time. It has been demonstrated in earlier studies that ZnO does not react with HMeIm and HEtIm to form ZIFs in conventional milling reactions [60,61]. However, Friščić et al. found that adding sub-stoichiometric amounts of salts like NH_4NO_3, $NH_4CH_3SO_3$, and $(NH_4)_2SO_4$ along with a modest amount of liquid additives such as EtOH, N,N-Diethylformamide (DEF), or DMF, improves the mechanochemical reactivity [62]. Friščić claimed that both the grinding liquid, as well as inorganic salt additives, have an impact on the final structure of ZIFs. In fact, a number of porous as well as non-porous ZIFs with various topologies were synthesized as a result of these alterations. It is interesting to note that the choice of salts used as additives greatly influences how the mechanochemical reaction turns out. Again, the appearance of intermediate phases and their subsequent conversion into the most stable form were also significantly influenced by the kind of added liquid, especially when HIm was utilized as the ligand. All the reactions of ZnO with HMeIm produced an extremely stable sodalite structure (ZIF-8). At the same time, they found that different additives and salts can give rise to ZIFs with different topologies for the same reactant. For instance, while

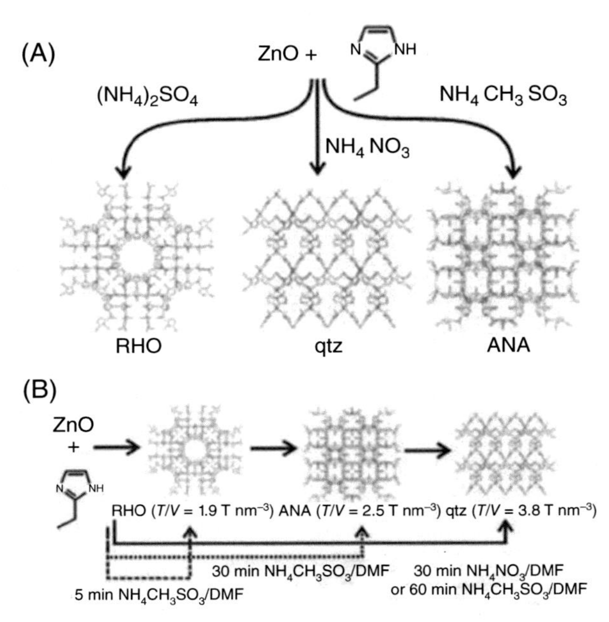

FIGURE 6.9 (A) An outline of mechanochemical reaction of ZnO with HEtIm: ion- and liquid-assisted grinding (ILAG) with $(NH_4)_2SO_4$; ILAG using NH_4NO_3 or $NH_4CH_3SO_3$ with EtOH as liquid additive; and ILAG with $NH_4CH_3SO_3$ and N,N-dimethylformamide or DEF as the liquid additives; (B) Structural alterations of zeolitic imidazolate framework with time during their synthesis by ILAG method (T/V denotes the amount of tetrahedral sites per nm^3) [62].

synthesizing ZIFs from HEtIm, various liquid and ionic salt additives result in the creation of three distinct topologies, RHO, qtz, and ANA (Fig. 6.9A).

6.3 Mechanochemical synthesis of MOF-based nano-composites

Composite refers to a solid substance constituting two or more components, where at least one substance functions as a matrix. In these types of material, each of the constituents retains their identities, while contributing desirable properties to the whole system. In most cases, uses of MOFs are restricted due to their poor electrical conductivity, fast photo-generated electron-hole recombination, etc. [63]. Therefore, high-porosity, structure-tunable MOF materials are combined with other functional materials to build a composite that can perform better than the constituent pure components. The synergistic effect produced by the material makes them more suitable for gas adsorption and storage, heterogeneous catalysis, and other wide range of applications. Mechanochemical reactions can be very helpful for synthesizing MOF-based nano-composites by grinding MOFs with other materials. Li et al. had prepared Cu-BTC and graphite oxide

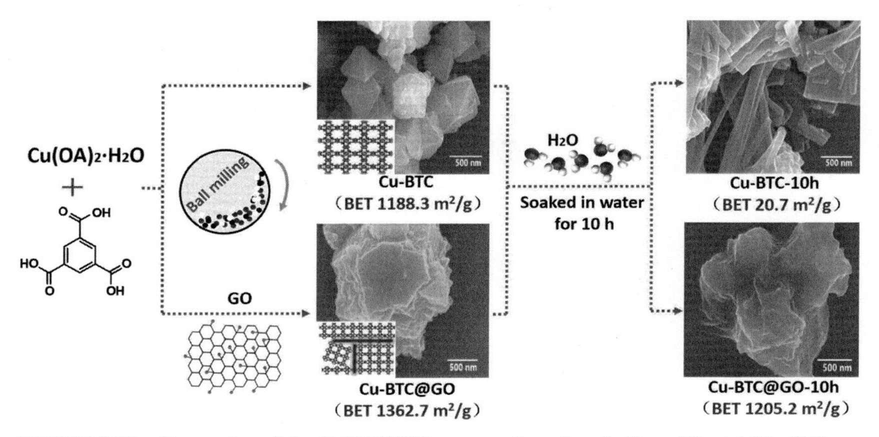

FIGURE 6.10 Preparation of the CuBTC@GOs composite and evaluation of its stability [64].

(Cu-BTC@GO) composite by ball milling of constituents under solvent-free conditions within 30 minutes [64]. This Cu-BTC@GO composite shows higher water stability, enhanced BET surface area (1362.7 m^2/g), and pore volume (0.87 cm^3/g), in comparison to the parent Cu-BTC MOF (Fig. 6.10). Furthermore, its toluene absorption capability increases by 47% when compared to pure Cu-BTC, making it a superior choice for use in VOC adsorption applications. By using a mechanochemical technique, Szczezsniak et al. created composite materials made of CuBTC, ordered mesoporous carbon, and graphene oxide (OMC/GO/CuBTC) [65]. In this method, triblock copolymer Pluronic F127 is used as a mesopore guiding agent together with tannin and graphene oxide as carbon precursors. In order to create a three-component composite with a distinct composition, the Cu-BTC MOF was synthesized using dry milling in the of presence of OMC/GO composite. The SSA of the as-prepared composite was 980 m^2/g, while the CO_2 uptake was 5.39 mmol/g.

MOF photocatalytic activity can be enhanced by forming a nanocomposite with g-C_3N_4. Since the majority of MOFs exhibit photocatalytic activity only within the ultraviolet range, their photocatalytic applications are quite limited. Similarly, the photocatalytic property of pure g-C_3N_4 is dominated by easy recombination of photo-induced electrons and holes. However, these two photocatalysts, MOF and g-C_3N_4 can be combined to conquer their shortcomings and unite their advantages, for instance the large SSA of the MOFs plus extraordinary chemical stability of g-C_3N_4. A good example is MIL-100(Fe)/g-C_3N_4 composite created by Du et al. using the ball milling and annealing process, which showed excellent photocatalytic activity for the conversion of Cr(VI) to Cr(III) and breakdown of diclofenac sodium when exposed to simulated sunlight (Fig. 6.11A) [66]. In addition, MOF composites are also widely used in the field of chemical sensing. Ko et al. employed a simple solvent-free ball milling approach to create a solid-state device for chemo resistive sensing of

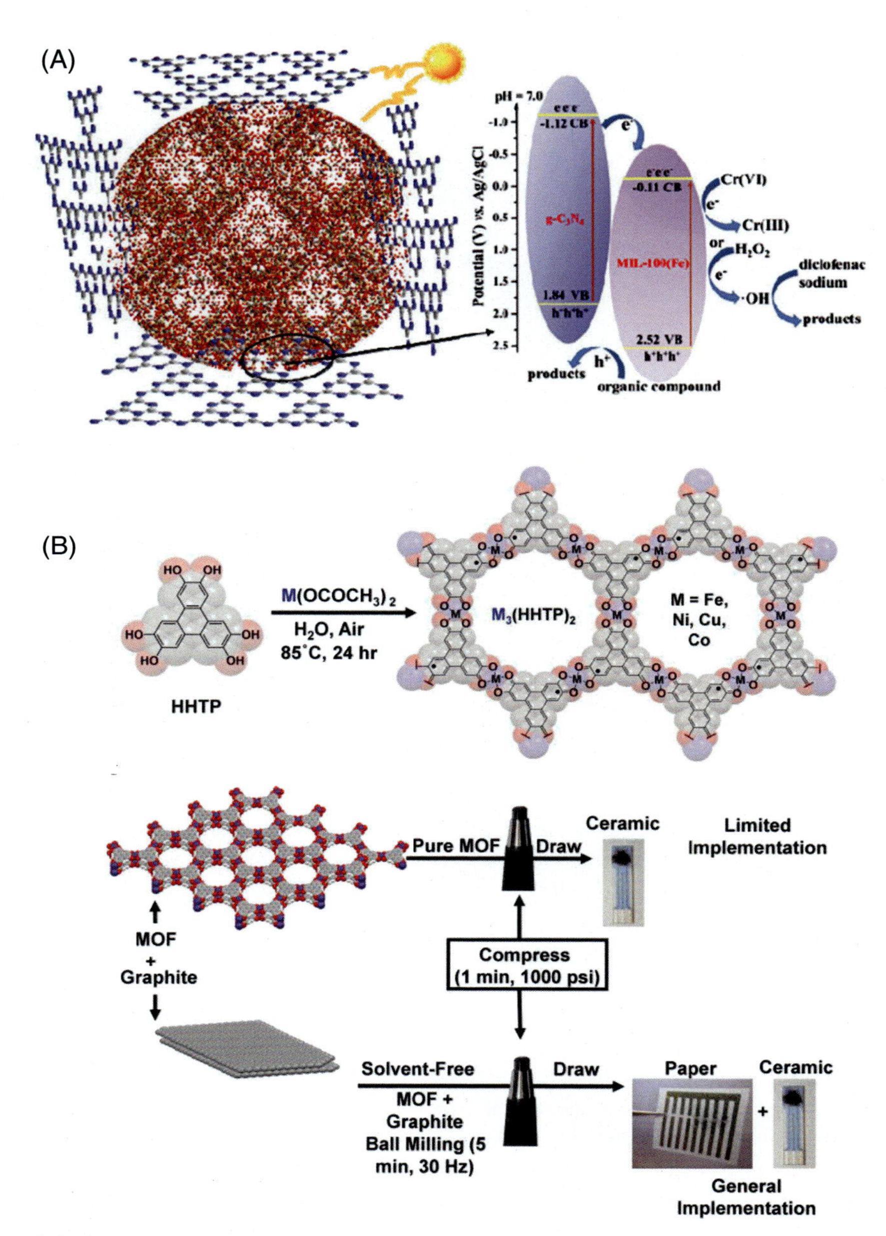

FIGURE 6.11 (A) A potential process for the photocatalytic removal of contaminants over MIL–100@g–C_3N_4 [66]; (B) An illustration showing the stepwise incorporation of M_3HHTP_2-MOF into chemiresistive sensor devices [67].

different gas molecules (Fig. 6.11B) [67]. The device is based on mixtures of triphenylene-based conductive MOFs (M_3HHTP_2) and graphite, which can distinguish between NH_3, H_2S, and NO at the ppm level.

6.4 Structural characterization of mechanochemically synthesized MOFs

Mechanochemistry has diverse applications in chemical synthesis, and the appropriate method for characterization of the product of mechanosynthesis might vary depending upon the sort of synthesis. Solution state ^1H and ^{13}C NMR spectroscopy is adequate in some fields, such as organic mechanochemical synthesis, to identify the products when the focus is more on molecular structure instead of crystal structure [22,57,68,69]. However, in the case of MOF, solution state structural characterization can be disregarded due to the possibility of these labile molecular products rearranging in a situation where crystal packing is the primary concern. In such instances, the powdered product obtained from mechanosynthesis must be characterized without dissolving or recrystallizing making it a challenging task. In general, PXRD, FTIR, and solid state NMR spectroscopy are employed to characterize majority of MOFs that are produced in powdered form using mechanochemical technique [22,33]. Although, FTIR and solid state NMR analyses give direct information on the product's chemical composition, they are not always adequate for complete structural characterization due to the creation of more than one polymorphic product. Consequently, the comprehensive structural characterization frequently relies upon the capacity to generate identical structure from solution via growing single crystals. In such circumstances, structural identification may be accomplished easily by solving the single crystal structure with the help of X-ray diffraction. However, if solution state reaction yields different product than mechanochemical reaction, it is theoretically feasible to produce single crystals by seeding it with finely ground grinding product and then allowing it to undergo heterogeneous nucleation from solution. Crystal structure analysis from PXRD data is another promising alternative to acquire the crystal structure of MOFs made by mechanochemical means. Indeed, certain new sophisticated laboratory equipment and software, when combined with other traditional spectroscopic information, have enabled structure to be determined from PXRD data [70]. However, it is undeniable that these techniques are still far more difficult to complete than determining structures using single-crystal XRD data. For instance, using PXRD data, it was possible to determine the structure of MOF $Zn_2(fum)_2(bipy)$, which was produced by the mechanochemical reaction of zinc acetate, H_2fum, and bipy [52]. In this example, the structural resemblance to its copper(II) analogue made it easier to determine the structure of $Zn_2(fum)_2(bpe)$ using PXRD data. The Rietveld refinement of the structural coordinates of copper(II) analogue to the

PXRD pattern produced for Zn-MOF yielded the structure of the Zn-based framework [22,71].

6.5 Mechanistic research on the synthesis of mechanically produced MOFs

6.5.1 Synchrotron PXRD for in situ and real-time monitoring of mechanochemical reactions

Although the mechanochemical reactions have appeared as emerging approach for MOF synthesis, the fundamental mechanism by which these types of reactions proceeds are not completely explored. This is owing to the inability to monitor the reaction in real time at the microscopic and molecular levels when the reactants turn into products. According to Drebuschak et al., understanding the processes that occur in a powder sample during milling or compacting is difficult since one cannot detect local temperature, pressure, shear stresses, or observe changes in diffraction patterns or vibrational spectra in situ [72]. However, the structural evolution of many intermediates that occur during mechanochemical syntheses may be investigated with the help of PXRD. In 2013, Friscic et al. employed the high energy powder X-ray diffraction technique to carry out in situ analyses of the mechanochemical reaction of ZnO with imidazole derivatives to generate ZIF structure [73]. The high penetration power of short-wavelength X-rays allows for real-time investigation and observation of crystalline materials without interrupting with the grinding process. These sort of real-time PRXD experiments give an outstanding dynamic image of mechanochemical processes and allow in-situ assessment of particle size growth whose life time of less than a minute. Julien et al. had used the same technique for mechanistic study of Zn-MOF-74 formation from ZnO and H_4dhta [51]. They have developed an unusual method of mechanochemical MOF synthesis that progresses through a densely packed reaction intermediate to generate a low-density metal-organic structure. LAG reaction of ZnO with H_4dhta in presence of water as grinding liquid gives MOF-74 of composition $Zn_2(H_2O)_2$(dhta) (Fig. 6.12A). The in situ monitoring experiment revealed that crystalline H_4dhta disappears as soon as the milling process begins, followed by the development of non-porous Zn-$(H_2O)_2(H_2$dhta) with a simultaneous decrease in amount of ZnO (Fig. 6.12B and E). As the reaction time rises, the reflection induced by residual ZnO and Zn-$(H_2O)_2(H_2$dhta) begins to fade due to the simultaneous development of Zn-MOF-74 and finally gives pure product after 70 minutes. The mechanism of Zn-MOF-74 creation is as follows: during the initial stage of milling, carboxylic acid groups on H_4dhta react promptly with ZnO to create $Zn(H_2O)_2(H_2$dhta) (Fig. 6.12C), and thereafter the less acidic phenol groups react with remaining ZnO on subsequent milling to give Zn-MOF-74 (Fig. 6.12A and D). The mechanochemical synthesis of Zn-MOF becomes a more complicated process if the milling liquid contains DMF solvent. In situ monitoring showed that

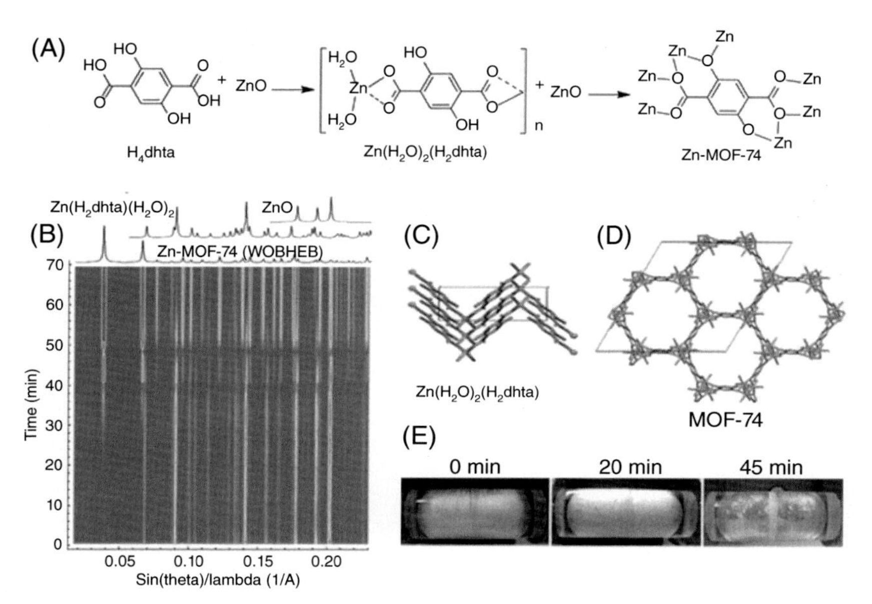

FIGURE 6.12 (A) Starting with ZnO and H_4dhta, a stepwise synthesis of Zn-MOF-74; (B) Time resolved in situ X-ray powder diffractogram for milling reaction of ZnO and H_4dhta (2:1 ratio) in presence of water (η = 0.625 μL/mg, λ = 0.207 Å); (C) Structure of $Zn(H_2O)_2(H_2dhta)$; (D) Structures of Zn-MOF-74and (E) The reaction mixture as seen at various milling periods. *(Reproduced with permission from [51]. Copyright (2016), American Chemical Society).*

the LAG reaction with a 4:1 (v/v) combination of DMF and H_2O yields $Zn(H_2O)_2(H_2dhta)$ and after about 20 minutes, a new and transient phase that is swiftly substituted by a new phase before Zn-MOF-74 reflections manifest at 45 minutes. The LAG approach based only on DMF was slower, taking nearly 3 hours to complete the conversion to Zn-MOF-74.

6.5.2 Raman spectroscopy for real-time monitoring of mechanochemical reactions

The foremost drawback of in situ X-ray diffraction studies of milling reaction is the need for a synchrotron source. Furthermore, PXRD does not provide any information about amorphous phase or molecular level transformations. However, infrared and Raman spectroscopy can be helpful for understanding molecular level complexation in LAG and ILAG reactions [74]. Raman spectroscopy, for instance, demonstrated conclusively that the mechanochemical reaction between $CdCl_2$ and cnge produces either $Cd(cnge)Cl_2$ with 3D structure or $Cd(cnge)_2Cl_2$ with 2D structure [75]. Changes in the adsorption bands roughly at 2200/cm for the nitrile group in cnge can be used to monitor the synthesis of each unique coordination structure and the consequent extinction of cnge.

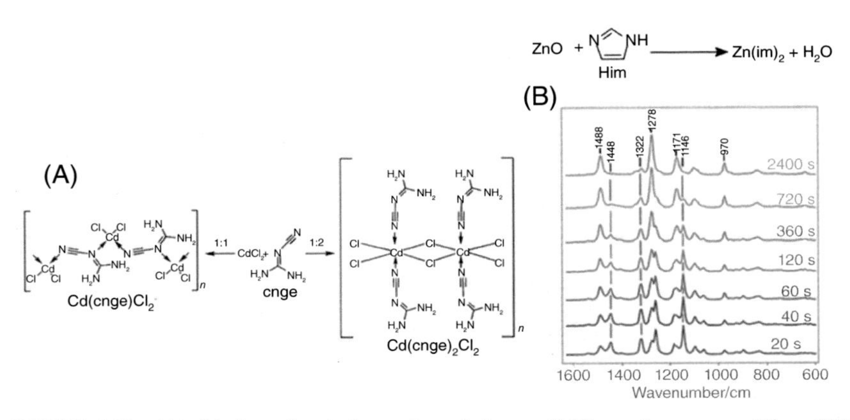

FIGURE 6.13 (A) Mechanochemical reactions between $CdCl_2$ and cyanoguanidine [75]; (B) ZnO and imidazole undergoing a mechanochemical reaction in the presence of anhydrous N,N-dimethylformamide (*top*), with time-dependent spectra demonstrating the changes seen throughout the reaction (*bottom*) [77].

Similar to this, variations in stretching band near 200/cm due to the alteration of coordination sphere of the metal ion might also be used to easily trace the disappearance of $CdCl_2$ and development of the product. However, in contrast to in situ monitoring, the ex-situ investigation of coordination polymer synthesis yielded a significantly different result. Raman reaction monitoring showed that $Cd(cnge)_2Cl_2$ is directly formed, in contrast to stepwise analysis, which showed that the 3D $Cd(cnge)Cl_2$ forms as a kinetic intermediate during the synthesis of 2D $Cd(cnge)_2Cl_2$ [76].

Ma et al. used Raman spectroscopy to investigate the LAG reaction between ZnO and HIm in the presence of anhydrous DMF [77]. They calculated the percentage of HIm that reacted using the change in band intensities at 1322/cm and 970/cm, which they had been monitoring (Fig. 6.13B). The reaction exhibits a very simple second-order kinetic model, with the rate controlled solely by the frequency of reactive collisions among reactant molecules. The frequency of reactive interactions between particles serves as the rate-determining factor, and as predicted, the reaction rate rises as the grinding frequency rises.

6.5.3 Combined X-ray diffraction and Raman spectroscopy for in-situ study of milling reactions

The integration of time-resolved XRD with Raman spectroscopy may offer a precise method for analyzing the critical steps of solid-state milling reactions. Batzdorf et al. had combined XRD and Raman spectroscopy to access complete information on the mechanochemical synthesis of MOFs ZIF-8, $(H_2Im)[Bi(1,4\text{-}bdc)_2]$ (Fig. 6.14) [78]. The benefits of this hybrid approach are as follows:

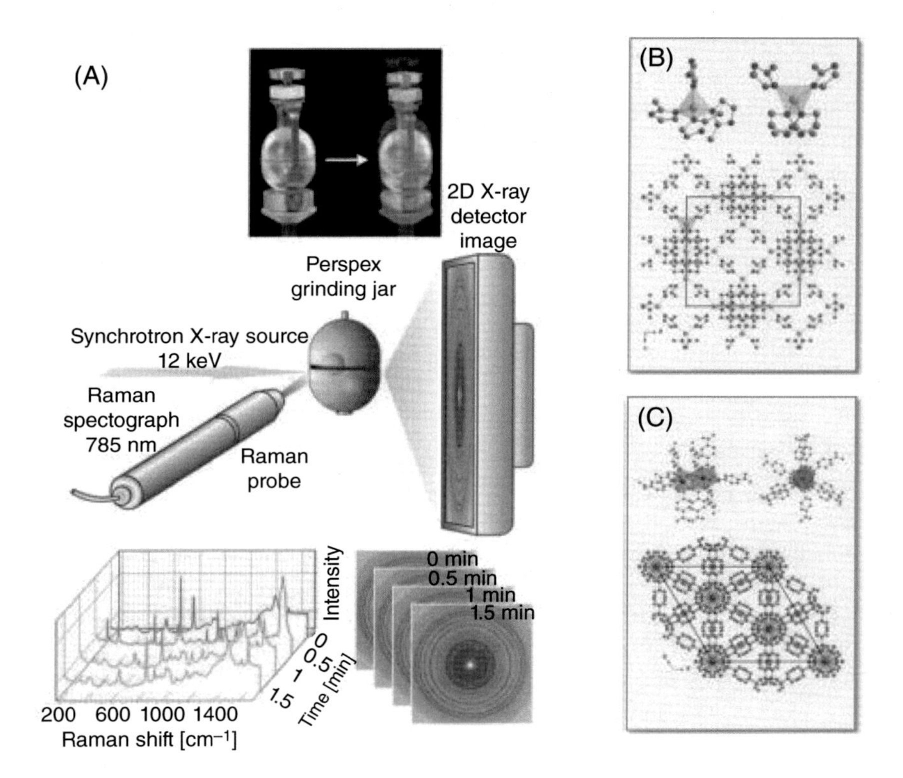

FIGURE 6.14 (A) Diagrammatic representation of the experimental equipment used to record powder X-ray diffraction (XRD) patterns and Raman spectra during the mechanochemical reaction. At the interval of every 30 seconds, XRD and Raman spectra were routinely collected; an illustration of the investigated metal–organic frameworks crystal structures (B) ZIF-8 [Zn(MeIm)$_2$] and (C) (H_2Im)[Bi(1,4-bdc)$_2$] [78].

Raman spectroscopy offers information on the molecular level irrespective of the physical state of materials, whereas high energy synchrotron PXRD offers details on the crystal's composition, its transition, and size of the crystallites. As a consequence, combining these two approaches in a single experiment allows for the simultaneous investigation of the entire product formation route at the molecular and crystalline levels. They had carried out in situ experiments on ILAG reaction of ZnO and HMeIm to generate ZIF-8 in the presence of trace amounts of NH_4NO_3 and DMF. The combined technique of investigation demonstrated that the final product develops within 2 minutes of milling time. The Raman spectra show that the HMeIm molecule is being adjusted at a slower rate in the crystal structure of ZIF-8. In contrast, (H_2im)[Bi(1,4-bdc)$_2$] was produced by grinding $Bi(NO_3)_3 \cdot 5H_2O$, terephthalic acid, and surplus amount of imidazole (HIm) for about 15 minutes [78]. The integration of XRD with Raman spectroscopy demonstrates rapid synthesis of the intermediate by reaction of $Bi(NO_3)_3{\cdot}5H_2O$ and HIm within 1 minute, which is characterized by significantly high reflection at $2\theta = 6.58$. Both studies support a reaction between the intermediate and terephthalic acid to form the final product. After a minute of milling, the PXRD pattern shows that $Bi(NO_3)_3{\cdot}5H_2O$ and HIm have completely converted to the intermediate, but terephthalic acid has not been affected by the reaction and is still present in the mixture. However, after milling the mixture for another 2.5 minutes, XRD pattern reveals that the intermediate reacted with the terephthalic acid to produce the final product $(H_2Im)[Bi(1,4\text{-}bdc)_2]$. The concurrently recorded Raman spectra provide further evidence in support of these findings. Furthermore, the Raman spectra also confirmed the protonation of imidazole ring during the formation of intermediate.

6.6 Challenges to the mainstream implementation of mechanochemical method

There are still some challenges that have to overcome before the industrial scale realization of mechanochemically synthesized MOFs. First of all, mechanochemical reactions may involve non-stoichiometric metal-to-ligand ratios that lead to impurities in the final products since they cannot always accomplish 100% conversion. Secondly, even though very little (a few drops) or no solvent is required in this method, the cleaning and activation procedure that follows unavoidably calls for the use of large amount of solvent, thus solvent free cannot be totally attained. Thirdly, because the various ions produced by the metal source and ligands might combine with the MOF products to make salts or they can bind within the pores of MOFs, there must be a separate cleaning and ion elimination step, which is a laborious procedure.

6.7 Conclusion and perspective

Because of its potential applicability in carrying out chemical reactions at ambient temperature (or somewhat elevated) as well as a considerable decrease in solvent usage, mechanochemical synthesis has become a very attractive approach. In addition, this approach is extremely rapid and has a high yield, making it far more efficient than traditional solvothermal techniques in terms of time, energy, and material use. This approach was used successfully to build large variety of MOFs, including well-known examples like as MOF-5, HKUST-1, ZIF-8, and MIL-100. In this chapter, we have highlighted some latest breakthroughs in the mechanochemical production of MOFs. The MOF-5, which was created by neatly grinding of the precursor materials, has the maximum SSA of all the MOFs that have been mechanochemically synthesized to far, reaching 3465 m^2/g. However, we must admit that it still has a lot of drawbacks that restrict it from being a widely used mainstream approach. In contrast to the great diversity of MOFs that have been reported up to this point, this approach is less researched and has only produced extremely small numbers of MOFs. In the near future, the way of governing various circumstances of ball milling reactions to accomplish cost efficient and industrial scale synthesis of functional MOFs might be an emerging research field. For this purpose, a depth understanding of the mechanism of MOF formation by mechanochemical reaction is required. The in-situ synchrotron X-ray diffraction study is a reliable method, and has been employed by a few research groups to understand this type of synthesis. However, they are still in their infancy, and the reaction mechanism underlying the synthesis of these MOFs is not well understood. Although the in-situ reaction analysis discloses several reaction routes and mesophases, these insights are insufficient to develop a strategy for the synthesis of novel MOFs. As a consequence, highly advanced research equipment are required for a superior understanding of the impacts of different parameters in mechanochemical production of MOFs. Finally, we anticipate that mechanochemistry research will expand further, resulting in more efficient and ecologically friendly techniques for obtaining novel functional MOFs.

Acknowledgment

The author is thankful to UGC-BSR for financial support for Start-up Research Grant [No.F.30-458/2019 (BSR)] and to A. M. School of Education, Assam University, Silchar, Assam, India for infrastructure facilities.

References

[1] H. Furukawa, K.E. Cordova, M. O′Keeffe, O.M. Yaghi, The chemistry and applications of metal-organic frameworks, Science 341 (2013) 1230444.

[2] U. Mueller, M. Schubert, F. Teich, H. Puetter, K. Schierle-Arndta, J. Pastré, Metal-organic frameworks-prospective industrial applications, J. Mater. Chem. 16 (2006) 626–636.

[3] G. Ferey, C. Mellot-Draznieks, C. Serre, F. Millange, J. Dutour, S. Surble, et al., A chromium terephthalate-based solid with unusually large pore volumes and surface area, Science 309 (2005) 2040–2042.

[4] J.R. Longa, O.M. Yaghi, The pervasive chemistry of metal-organic frameworks, Chem. Soc. Rev. 38 (2009) 1213–1214.

[5] F. Meng, Z. Fang, Z. Li, W. Xu, M. Wang, Y. Liu, et al., Porous Co_3O_4 materials prepared by solid-state thermolysis of a novel Co-MOF crystal and their superior energy storage performances for supercapacitors, J. Mater. Chem. A 1 (2013) 7235–7241.

[6] W.H. Li, K. Ding, H.R. Tian, M.S. Yao, B. Nath, W.H. Deng, et al., Conductive metal-organic framework nanowire array electrodes for high-performance solid-state supercapacitors, Adv. Energy Mater. 27 (2017) 1702067.

[7] D.P. Dubal, N.R. Chodankar, D.H. Kim, P. Gomez-Romero, Towards flexible solid-state supercapacitors for smart and wearable electronics, Chem. Soc. Rev. 47 (2018) 2065–6129.

[8] N. Balahmar, A.C. Mitchell, R. Mokaya, Generalized mechanochemical synthesis of biomass-derived sustainable carbons for high performance CO_2 storage, Adv. Energy Mater. 5 (2015) 1500867.

[9] J.R. Li, R.J. Kupplera, H.C. Zhou, Selective gas adsorption and separation in metal-organic frameworks, Chem. Soc. Rev. 38 (2009) 1477–1504.

[10] P.D. Zhang, X.Q. Wu, T. He, L.H. Xie, Q. Chen, J.R. Li, Selective adsorption and separation of C_2 hydrocarbons in a "flexible-robust" metal-organic framework based on a guest-dependent gate-opening effect, Chem. Commun. 56 (2020) 5520–5523.
[11] V. Pascanu, G.G. Miera, A.K. Inge, B. Martín-Matute, Metal–organic frameworks as catalysts for organic synthesis: a critical perspective, J. Am. Chem. Soc. 141 (2019) 7223–7234.
[12] M. Ranocchiari, J.A. van Bokhoven, Catalysis by metal-organic frameworks: fundamentals and opportunities, Phys. Chem. Chem. Phys. 13 (2011) 6388–6396.
[13] M.S. Yao, W.X. Tang, G.E. Wang, B. Nath, G. Xu, MOF thin film-coated metal oxide nanowire array: significantly improved chemiresistor sensor performance, Adv. Mater. 28 (2016) 5229–5234.
[14] G.M. Espallargas, E. Coronado, Magnetic functionalities in MOFs: from the framework to the pore, Chem. Soc. Rev. 47 (2018) 533–557.
[15] H.D. Lawson, S.P. Walton, C. Chan, Metal-organic frameworks for drug delivery: a design perspective, ACS Appl. Mater. Interfaces 13 (2021) 7004–7020.
[16] Y. Cui, Y. Yue, G. Qian, B. Chen, Luminescent functional metal-organic frameworks, Chem. Rev. 112 (2012) 1126–1162.
[17] J.D. Rocca, D. Liu, W. Lin, Nanoscale metal-organic frameworks for biomedical imaging and drug delivery, Acc. Chem. Res. 44 (2011) 957–968.
[18] M. Kalaj, S.M. Cohen, Postsynthetic modification: an enabling technology for the advancement of metal-organic frameworks, ACS Cent. Sci. 6 (2020) 1046–1057.
[19] H.C. Zhou, J.R. Long, O.M. Yaghi, Introduction to metal-organic frameworks, Chem. Rev. 112 (2012) 673–674.
[20] N. Stock, S. Biswas, synthesis of metal-organic frameworks (MOFs): routes to various MOF topologies, morphologies, and composites, Chem. Rev. 112 (2012) 933–969.
[21] M. Rubio-Martinez, C. Avci-Camur, A.W. Thornton, I. Imaz, D. Maspoch, M.R. Hill, New synthetic routes towards MOF production at scale, Chem. Soc. Rev. 46 (2017) 3453–3480.
[22] S.L. James, C.J. Adams, C. Bolm, D. Braga, P. Collier, T. Friscic, et al., Mechanochemistry: opportunities for new and cleaner synthesis, Chem. Soc. Rev. 41 (2012) 413–447.
[23] A.L. Garay, A. Pichon, S.L. James, Solvent-free synthesis of metal complexes, Chem. Soc. Rev. 36 (2007) 846–855.
[24] G. Kaupp, Solid-state molecular syntheses: complete reactions without auxiliaries based on the new solid-state mechanism, CrystEngComm 5 (2003) 117–133.
[25] T. Friscic, Supramolecular concepts and new techniques in mechanochemistry: cocrystals, cages, rotaxanes, open metal-organic frameworks, Chem. Soc. Rev. 41 (2012) 3493–3510.
[26] P. Balaz, Mechanochemistry in Nanoscience and Minerals Engineering, Springer, 2008.
[27] S. Głowniak, B. Szczezesniak, J. Choma, M. Jaroniec, Mechanochemistry: toward green synthesis of metal-organic frameworks, Mater. Today 46 (2021) 109–124.
[28] C.A. Tao, J.F. Wang, Synthesis of metal organic frameworks by ball-milling, Crystals 11 (2021) 15.
[29] T. Friščić, Metal-organic frameworks: mechanochemical synthesis strategies, in: R.A. Scott (Ed.), Encyclopedia of Inorganic and Bioinorganic Chemistry, John Wiley & Sons, 2014, pp. 1–19.
[30] A. Pichon, A. Lazuen-Garaya, S.L. James, Solvent-free synthesis of a microporous metal-organic framework, CrystEngComm 8 (2006) 211–214.
[31] W. Yuan, A.L. Garay, A. Pichon, R. Clowes, C.D. Wood, A.I. Cooper, et al., Study of the mechanochemical formation and resulting properties of an archetypal MOF: $Cu_3(BTC)_2$ (BTC= 1,3,5-benzenetricarboxylate), CrystEngComm 12 (2010) 4063–4065.

[32] M. Klimakow, P. Klobes, F. Thünemann, K. Rademann, F. Emmerling, Mechanochemical synthesis of metal-organic frameworks: a fast and facile approach toward quantitative yields and high specific surface areas, Chem. Mater. 22 (2010) 5216–5221.
[33] M. Klimakow, P. Klobes, K. Rademann, F. Emmerling, Characterization of mechanochemically synthesized MOFs, Microporous Mesoporous Mater. 154 (2013) 113–118.
[34] N.K. Singh, M. Hardi, V.P. Balema, Mechanochemical synthesis of an Yttrium based metal-organic framework, Chem. Commun. 49 (2013) 972–974.
[35] T. Alammar, I.Z. Hlova, S. Gupta, V. Balema, V.K. Pecharskyab, A.V. Mudring, Luminescence properties of mechanochemically synthesized lanthanide containing MIL-78 MOFs, Dalton Trans. 47 (2018) 7594–7601.
[36] Z. Wang, Z. Li, M. Ng, P.J. Milner, Rapid mechanochemical synthesis of metal-organic frameworks using exogenous organic base, Dalton Trans. 49 (2020) 16238–16244.
[37] N.L. Rosi, J. Kim, M. Eddaoudi, B. Chen, M. O'Keeffe, O.M. Yaghi, Rod packings and metal-organic frameworks constructed from rod-shaped secondary building units, J. Am. Chem. Soc. 127 (2005) 1504–1518.
[38] J.E. Bachman, Z.P. Smith, T. Li, T. Xu, J.R. Long, Enhanced ethylene separation and plasticization resistance in polymer membranes incorporating metal-organic framework nanocrystals, Nat. Mater. 15 (2016) 845–849.
[39] D. Chen, J. Zhao, P. Zhang, S. Dai, Mechanochemical synthesis of metal-organic frameworks, Polyhedron 162 (2019) 59–64.
[40] W. Yuan, J. O'Connor, S.L. James, Mechanochemical synthesis of homo- and hetero-rare-earth(III) metal-organic frameworks by ball milling, CrystEngComm 12 (2010) 3515–3517.
[41] M. Schlesinger, S. Schulze, M. Hietschold, M.E. Mehring, Evaluation of synthetic methods for microporous metal-organic frameworks exemplified by the competitive formation of $[Cu_2(btc)_3(H_2O)_3]$ and $[Cu_2(btc)(OH)(H_2O)]$, Microporous Mesoporous Mater. 132 (2010) 121–127.
[42] H. Yang, S. Orefuwa, A. Goudy, Study of mechanochemical synthesis in the formation of the metal-organic framework $Cu_3(BTC)_2$ for hydrogen storage, Microporous Mesoporous Mater. 143 (2011) 37–45.
[43] T. Stolar, L. Batzdorf, S. Lukin, D. Žilic, C. Motillo, T. Friščic, et al., In situ monitoring of the mechanosynthesis of the archetypal metal-organic framework HKUST-1: effect of liquid additives on the milling reactivity, Inorg. Chem. 56 (2017) 6599–6608.
[44] D. Braga, M. Curzi, A. Johansson, M. Polito, K. Rubini, F. Grepioni, Simple and quantitative mechanochemical preparation of a porous crystalline material based on a 1D coordination network for uptake of small molecules, Angew. Chem. Int. Ed. 45 (2006) 142–146.
[45] M. Pilloni, F. Padella, G. Ennas, S. Lai, M. Bellusci, E. Rombi, et al., Liquid-assisted mechanochemical synthesis of an iron carboxylate metal organic framework and its evaluation in diesel fuel desulfurization, Microporous Mesoporous Mater. 213 (2015) 14–21.
[46] B. Karadeniz, D. Žilić, I. Huskić, L.S. Germann, A.M. Fidelli, S. Muratović, et al., Controlling the polymorphism and topology transformation in porphyrinic zirconium metal-organic frameworks via mechanochemistry, J. Am. Chem. Soc. 141 (2019) 19214–19220.
[47] K. Užarević, T.C. Wang, S.Y. Moon, A.M. Fidelli, J.T. Hupp, O.K. Farha, et al., Mechanochemical and solvent-free assembly of zirconium-based metal-organic frameworks, Chem. Commun. 52 (2016) 2133–2136.
[48] L.S. Germann, A.D. Katsenis, I. Huskić, P.A. Julien, K. Užarević, M. Etter, et al., Real-time in situ monitoring of particle and structure evolution in the mechanochemical synthesis of UiO-66 metal-organic frameworks, Cryst. Growth Des. 20 (2020) 49–54.

[49] G. Ayoub, B. Karadeniz, A.J. Howarth, O.K. Farha, I. Dilović, L.S. Germann, et al., Rational synthesis of mixed-metal microporous metal-organic frameworks with controlled composition using mechanochemistry, Chem. Mater. 31 (2019) 5494–5501.
[50] D. Yan, R. Gao, M. Wei, S. Li, J. Lu, D.G. Evans, et al., Mechanochemical synthesis of a fluorenone-based metal organic framework with polarized fluorescence: an experimental and computational study, J. Mater. Chem. C 1 (2013) 997–1004.
[51] P.A. Julien, K. Užarević, A.D. Katsenis, S.A.J. Kimber, T. Wang, O.K. Farha, et al., In situ monitoring and mechanism of the mechanochemical formation of a microporous MOF-74 framework, J. Am. Chem. Soc. 138 (2016) 2929–2932.
[52] T. Friščić, L. Fábián, Mechanochemical conversion of a metal oxide into coordination polymers and porous frameworks using liquid-assisted grinding (LAG), CrystEngComm 11 (2009) 743–745.
[53] J.L. Howard, Q. Cao, D.L. Browne, Mechanochemistry as an emerging tool for molecular synthesis: what can it offer? Chem. Sci. 9 (2018) 3080–3094.
[54] D. Tan, F. García, Main group mechanochemistry: from curiosity to established protocols, Chem. Soc. Rev. 48 (2019) 2274–2292.
[55] C. Xu, S. De, A.M. Balu, M. Ojeda, R. Luque, Mechanochemical synthesis of advanced nanomaterials for catalytic applications, Chem. Commun. 51 (2015) 6698–6713.
[56] G.A. Bowmaker, Solvent-assisted mechanochemistry, Chem. Commun. 49 (2013) 334–348.
[57] K.K. Sarmah, T. Rajbongshi, A. Bhuyana, R. Thakuria, Effect of solvent polarity in mechanochemistry: preparation of a conglomerate vs. racemate, Chem. Commun. 55 (2019) 10900–10903.
[58] V.R. Remya, M. Kurian, Synthesis and catalytic applications of metal-organic frameworks: a review on recent literature, Int. Nano Lett. 9 (2019) 17–29.
[59] T. Friščić, D.G. Reid, I. Halasz, R.S. Stein, R.E. Dinnebier, M.J. Duer, Ion- and liquid-assisted grinding: improved mechanochemical synthesis of metal-organic frameworks reveals salt inclusion and anion templating, Angew. Chem. Int. Ed. 49 (2010) 712–715.
[60] J.F. Fernndez-Bertrn, M.P. Hernndez, E. Reguera, H. Yee-Madeira, J. Rodriguez, A. Paneque, et al., Characterization of mechanochemically synthesized imidazolates of Ag^{+1}, Zn^{+2}, Cd^{+2}, and Hg^{+2}: solid state reactivity of nd^{10} cations, J. Phys. Chem. Solids 67 (2006) 1612–1617.
[61] C.J. Adams, H.M. Colquhoun, P.C. Crawford, M. Lusi, A.G. Orpen, Solid-state interconversions of coordination networks and hydrogen-bonded salts, Angew. Chem. Int. Ed. 46 (2007) 1124–1128.
[62] P.J. Beldon, L. Fábián, R.S. Stein, A. Thirumurugan, A.K. Cheetham, T. Friščić, Rapid room-temperature synthesis of zeolitic imidazolate frameworks by using mechanochemistry, Angew. Chem. Int. Ed. 49 (2010) 9640–9643.
[63] C.C. Wang, X.H. Yi, P. Wang, Powerful combination of MOFs and C_3N_4 for enhanced photocatalytic performance, Appl. Catal. B 247 (2019) 24–48.
[64] Y. Li, J. Miao, X. Sun, J. Xiao, Y. Li, H. Wang, et al., Mechanochemical synthesis of Cu-BTC@GO with enhanced water stability and toluene adsorption capacity, Chem. Eng. J. 298 (2016) 191–197.
[65] B. Szczęśniak, J. Phuriragpitikhon, J. Choma, M. Jaroniec, Mechanochemical synthesis of three-component graphene oxide/ordered mesoporous carbon/metal-organic framework composites, J. Colloid Interface Sci. 577 (2020) 163–172.
[66] X. Du, X. Yi, P. Wang, J. Deng, C.C. Wang, Enhanced photocatalytic Cr(VI) reduction and diclofenac sodium degradation under simulated sunlight irradiation over MIL-100(Fe)/g-C_3N_4 heterojunctions, Chinese J. Catal. 40 (2019) 70–79.

[67] M. Ko, A. Aykanat, M.K. Smith, K.A. Mirica, Drawing sensors with ball-milled blends of metal-organic frameworks and graphite, Sensors 17 (2017) 2192.

[68] D. Braga, S.L. Giaffreda, F. Grepioni, A. Pettersen, L. Maini, M. Curzi, et al., Mechanochemical preparation of molecular and supramolecular organometallic materials and coordination networks, Dalton Trans. (2006) 1249–1263.

[69] C. Espro, D. Rodríguez-Padrón, Re-thinking organic synthesis: mechanochemistry as a greener approach, Curr. Opin. Green Sustain. 30 (2021) 100478.

[70] K.D.M. Harris, E.Y. Cheung, How to determine structures when single crystals cannot be grown: opportunities for structure determination of molecular materials using powder diffraction data, Chem. Soc. Rev. 33 (2004) 526–538.

[71] T. Friščić, Towards mechanochemical synthesis of metal-organic frameworks (MOFs): from coordination polymers and lattice inclusion compounds to porous materials, in: L.R. MacGillivray (Ed.), Metal-Organic Frameworks: Design and Application, John Wiley & Sons, 2010.

[72] T.N. Drebuschak, A.A. Ogienko, E.V. Boldyreva, 'Hedvall effect' in cryogrinding of molecular crystals. A case study of a polymorphic transition in chlorpropamide, CrystEngComm 13 (2011) 4405–4410.

[73] T. Friščić, I. Halasz, P.J. Beldon, A.M. Belenguer, F. Adams, S.A.J. Kimber, et al., Real-time and in situ monitoring of mechanochemical milling reactions, Nat. Chem. 5 (2013) 66–73.

[74] P. Vandenabeele, H.G.M. Edwardsb, J. Jehlička, The role of mobile instrumentation in novel applications of Raman spectroscopy: archaeometry, geosciences, and forensics, Chem. Soc. Rev. 43 (2014) 2628–2649.

[75] D. Gracin, V. Strukil, T. Friscic, I. Halasz, K. Uzarevic, Laboratory real-time and in situ monitoring of mechanochemical milling reactions by Raman spectroscopy, Angew. Chem. Int. Ed. 53 (2014) 6193–6197.

[76] V. Strukil, L. Fabian, D.G. Reid, M.J. Duer, G.J. Jackson, M. Eckert-Maksic, et al., Towards an environmentally-friendly laboratory: dimensionality and reactivity in the mechanosynthesis of metal-organic compounds, Chem. Commun. 46 (2010) 9191–9193.

[77] X. Ma, W. Yuan, S.E.J. Bell, S.L. James, Better understanding of mechanochemical reactions: Raman monitoring reveals surprisingly simple 'pseudo-fluid' model for a ball milling reaction, Chem. Commun. 50 (2014) 1585–1587.

[78] L. Batzdorf, F. Fischer, M. Wilke, K.J. Wenzel, F. Emmerling, Direct in situ investigation of milling reactions using combined X-ray diffraction and Raman spectroscopy, Angew. Chem. Int. Ed. 54 (2015) 1799–1802.

Chapter 7

Sonochemical synthesis of metal–organic frameworks

Taposi Chatterjee[a,b], SK Khalid Rahaman[a] and Seikh Mafiz Alam[a]
[a] *Department of Chemistry, Aliah University, Kolkata, West Bengal, India,* [b] *Department of Basic Science & Humanities, Techno International New Town, Kolkata, West Bengal, India*

7.1 Introduction

7.1.1 Sonochemistry: a brief overview

Sonochemistry uses ultrasonic waves to facilitate ultrasound-initiated chemical processes in fluids. Ultrasound belongs to the acoustic spectrum and its frequency range extends from 20 kHz to 10 MHz [1–4]. Sound is transmitted by inducing vibrational motions of the molecules present in the medium through which it travels. This behavior can be simulated with waves generated when gravel is thrown into still water. The waves propagate however the water molecules constitute the waves and regain their normal positions once the waves are successfully traveled. It is likely to feel the physical effects of sound vibrations by standing in front of a loudspeaker playing music at a high pitch. Ultrasound has the same physical properties as normally audible sound, except that it is inaudible to the human ear. It can make a much louder roar than a jet engine, but it is completely imperceptible to our ears. Therefore, it is sometimes called silent sound. Ultrasonic frequencies above 100 kHz do not occur in nature. Only artificial devices can both generate and detect these frequencies. Ultrasonic waves can be classified into three different ranges: 20–100 kHz (low frequency), 100 kHz–1 MHz (intermediate frequency), and 1–10 MHz (high frequency) [5]. When these ranges of frequencies are allowed to interact with the molecules in the solution, it causes the production, growth, and implosive collapse of gaseous microbubbles in the liquid phase, which is called acoustic cavitation. It is followed by the generation of extreme circumstances (high temperature, pressure, etc.), which is the fundamental aspect of the entire sonochemical process [6–8]. All of these phenomena cause to produce radical shear forces, shock waves, light emission–sonoluminescence (SL), microjets, microstreaming, and turbulence [9]. This process can be sub-divided into three categories as

Synthesis of Metal–Organic Frameworks via Water-Based Routes: A Green and Sustainable Approach.
DOI: https://doi.org/10.1016/B978-0-323-95939-1.00015-0

primary sonochemistry (gas-phase chemistry occurring inside collapsing bubbles), secondary sonochemistry (solution-phase chemistry occurring outside the bubbles), and physical modifications (caused by high-speed jets or shock waves derived from bubble collapse). Interestingly, if experimental conditions like solvent compatibility, better waste management, proper use of the raw material, low energy consumptions, and suitable reactivity time are optimized during the sonochemical processes, it can be considered under the domain of green chemistry [10]. Both green chemistry and sonochemistry are multidisciplinary in nature, and the environment-friendly techniques based on ultrasound have been reported in diverse fields such as preparation of organic/inorganic materials, sonocatalysis, biomass conversion, extraction, electrochemistry, enzymatic catalysis, environmental remediation, etc. [11]. The goal of this chapter is to provide readers with an understanding of the fundamentals of sonochemistry, as well as its historical development, the instruments involved, and its potential use in the synthesis of inorganic–organic hybrids, or metal–organic frameworks (MOF), using a water-based, environmentally friendly, and sustainable method [12–15]. It has been established that the primary mechanism underlying this technique is "acoustic cavitation." Sonochemistry has become a rapidly expanding technique, which takes the lead in the ultrasound power for MOF synthesis. This technique has made itself superior to the existing methodologies in terms of reduced reaction time, lower consumption of energy, as well as simplicity. Further, this article offers examples of how the physical and chemical effects of high-intensity ultrasound can be exploited for the fabrication of MOF. Advancement of research in this field led to the surfacing of a new twig of chemical science. Ultrasonic irradiation significantly differs from other existing energy sources like heat, light, or ionizing radiation, in terms of duration, pressure, as well as energy per molecule. The reason of the gigantic temperatures, mammoth pressures, and astounding heating and cooling rates created by cavitation bubble collapse, ultrasound offers an extraordinary means for generating high-energy chemistry. In addition, as sonochemistry holds a high-pressure element, it results in forming similar conditions created during detonations or by shock waves. Sonochemistry finds extensive applications in industries for the physical processing of liquids, such as emulsification, solvent degassing, solid dispersion, and soil formation. Along with these, it also offers numerous applications in solids processing as well, which includes cutting, welding, cleaning, precipitation, and many more.

7.1.2 Historical developments of sonochemistry

The phenomenal journey of sonochemistry started in 1893 by F. Galton with the invention of "Galton whistle," which was used to measure the hearing range of humans and other animals, demonstrating that many animals could hear sounds above the hearing range of humans. "Galton whistle" was an adjustable whistle that can produce ultrasound [16]. However, the first successful technological innovation of ultrasound was made with the invention of the quartz sandwich transducer for underwater sound transmission in submarine detection via the use of piezoelectricity by Professor Paul Langevin (1872–1946) during World War I. Professor Langevin suggested the use of ultrasound to develop a device named "sonar" for the detection of icebergs after the Titanic tragedy of 1912. These remarkable inventions paved the way for developing a wide range of complicated techniques for non-destructive testing and medical imaging [17]. Medical diagnostic tools use low powers and very high frequencies of ultrasound thereby not affecting the physical or chemical character of the medium under investigation. Contrarily, if a lower frequency and a higher power are applied to a fluid, it suffers considerable physical and chemical changes in the medium [18]. In 1927, Professor Robert Wood carried out collaborative research with Professor Paul Langevin which made sonochemistry a viable research field in physics, chemistry, and biology. In 1920, Alfred Lee Loomis, a business magnet from the United States, did experiments in his private Manhattan mansion in New York along with group members on sound waves beyond the range of hearing frequency. The first radiation force balance instrument was invented by Alfred Lee Loomis utilizing ultrasound radiation force. It has become the first prototype of a modern device for the measurement of ultrasonic power in liquids. W.T. Richards, in 1929, in his paper entitled "an intensity gauge for supersonic radiation in liquids" illustrated a number of methods for estimating high sound intensities and improved technical difficulties faced by ultrasonic devices for precise measurement [19]. By the end of 1929, the detection of metal devices using ultrasonic vibrations was developed. In 1931, Malhauser published a patent on "ultrasonic transducers" for the detection of solid flaws. Ultrasound made a late but robust entry into the medical diagnostic domain, which began during World War II. It made an entry for cleaning applications in textile and food industry technology in 1950, although it still remains appreciated globally as a surface cleaning tool [20]. The widespread application of sonochemistry in different disciplines of chemistry, along with the terrific boost in MOF research, has made this technique one step closer, with the intention of attaining lesser reaction times, phase-selectivity, and smaller particle sizes. The new beginning of sonochemistry took place in the 1980s, right after the arrival of low-cost and trustworthy science laboratory generators of high-intensity ultrasound.

7.1.3 Acoustic cavitation

7.1.3.1 What is acoustic cavitation?

In general terms, cavitations can be described as the generation of one or more pockets of gas in a liquid. Sufficient reduction of the pressure below the vapor pressure of the liquid or the elevation of the temperature above the boiling point leads to the formation of cavitations. However, chemical, electrical, and radiation-induced phenomena can also play a big role in the cavitation process. Cavities formed in such ways can be termed "bubbles." If such cavities are formed when a sample material is irradiated with acoustic sound, the phenomenon is known as acoustic cavitations. Acoustic cavitation accounts for the foremost reason for reaction intensification in sonochemistry. When ultrasound waves propagate in a liquid medium, they generate a repeating pattern of compressions and rarefactions to supply energy to the liquid phase which leads to the consecutive formation, maturation, and disintegration of bubbles, popularized as "hot spot." Destruction of these hot spots results in the production of enormous energy due to the conversion of the kinetic energy of liquid motion into heat energy of bubble contents. This energy sets the path of enormously higher local temperatures and pressures, followed by extremely rapid cooling, which serves the so-called "hot spot" as a unique source for driving chemical reactions under such conditions. More explicitly, the effective temperature reached in hot spots is about an astounding 5000°C, pressures of mammoth 1000 atm, and heating/cooling at an ultra-fast rate of above 1010 K/s. Due to these extreme conditions, reactions that were previously difficult to execute by other methods were easily accomplished by using the sonochemical technique. Further, it has been observed that cavitation processes are an effective tool for integrating diffuse acoustic energy in a specific set of environments to produce unusual materials from dissolved (and generally volatile) precursors. Chemical reactions are not usually seen when ultrasound is irradiated on solid systems. It is because of the formation of cavitation bubbles and frontier space around cavitation bubbles, which has enormous temperature, pressure, and electric field gradients. Such cavitation bubbles are missing in solid systems. Apart from that, liquid motion in the cavitation bubble also generates very large shear and strain gradients; these are caused by the very rapid streaming of solvent molecules around the cavitation bubble, as well as the intense shock waves emanating from the collapse. Consequently, water vapor and oxygen, if present, undergo dissociation inside a bubble to produce oxidants such as hydroxyl (OH), oxygen radical (O), and hydrogen peroxide (H_2O_2) [21]. Thus, the phenomenon which includes the development of multiple bubbles to the formation of chemical products via the utilization of ultrasonic waves is called a sonochemical reaction [22]. The physical and chemical consequences of ultrasonic cavitation can be explored to carry out many assignments, for example, starting off and speeding up chemical reactions [23,24], emulsification [25,26] in medical diagnostics [27,28], sterilization of liquids [29], isolation of biologically active compounds from plant cells [30], and many more.

7.1.3.2 Mechanism involved in acoustic cavitation

An acoustic wave is basically the transmission of pressure oscillation with the sonic speed in a medium such as liquid, gas, or solid [31]. Ultrasound can be described as an inaudible sound with a frequency of pressure oscillation higher than 20 kHz. The distance for a single pressure oscillation of an acoustic wave can be defined as the wavelength (λ), while the number of pressure oscillations per unit time (second) is termed as frequency ($f = 1/T_a$). The sonic velocity (c) is defined as the distance for a pressure disturbance propagating per unit of time. The acoustic period (T_a) is defined as the time required for one pressure oscillation. The entire process of acoustic cavitation is comprised of three steps [32,33], which are (1) formation of cavitation nuclei, (2) growth (radial motion) of cavitation bubbles, and (3) collapse or the next oscillation of bubbles. During the acoustic cavitations process, when acoustic pressure reaches a threshold value in the liquid, bubbles make it appear. The ultrasonic field is then submitted to the action of the bubbles. A local decrement in the pressure of a liquid due to the transmission of an acoustic wave can lead to the development of vapor–gas bubbles. These processes go hand in hand with powerful hydrodynamic turbulence with microflows and micro-shock waves [34,35]. The tiny bubble thus kicks off its motion with repeating expansion and compression. In the course of compression, the acoustic waves deploy a surged pressure on the liquid thereby reducing the average distances between the molecules [36–38]. Contrarily, during the rarefaction cycle, a negative pressure causes the distances to be amplified. If the negative pressure during the rarefaction cycle is abundant, so that the average distance between molecules surpasses the critical molecular distance necessary to hold the liquid unharmed, the liquid breaks down and consequently cavities are created. The negative pressure developed in the rarefaction cycle has to necessarily surmount the tensile strength of the liquid. Pure liquids usually possess tremendous tensile strengths and require significant negative pressures to the creation of cavities. However, the occupancy of trapped gases in crevices of small solid particles largely lowers the liquid's tensile strength. When a gas-filled crevice is uncovered to the rarefaction cycle of a sonic wave, the gas inflates until a tiny bubble is unleashed into the solution. The threshold value for such a mechanism depends on several key factors, such as hydrostatic pressure, temperature of the liquid, and rate of change of pressure. Acoustic cavitation can be categorized into stable and transient cavitation [35,39]. Stable cavitation can be distinguished by sustained small amplitude oscillations of the bubble around its equilibrium. The bubble's oscillations diffuse pressure to the surrounding fluid thereby generating motion around the bubble termed microstreaming [40]. Contrarily, transient bubbles only survive for one or a few acoustic cycles, in which their radius inflates to at least double the inceptive size. Because of this restricted time, it is thought that there exists no mass flow of permanent gas through the bubble–liquid interface [41]. This shortage of gas cushioning results in the formation of transient bubbles to collapse vehemently resulting in the dissolution of minor bubbles. As per the report, the number of bubbles produced in an actual sonochemical reaction was found to exceed an astronomical figure of 10,000 per mL of solution at a sonic frequency of 20 kHz. Bubbles formed at various locations of the acoustic field tend to follow different cavitation behavior and therefore have different cavitation energy on account of the spatial diversity [42]. The acoustic cavitation phenomenon can be schematically depicted in Fig. 7.1.

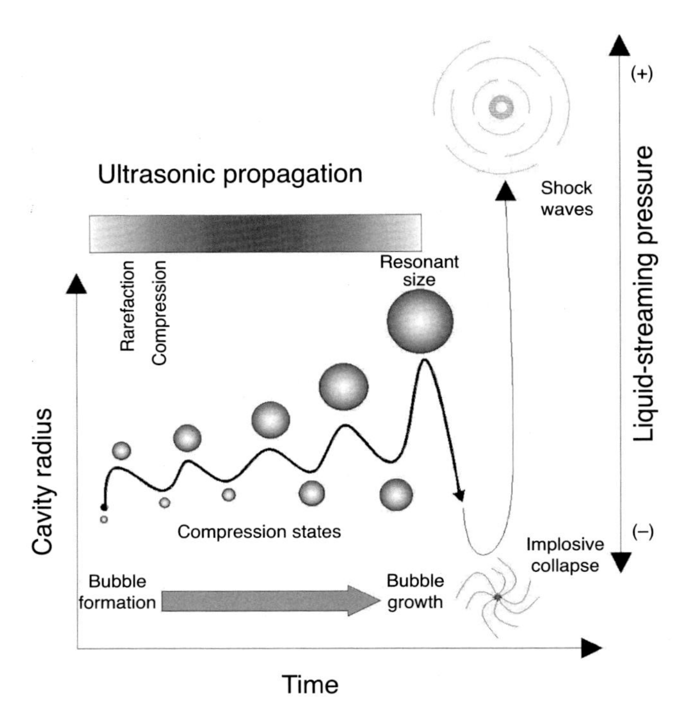

FIGURE 7.1 Schematic illustration of acoustic cavitation.

7.2 Instruments employed and experimental variables

7.2.1 Types of instruments used in sonochemistry

Depending upon the type of application, ultrasonic devices can have different designs and shapes, which can be broadly classified into two categories:

(1) Bath-type sonicator

The principle of bath-type sonicator works on indirect interaction between the sample and ultrasonic waves. This process is conducted in an energized water bath filled with degassed water in order to avoid the generation of bubbles [43]. Firstly, ultrasonic energy is irradiated through a liquid such as water, taken inside the device followed by the walls of the sealed sample container, and finally interacts with the sample. Therefore, the intensity of waves inside the sample container becomes lower than usual [44]. Sample containers are basically made up of glass materials. For this reason, a method is found to be effective for all sample sizes with a low risk of cross-contamination. Ultrasonic baths can be broadly classified into three types [45]. The first one happens to be the classic ultrasonic bath, which has been used extensively in laboratories. The bath operates at a fixed frequency, typically at 40 kHz, supplied with a temperature-control unit. A second type is constituted by a multi-frequency unit, which basically operates in the range of 25–40 kHz. Moreover, it delivers a uniform ultrasonic power supply. The most sophisticated bath-type ultra-sonicator is found to be operated in the range of 25/45 kHz or 35/130 kHz. These baths are so designed that they can function with one of the two frequencies at a time and sanitation intensity can be altered through amplitude control of about 10–100%. There are three operation modes which are available in modern baths which are as follows: (1) Sweep: the frequency varies within a fixed range. Therefore, ultrasonic efficiency will be more homogeneously distributed in the bath. (2) Standard: in this mode, the ultrasonic efficiency will be less homogeneously distributed in the bath. (3) Degas: in this mode, power is disrupted for a short span to prohibit bubbles production due to ultrasonic forces. However, this process needs significantly more energy and often, overheats the sample [46,47]. When bubbles are generated inside the water contained in an ultrasonic bath, degassing takes place by the bubbles, and subsequently the gas concentration in the water falls down. As the time passes by, the number of bubbles gradually decreases. When these bubbles powerfully soothe the ultrasonic wave, the acoustic intensity in a small liquid container is altered by the occurrence of bubbles in the surrounding bath. The decline in the quantity of bubbles in an ultrasonic bath consequently shoots up the acoustic intensity in a small liquid container. Therefore, for the indirect irradiation technique, it is desirable that an ultrasonic bath should be filled with degassed water. Conventionally, degassed water is prepared by reducing the normal atmospheric pressure using a vacuum pump or by simply boiling it. In order to keep the identical irradiation condition, the liquid surface in a small liquid container is frequently fixed at the same level as the water surface in a bath. The amount of liquid in a bath remains an important parameter, especially for relatively low ultrasonic frequencies. A typical bath-type system has been shown in Scheme 7.1. A signal generator generates an alternating current (AC) voltage of a sinusoidal waveform of a desired frequency. A power amplifier amplifies the AC power. A suitable matching circuit is inserted between an ultrasonic transducer and a power amplifier in order to prevent a reflection of the AC power. The electrical drive system required here are identical with the probe type sonicator.

(2) Probe-type sonicator

The mechanism of probe-type sonicator works on a sort of direct interaction between the sample and ultrasonic waves. Consequently, more concentrated energy can be focused directly onto the sample. The ultrasonic probe is placed straightway into the solution containing sample, where the sonication process is underway. The ultrasonic power furnished by the probe is not less than up to 100 times more intense than that of supplied by the bath type. Such differences make each system suitable for a diverse set of desired uses. In order to achieve cavitation, it is not always recommended to utilize high magnitude levels; or else there could be a possibility of potential deterioration of the tip. Probes made of titanium alloy can adulterate the reaction mixture [48]. Thus, there has been an urgency for the development of a metal-free probe to be

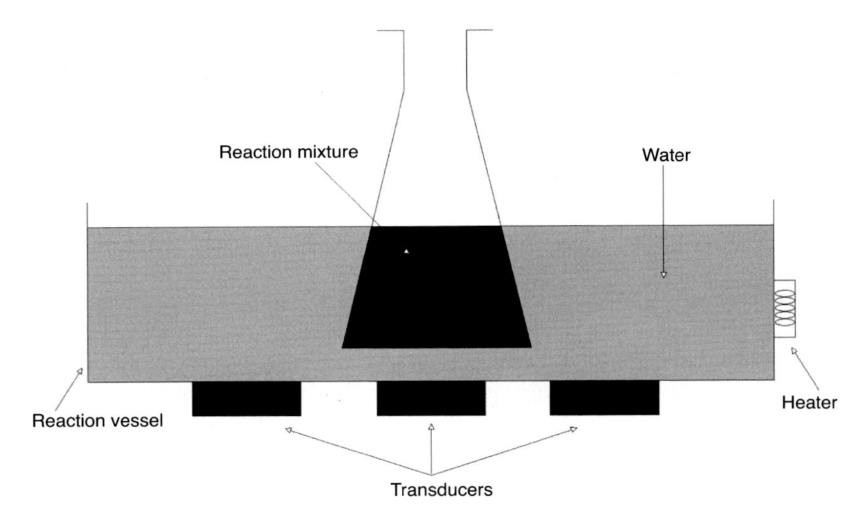

SCHEME 7.1 Schematic representation of bath-type sonicator.

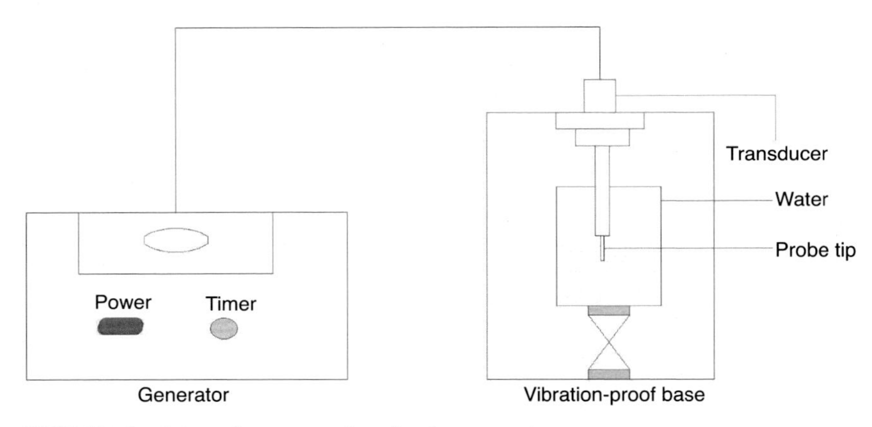

SCHEME 7.2 Schematic representation of probe-type sonicator.

implemented for long. In order to address such shortcomings, the utilization of silica for the fabrication of such probes has been developed [49]. It offers high chemical resistance along with significant temperature shock resistance. A typical probe-type sonicator is schematically represented in Scheme 7.2. The function of the generator is to convert mains voltage into high-frequency electrical energy. The most vital component of the entire assembly is the probe or the detachable horn, which empowers the tremors of the horn to be transmitted through an additional length of metal in such a fashion that the power transported gets magnified. The appropriate designing of devoted probes, therefore, remains vital for such techniques. Exploration of small probes for large sample volumes or bigger probes for lesser sample volumes must be kept away from. Presently, multiple probes have also been manufactured to permit two or more probes simultaneously at play. Consequently, enhanced throughput by the sonicator is achieved. A probe-type sonicator has been schematically shown in Scheme 7.2.

7.2.2 Factors influencing sonochemical processes

- Effect of sonic intensity: The intensity of ultrasound wave is directly proportional to the amplitude of vibration of the ultrasonic source; therefore, any increase in the amplitude of vibration will consequently give rise to the enhanced intensity in the vibration, thereby substantially influencing sonochemical effects to a large extent. The minimum intensity required to initiate a cavitations process is termed as cavitations threshold. But higher amplitude may be necessary to working samples with higher viscosity. This is because of the fact that largely viscous samples impede the movement of ultrasonic waves to pass through the liquid media. Extensive studies have been made by various scientists in order to check the effect of sonication amplitude on the efficiency of the solid–liquid extraction of metals. Generally, greater amplitude results in larger extraction of the samples until an optimum is reached for a given set of conditions and sonication duration [50,51]. The concept of optimum can be ascribed to the fact that at significantly higher power dissipation levels, cushioning effects are observed due to the generation of a large number of cavities that collapse in a cloud-like structure. Also, the decoupling losses decrease the acoustic energy available for the reaction mixture [52].
- Effect of temperature: The exploitation of the ultrasonic probe leads to the increment in temperature of the entire reaction assembly. The higher the value of amplitude results in a sharp increase in the temperature. If the temperature is not properly regulated, the potential degradation of compounds of interest may eventually take place [53]. Additionally, a rise in temperature may severely affect the physical characteristics of the liquid media so that ultrasonic transmission of the ultrasonic waves gets hurt as a result no cavitation is attained. This condition is referred to as the decoupling phenomenon [54]. Amongst various approaches to control the temperature, the utilization of an ice bath happens to be the first and simplest one. The entire vessel is inserted into the ice bath. This facilitates the rapid dissociation of the produced heat. However, this technique fits well for the short-duration synthesis routes. In next, the utilization of specially designed dedicated vessels to dissipate warming is encouraged. In the third process, and for prolonged sonication, "pulse" mode can be adopted where the amplifier periodically switches on and off in order to avert the development of high reaction temperature. Some moments come when the principal reason of sonochemical effects happens to be a cavitational collapse, then a bubbled gas should be utilized to generate abundant numbers of nucleation sites for the process to continue.

- Effect of frequency: The frequency of the ultrasonic waves plays a significant role in the sonication process. Three ranges of ultrasonic frequency have been reported: (1) low-frequency range (20–100 kHz), (2) medium frequency range (300–1000 kHz), and (3) high-frequency range (2–10 MHz) [55]. Sonication process conducted in the low-frequency range, a large number of compressions and rarefactions take place, which delays the collapse stage and allows volatile compounds to move into the gas phase. Hydrophobic compounds are found to be successfully decomposed by low-frequency ultrasound methods.
- Contour and material of reaction vessel: The size and shape of the reaction container is vital for the accurate application in sonochemical synthesis. When the ultrasonic wave is made to fall against any solid surface, a quantity of energy gets reflected. If the base of the vessel becomes flat, the reflection becomes minimum. Contrarily, when the base of the container is spherical the ultrasonic wave hits the container at an angle, and a huge proportion of the ultrasonic wave gets reflected away. The intensity of ultrasound is gradually attenuated as it transmits via a medium. The degree of such attenuation is inversely proportional to the frequency of the ultrasonic wave. Thus, the width of the wall of the container should be maintained at a minimum value in order to avoid such a phenomenon.
- Effect of reaction time: The typical durations of sonochemical synthesis are up to 2 hours, while durations lesser than 1 hour are not uncommon, depending on the ultrasound intensity and temperature, something that makes this technique ideal compared to prolonged reaction times (12–24 hours) of the solvothermal method.

7.2.3 Safety measures during the process

The main aim of ultrasound power is to produce high-power vibrational energy of sufficient intensity for the liquid sample to allow permanent physical change. Hence, the major hazard is associated with unintentional exposure to ultrasonic waves. In many industrial and commercial applications, sometimes ultrasound can also incidentally originate and transmit high sound–pressure magnitude in the air in the sonic and ultrasonic range. As a result of this, a hazard may also produce from the ear's reception of the air-borne ultrasound. Therefore, the following steps are suggested to ensure the safe handling of high-power ultrasound:

- Precautionary measures: Warning signs should be put in any space which contains high-power ultrasound equipment. As well as, there should be a list of instructions, which should be followed while the ultrasound power is on.
- Restricted handling: Only skilled operators should be allowed within the boundaries of the controlled area while the equipment is functioning.
- Device handling: Ultrasonic cleaning tanks should have labels on them to warn nearby personnel not to touch the body in the tank while it is operating.

7.3 Synthesis of MOFs

7.3.1 MOFs: a brief overview

Metal–organic framework is a class of inorganic–organic hybrid materials exhibiting extremely high porosity that extends up to 90% free volume and enormous internal surface areas of 6000 m^2/g [56]. MOFs are popular for their extraordinarily tunable pore size, and adjustable internal surface properties. Each unit assemblies consist of inorganic node or cluster with organic moieties to produce one-, two-, or three-dimensional material with desired morphologies like micro- or nano-spheres, -cubes, -sheets, -rods, etc. These hybrid materials have attracted a lot of attention in the past few decades owing to their abundant functionalities and unique characteristics [57]. In terms of applications, MOFs have been very promising in a wide spectrum, ranging from the well-known gas storage/adsorption and separation, catalysis, sensing, and dye/toxic material removal to recently rising fields, such as magnetism, luminescence, electrode materials, membranes, solid-phase extraction, and drug delivery. It has been observed that the preparation method of MOFs has a great influence in an effective manner for the unique and desirable properties of MOF. Therefore, variable synthetic processes were tried and compared on the basis of reaction parameters from the last decade. The variable conventional synthesis process includes hydrothermal, solvothermal, diffusion method, a slow evaporation method, and using an autoclave at high temperatures, but all these processes take a longer time period of hours or even days [58]. In this regard, sonochemistry provides a simple route to MOF synthesis utilizing a non-audible frequency range of sound waves, i.e., ultrasound specifically 10 kHz–20 MHz. Sonochemistry synthetic route has made the MOF synthesis process much faster and shape controllable with more choices of phase selectivity [59].

7.3.2 Conventional synthesis of MOFs

Morphology, porosity, and crystallinity of synthesized greatly rely upon various experimental parameters. Therefore, the suitable choice of synthesis approaches plays an important role in the physicochemical properties of the products. Further, large-scale synthesis of such compounds may also require economic and environmental aspects. Among diversified synthetic approaches studied so far for the synthesis of MOFs, solvothermal synthesis remains one of the most popular choices. This particular technique utilizes a solvent-based reaction of metal nodes with organic spacers and their subsequent crystallization in a sealed container where high temperature and pressure promote self assembly and crystal formation. In a similar fashion, hydrothermal synthesis is conducted in aqueous solutions. In the conventional solvothermal/hydrothermal processes electric heating provides the requisite energy to bring about and accelerate the crystallization process for MOF synthesis [60]. Another conventional popular technique is the layer-by-layer method which is based on surface chemistry, where a self-assembled monolayer is used as a template for the growth of MOF crystals. In this case, MOFs grow as a result of sequential reactions between the organic ligand and the inorganic nodes [61,62]. Alternatively, the electrochemical (EC) synthetic method provides an option, which is precisely controlled, simple, and viable. The major advantage of such techniques is their fastness and high purity due to the absence of counter ions. In this method, metal ions rather than the metal salt are introduced in the anode whereas the organic linker is placed at the cathode and the EC cell is filled with conducting salt [63]. In the slow diffusion method, crystals of appropriate sizes are obtained owing to a slow and steady diffusion mechanism. The solvent–liquid diffusion occurs in three discrete layers, the first layer consists of precipitant solvent, the second layer is comprised of product solvent, and the third layer basically, separates the first and second layers and allows diffusion at a lower rate [64]. Mechanochemical synthesis is the simplest one, where synthesis of MOF, is carried out by grinding the MOF precursors together manually in a mortar or by utilizing a ball mill [65,66].

7.3.3 Sonochemical synthesis of MOF

Sonochemistry finds itself as a simple method for the synthesis of MOF with the exploration of high-intensity ultrasound. When ultrasound is made to transmit via a solution, it leads to the generation of high and low-pressure regions inside the solution, as a result of subsequent compression and expansion taking place [67,68]. Such changes in pressure mark the foundation of sonochemistry, as it sets the path for the significant process of acoustic cavitation, i.e., generation, growth, and subsequent collapse of the acoustic bubble. Air molecules dissolved in the solution tend to diffuse to create bubbles at low pressure. On advancing to the next cycle, the high external pressure compresses the bubble viciously. Such phenomenon of bubble growth and compression goes on until the external pressure vehemently collapses the bubble. Pressure and temperature within the bubble have been found to have crossed the astounding mark of 1000 atm and 5000 K, respectively. Such conditions can bring about unusual physical and chemical changes and help in the reaction between atoms and molecules to turn out an amazing set of materials [69–72]. The benign technique paves the direction of synthetic routes without any bias towards a certain set of materials. Consequently, new and advanced applications have emerged to address synthetic challenges involved in various fields of science and technology.

Over the last decade, sonochemistry has made its profound entry as a synthetic route for numerous organic compounds. The popularity of this technique is largely due to its convenience, shorter reaction times, along with high yields [73]. Sonochemical methods can induce homogeneous nucleation and a significant reduction in the duration of the crystallization process compared with conventional oven heating. The utilization of high-intensity ultrasound to increase the reactivity of metals as a stoichiometric reagent has become a synthetic route for many heterogeneous organic and organometallic reactions [74–78].

Won et al. in 2008, synthesized high-quality MOF-5 crystals of 5–25 μm in size for the first time using a sonochemical method in a substantially low synthesis time of 30 minutes compared with conventional solvothermal synthesis (24 hours). The device was SONOPLUS HD 2200 sonicator with an adjustable power output having a 200 W maximum limit at 20 kHz fitted by a custom-made horn-type tube pyrex reactor. Firstly, the reaction mixture was vigorously agitated under nitrogen until a transparent solution was achieved. The resulting solution was moved to a 50 mL transparent horn-type tube reactor and subjected to an ultrasonic irradiation for 10–75 minutes at various power levels to obtain the MOF-5 crystals. The resulting solution became increasingly dark with sonication, and white crystals started to precipitate in 8 minutes to 30 minutes. With the power level intensified from 10% to 50% of the maximum power, the actual synthesis temperature attained increased from 129°C to 164°C with faster crystallization [79].

In the same year in 2008, Qiu et al. discovered a sonochemical synthetic route for the fast synthesis of fluorescent microporous MOF $Zn_3(BTC)_2 \cdot 12H_2O$ and utilized this material for selective sensing of organoamines. The outcomes show that ultrasonic synthesis is a straightforward, effective, affordable, and environmentally friendly method for producing nanoscale MOFs. During the synthesis, zinc acetate dihydrate was allowed to react with benzene-1,3,5-tricarboxylic acid (H3BTC) in 20% of ethanol in water (v/v). When the ultrasonic method was applied, reaction time was varied from 5 minutes, 10 minutes, 30 minutes, to 90 minutes under ambient temperature and atmospheric pressure simultaneously, the hydrothermal method was also carried out at 140°C at high pressure for 24 hours to synthesize the same compound. The structures of the ultrasonic product and hydrothermal product were confirmed by IR, single, and powder X-ray diffraction (XRD) patterns and found to be identical. Surprisingly, irradiation at ambient temperature and pressure for 5 minutes produces the compound in a remarkably high yield (75.3%, based on H3BTC). Also, the yield of the compound increased gradually from 78.2% to 85.3% with increasing the reaction time from 10 minutes to 90 minutes. This result suggests that rapid synthesis of MOF can be realized in a significantly high yield using the ultrasonic method. Compared with the hydrothermal synthesis of the same compound, which is carried out for a longer time period, ultrasonic synthesis is found to be a highly efficient method with high yield and low cost [80].

After four years, in 2012 Norbert Stock and Shyam Biswas published a review on various routes of MOF synthesis and their aspects where he mentioned the sonochemical MOF synthesis procedure in detail. He has stated that ultrasound is a cyclic mechanical vibration having a frequency range from 20 kHz (the upper limit of human hearing) to 10 MHz and this range of energy is very low to cause interaction between ultrasound and molecules to initiate chemical reactions, however when high-frequency ultrasound comes in contact with liquids, cyclic alternating areas of compression and rarefaction are produced. In the rarefaction area, the pressure falls below the vapor pressure of the solvent and reactants, resulting in the formation of small bubbles, i.e., cavities. This way, bubbles continue to form in the range of tens of micrometers through the diffusion of solute vapor into the volume of the bubble. Hence, ultrasonic energy is started to be stored in the vicinity. The moment bubbles attend their maximum size limit, become unstable and collapse. This progression of bubble formation, development, and breakdown is called cavitation which leads to the dissipation of energy into a high temperature almost equivalent to 5000 K, pressure around 1000 bars, intense shear forces, and heating and cooling rates of greater than 10^{10} K/s. This "hot spot" provides a platform for chemical reactions, i.e., causing molecules to achieve the excited state, bond cleavage, and the formation of radicals to react and furnish the products. There are a number of factors to decide the hot spot temperature and pressure. For example, the choice of organic solvents can alter the intensity of cavitational collapse and hence can change the resulting temperature and pressure. The presence of a solid surface in a hot spot, causes cavitation to produce microjets that clean or activate the surface, and accumulations of smaller particles are distributed while for a homogeneous liquid, the formation of extreme conditions that led to the chemical reactions can happen within the cavity at the interface or in the bulk media, where intense shear forces are present [81].

Thus, it can be said that when a reactant mixture of the solution is put in place to a reactor fitted to a sonicator bar with a regulating power output as depicted in Scheme 7.3, the formation and collapse of bubbles formed in the solution after sonication. This results in enormously speedy heating and cooling rates producing fine crystals of MOF.

Most recently in 2018, four MOF films (Cu-BTC, Cu-BDC, ZIF-8, and MOF-5) were effectively synthesized by Abuzalat et al. exploring sonochemical techniques. In the first step, the substrates were treated with a strong oxidizing agent in order to convert the metal to the corresponding metal hydroxide. Ultrasonic irradiation provided the energy to drive the reaction between the metal ion sources and organic linkers [82].

In a similar fashion, ZIF-8 (Fig. 7.2) was prepared by Cho et al. utilizing a sonochemical approach under pH-adjusted synthesis conditions. After optimizing the synthesis parameters, it was extended successfully to 1 L-scale synthesis with an 85% product yield at a high substrate concentration in 2 hours without compromising the textural properties [83].

Li et al. have demonstrated a rapid synthetic procedure for the synthesis of $[Cu_3(BTC)_2(H_2O)_3]_n$ by the reaction of cupric acetate and H_3BTC with the help of ultrasonic irradiation.

The synthesis of a 3D MOF, $Cu_3(BTC)_2$ (HKUST-1, BTC = benzene-1,3,5-tricarboxylate), has been reported by using the ultrasonic method. The reaction mixture of cupric acetate and H_3BTC in DMF/EtOH/H_2O (3:1:2, v/v) was subjected to ultrasonic irradiation at ambient temperature and atmospheric

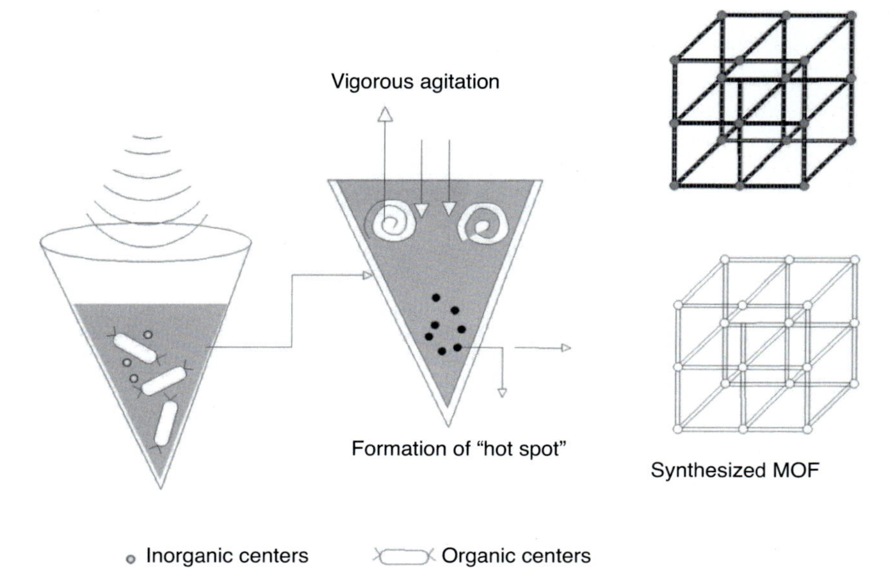

SCHEME 7.3 Sonochemical fabrication of metal–organic framework structures.

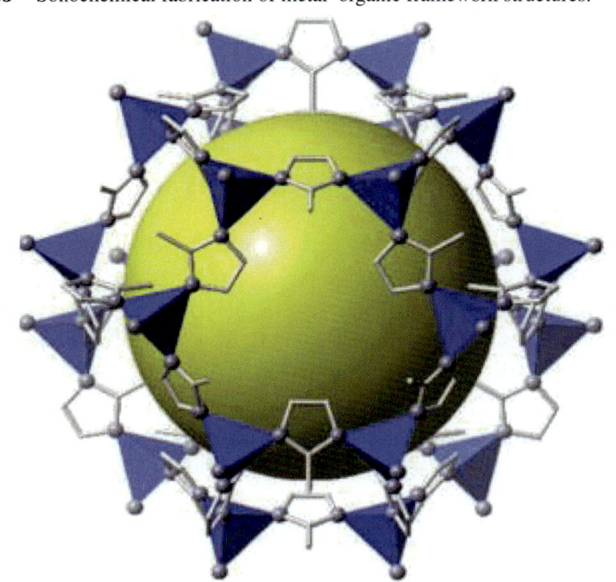

FIGURE 7.2 Crystal structures of ZIF-8: Zn (*polyhedra*), N (*sphere*), and C (*line*). The large spheres represent the largest van der Waals spheres that would fit in the cavities without touching the framework. All hydrogen atoms were omitted for clarity. (*Reproduced from H.Y. Cho, J. Kim, S.N. Kim, W.S. Ahn, High yield 1-L scale synthesis of ZIF-8 via a sonochemical route, Microporous Mesoporous Mater. 169 (2013) 180–184*).

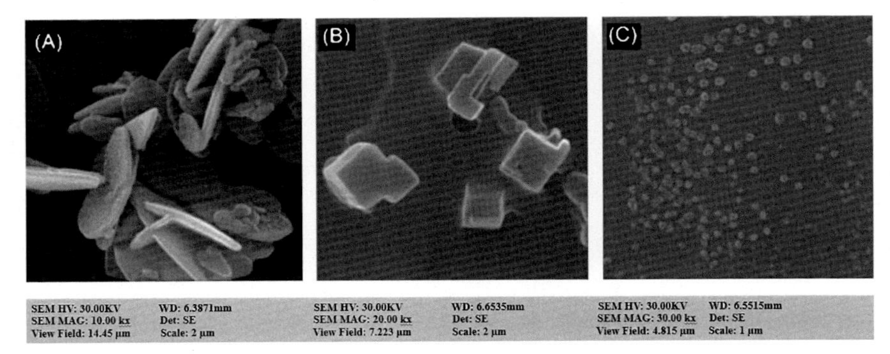

FIGURE 7.3 Scanning electron microscope (SEM) images of sonochemically synthesized TMU-31 (30 minutes) with concentrations of (A) 0.01 M, (B) 0.005 M, and (C) 0.001 M. (*Reproduced from L. Esrafili, A.A. Tehrani, A. Morsali, Ultrasonic assisted synthesis of two urea functionalized metal organic frameworks for phenol sensing: a comparative study, Ultrason. Sonochem. 39 (2017) 307–312*).

pressure. A high yield of 62.6–85.1% was observed in a short reaction time of 5–60 minutes. There were no noteworthy differences in physicochemical properties were found. Compared with other existing conventional techniques, such as solvent diffusion technique, hydrothermal and solvothermal methods, the ultrasound irradiated technique for the construction of such compounds was found to be highly efficient [84].

Esrafili et al. have successfully synthesized two MOF, TMU-31, [Zn(L1)(L2)]·DMF and TMU-32, [Zn(oba)(L2)]·2DMF·H_2O by sonochemical irradiation. The effects of various parameters such as various concentrations of initial reagents and different times of irradiation were also tested for obtaining homogeneous morphology [85]. Scanning electron microscope images of TMU-31 with different morphologies due to changes in concentration of reagent have been shown in Fig. 7.3.

Tahmasian et al. utilized an ultrasonically-assisted route and reported an efficient, low-cost, and environmental-friendly path. A 3D supramolecular nanostructured MOF {[Mg(HIDC)(H_2O)2]·1.5H_2O}n (H3L = 4,5-imidazole-dicarboxylic acid) was synthesized. Firstly, 20 mL solution of the ligand H_3IDC (0.05M) and potassium hydroxide (0.1 M) was placed in a high-density ultrasonic probe with a maximum power output of 305W followed by, dropwise addition of 20 mL aqueous solution of magnesium nitrate (0.05 M). The resultant precipitates were filtered off, washed with water and ethanol, and air dried [86].

7.3.4 Advantages sonochemical synthesis over conventional methods

Ultrasound has proven itself as vital technique to improve the quality of synthetic approaches in versatile field of material chemistry owing to following points:

- The high reaction rates can be achieved through sonochemical synthesis.

- Many enhanced properties are obtained in the field of kinetics, selectivity, extraction, dissolution, filtration, and crystallinity.
- It requires no reaction catalyst and therefore, does not have a need to change or replace on a regular interval.
- It is an environment-friendly chemical synthesis process.
- It is highly versatile by making slight modifications in the reaction.
- It is a time-efficient process.
- The quality of yield is relatively higher than other conventional processes.

7.4 Conclusion

Ultrasonic irradiation offers exceptional reaction state of affairs via acoustic cavitation. Bubbles originated during sonication can efficiently build up the diffuse energy of ultrasound, and on collapse, a massive amount of energy is on the rampage to heat up the contents of the bubble. These short lived, localized hot spots with enormously elevated temperatures and pressures are principally held accountable for the chemical effects because of ultrasound. The worth of sonochemical method as a synthetic tool resides in its versatility. With a simple adjustment in reaction parameters, different morphologies of MOFs can be successfully fabricated. The exploration of ultrasound in supramolecular chemistry is manifold, and there still remains a lot to explore. However, the significant challenges that face a wider application of sonochemistry include issues of scale up and energy consumption. While small-scale equipment for sonochemical synthesis is commercially available, larger-scale equipment remains relatively uncommon. Simply, ultrasound does not induce chemical reactions by itself. Instead, the ultrasound serves itself as a platform to nebulize precursor solutions producing micron-sized droplets that confine chemical reactions within their periphery. Therefore, such micro-reactors allow for easy maneuvering over chemical composition. In the present chapter, we have compiled the ultrasound-assisted preparation of MOFs and the synthesis parameters that can directly affect the properties of the final product. The intensified nucleation and crystallization duration of the process have made sonochemical synthesis very appealing, but appropriate handling is necessary in order to obtain desired morphology and particle dimension. Depending on the target application, there should be several parameters put in place for the selection of the appropriate reagents and solvents, appropriate techniques, as well as post-synthetic modifications. The current endeavors have been systematically considered and the benefits of the use of ultrasounds in all these areas of MOF preparation and MOF applications have been elucidated.

References

[1] T.J. Mason, Sonochemistry, first ed., Oxford University Press, Oxford, 1999.
[2] T.J. Mason, Advances in Sonochemistry, first ed., Jai Press Inc., Stamford, 1999.

[3] M. Draye, J.P. Bazureau, Ultrasound and Microwaves: Recent Advances in Organic Chemistry, Transworld Research Network, Trivandrum, 2011.
[4] N.A. Khan, S.H. Jhung, Synthesis of metal-organic frameworks (MOFs) with microwave or ultrasound: rapid reaction, phase-selectivity, and size reduction, Coord. Chem. Rev. 285 (2015) 11–23.
[5] K.S. Suslick, Kirk-Othmer Encyclopedia of Chemical Technology, vol. 26, John Wiley & Sons, Inc., 1998, pp. 517–541.
[6] G. Chatel, How sonochemistry contributes to green chemistry? Ultrason. Sonochem. 40 (2018) 117–122.
[7] K.S. Suslick, D.A. Hammerton, R.E. Cline, Sonochemical hot spot, J. Am. Chem. Soc. 108 (1986) 5641–5642.
[8] T.J. Mason, D. Peters, Practical Sonochemistry: Power Ultrasound Uses and Applications, second ed., Woodhead Publishing, Cambridge, UK, 2002.
[9] Bhangu Kaur Sukhvir, Ashokkumar Muthupandian, Theory of Sonochemistry, Sonochemistry: From Basic Principles to Innovative Applications, in: Juan Colmenares, Gregory Chatel (Eds.), Topics in Current Chemistry Collections, Springer International Publishing, Cham, 2016, pp. 1–28. https://link.springer.com/chapter/10.1007/978-3-319-54271-3_1.
[10] G. Chatel, Sonochemistry: New Opportunities for Green Chemistry, World Scientific Publishing Company, New York, 2016.
[11] N. Pokhrel, P.K. Vabbina, N. Pala, Sonochemistry: science and engineering, Ultrason. Sonochem. 29 (2016) 104–128.
[12] M. Kamali, R. Dewil, L. Appels, T.M. Aminabhavi, Nanostructured materials via green sonochemical routes–sustainability aspects, Chemosphere 276 (2021) 130146.
[13] M. Tanhaei, A.R. Mahjoub, V. Safarifard, Energy-efficient sonochemical approach for the preparation of nanohybrid composites from graphene oxide and metal-organic framework, Inorg. Chem. Commun. 102 (2019) 185–191.
[14] R. Al-Attri, R. Halladj, S. Askari, Green route of flexible Al-MOF synthesis with superior properties at low energy consumption assisted by ultrasound waves, Solid State Sci. 123 (2022) 106782.
[15] B. Souri, P. Hayati, A.R. Rezvani, J. Janczak, The effects of modifying reaction conditions in green sonochemical synthesis of a copper (II) coordination polymer as well as in achieving to different morphologies of copper (II) oxide micro crystals via solid-state process, Inorganica Chim. Acta. 483 (2018) 516–526.
[16] L. Tait, Galton's whistles, Nature 15 (1877) 294.
[17] J.H. Clark, D.J. Macquarrie, Handbook of Green Chemistry and Technology, John Wiley & Sons, Ireland, 2008.
[18] K.F. Graff, A history of ultrasonics, Phys. Acoust. 15 (1981) 1–97.
[19] W.T. Richards, An intensity gauge for "supersonic" radiation in liquids, Proc. Natl Acad. Sci. 15 (1929) 310–314.
[20] T.J. Mason, Ultrasonic cleaning: an historical perspective, Ultrason. Sonochem. 29 (2016) 519–523.
[21] C. Vaitsis, G. Sourkouni, C. Argirusis, Metal organic frameworks (MOFs) and ultrasound: a review, Ultrason. Sonochem. 52 (2019) 106–119.
[22] C. Zhang, L. Xin, J. Li, J. Cao, Y. Sun, X. Wang, J. Luo, Y. Zeng, Q. Li, Y. Zhang, T. Zhang, Metal–organic framework (MOF)-based ultrasound-responsive dual-sonosensitizer nanoplatform for hypoxic cancer therapy, Adv. Healthc. Mater. 11 (2022) 2101946.
[23] K.S. Suslick, N.C. Eddingsaas, D.J. Flannigan, S.D. Hopkins, H. Xu, The chemical history of a bubble, Acc. Chem. Res. 51 (2018) 2169–2178.

[24] S.S. Rashwan, I. Dincer, A. Mohany, An investigation of ultrasonic based hydrogen production, Energy 205 (2020) 118006.

[25] S. Zhao, C. Yao, Z. Dong, Y. Liu, G. Chen, Q. Yuan, Intensification of liquid-liquid two-phase mass transfer by oscillating bubbles in ultrasonic microreactor, Chem. Eng. Sci. 186 (2018) 122–134.

[26] T.J. Tiong, J.K. Chu, L.Y. Lim, K.W. Tan, Y.H. Yap, U.A. Asli, A computational and experimental study on acoustic pressure for ultrasonically formed oil-in-water emulsion, Ultrason. Sonochem. 56 (2019) 46–54.

[27] C.E. Brennen, Cavitation in medicine, Interface Focus 5 (2015) 20150022.

[28] E. Stride, T. Segers, G. Lajoinie, S. Cherkaoui, T. Bettinger, M. Versluis, M. Borden, Microbubble agents: new directions, Ultrasound Med. Biol. 46 (6) (2020) 1326–1343.

[29] G. Matafonova, V. Batoev, Dual-frequency ultrasound: strengths and shortcomings to water treatment and disinfection, Water Res. 182 (2020) 116016.

[30] A.D. Alarcon-Rojo, L.M. Carrillo-Lopez, R. Reyes-Villagrana, M. Huerta-Jiménez, I.A. Garcia-Galicia, Ultrasound and meat quality: a review, Ultrason. Sonochem. 55 (2019) 369–382.

[31] F.J. Fahy, Foundations of Engineering Acoustics, Academic Press, Elsevier, San Diego, 2000.

[32] Y.T. Shah, A.B. Pandit, V.S. Moholkar, Cavitation Reaction Engineering, Plenum Press, Springer Science & Business Media, New York, 1999.

[33] L.K. Doraiswamy, L.H. Thompson, Sonochemistry: science and engineering, Ind. Eng. Chem. Res. 38 (1999) 1215–1249.

[34] C. Vanhille, C. Campos-Pozuelo, Acoustic cavitation mechanism: a nonlinear model, Ultrason. Sonochem. 19 (2) (2012) 217–220.

[35] M. Saclier, R. Peczalski, J. Andrieu, A theoretical model for ice primary nucleation induced by acoustic cavitation, Ultrason. Sonochem. 17 (1) (2010) 98–105.

[36] A. Eller, H.G. Flynn, Rectified diffusion during nonlinear pulsations of cavitation bubbles, J. Acoust. Soc. Am. 37 (1965) 493–503.

[37] E.A. Neppiras, Acoustic cavitation series: part one: acoustic cavitation: an introduction, Ultrasonics 22 (1984) 25–28.

[38] H. Ji, H. Chen, M. Li, Effect of ultrasonic transmission rate on microstructure and properties of the ultrasonic-assisted brazing of Cu to alumina, Ultrason. Sonochem. 34 (2017) 491–495.

[39] C. Vanhille, C. Campos-Pozuelo, Acoustic cavitation mechanism: a nonlinear model, Ultrason. Sonochem. 19 (2012) 217–220.

[40] K.S. Suslick, Y. Didenko, M.M. Fang, T. Hyeon, K.J. Kolbeck, W.B. McNamara III, M.M. Mdleleni, M. Wong, Acoustic cavitation and its chemical consequences, Philos. Trans. R. Soc. Series A Mathemat. Phys. Eng. Sci. 357 (1999) 335–353.

[41] R.P. Taleyarkhan, C.D. West, J.S. Cho, R.T. Lahey Jr, R.I. Nigmatulin, R.C. Block, Evidence for nuclear emissions during acoustic cavitation, Science 295 (2002) 1868–1873.

[42] D.G. Shchukin, E. Skorb, V. Belova, H. Möhwald, Ultrasonic cavitation at solid surfaces, Adv. Mater. 23 (2011) 1922–1934.

[43] T. Tuziuti, K. Yasui, M. Sivakumar, Y. Iida, N. Miyoshi, Correlation between acoustic cavitation noise and yield enhancement of sonochemical reaction by particle addition, J. Phys. Chem. A 109 (2005) 4869–4872.

[44] H.M. Santos, C. Lodeiro, J.L. Capelo-Martinez, The power of ultrasound, Ultrason. Chem. Analyt. App. Wiley-VCH (2009) 1–6.

[45] H.M. Santos, J.L. Capelo, Trends in ultrasonic-based equipment for analytical sample treatment, Talanta 73 (2007) 795–802.

[46] Y.G. Adewuyi, Sonochemistry: environmental science and engineering applications, Ind. Eng. Chem. Res. 40 (2001) 4681–4715.

[47] R.F. Contamine, A.M. Wilhelm, J. Berlan, H. Delmas, Power measurement in sonochemistry, Ultrason. Sonochem. 2 (1995) S43–S47.

[48] G. Wibetoe, D.T. Takuwa, W. Lund, G. Sawula, Coulter particle analysis used for studying the effect of sample treatment in slurry sampling electrothermal atomic absorption spectrometry, Fresenius J. Anal. Chem. 363 (1999) 46–54.

[49] L.A. Crum, Comments on the evolving field of sonochemistry by a cavitation physicist, Ultrason. Sonochem. 2 (1995) S147–S152.

[50] L. Amoedo, J.L. Capelo, I. Lavilla, C. Bendicho, Ultrasound-assisted extraction of lead from solid samples: a new perspective on the slurry-based sample preparation methods for electrothermal atomic absorption spectrometry, J. Anal. At. Spectrom. 14 (1999) 1221–1226.

[51] É.C. Lima, F. Barbosa Jr, F.J. Krug, M.M. Silva, M.G. Vale, Comparison of ultrasound-assisted extraction, slurry sampling and microwave-assisted digestion for cadmium, copper and lead determination in biological and sediment samples by electrothermal atomic absorption spectrometry, J. Anal. At. Spectrom. 15 (2000) 995–1000.

[52] P.R. Gogate, A.M. Wilhelm, A.B. Pandit, Some aspects of the design of sonochemical reactors, Ultrason. Sonochem. 10 (2003) 325–330.

[53] J.L. Capelo, M.M. Galesio, G.M. Felisberto, C. Vaz, J.C. Pessoa, Micro-focused ultrasonic solid–liquid extraction (μFUSLE) combined with HPLC and fluorescence detection for PAHs determination in sediments: optimization and linking with the analytical minimalism concept, Talanta 66 (2005) 1272–1280.

[54] B. Savun-Hekimoğlu, A review on sonochemistry and its environmental applications, Acoustics 2 (2020) 766–775.

[55] M.A. Margulis, Fundamental aspects of sonochemistry, Ultrasonics 30 (1992) 152–155.

[56] P.J. Perez, Advances in Organometallic Chemistry, first ed., Academic Press, Elsevier, 2018.

[57] M.A. Beckett, I. Hua, Impact of ultrasonic frequency on aqueous sonoluminescence and sonochemistry, Phys. Chem. A 105 (2001) 3796–3802.

[58] D.J. Tranchemontagne, J.R. Hunt, O.M. Yaghi, Room temperature synthesis of metal-organic frameworks: MOF-5, MOF-74, MOF-177, MOF-199, and IRMOF-0, Tetrahedron 64 (36) (2008) 8553–8557.

[59] M.A. Margulis, Sonochemistry as a new promising area of high-energy chemistry, High Energy Chem. 38 (2004) 135–142.

[60] Y.R. Lee, J. Kim, W.S. Ahn, Synthesis of metal-organic frameworks: a mini review, Korean J. Chem. Eng. 30 (2013) 1667–1680.

[61] S. Natarajan, S. Mandal, P. Mahata, V.K. Rao, P. Ramaswamy, A. Banerjee, A.K. Paul, K.V. Ramya, The use of hydrothermal methods in the synthesis of novel open-framework materials, J. Chem. Sci. 118 (2006) 525–536.

[62] G. Cravotto, P. Cintas, Power ultrasound in organic synthesis: moving cavitational chemistry from academia to innovative and large-scale applications, Chem. Soc. Rev. 35 (2006) 180–196.

[63] U. Mueller, M. Schubert, F. Teich, H. Puetter, K. Schierle-Arndt, J. Pastre, Metal–organic frameworks—prospective industrial applications, J. Mater. Chem. 16 (2006) 626–636.

[64] Q.R. Fang, G.S. Zhu, Z. Jin, Y.Y. Ji, J.W. Ye, M. Xue, H. Yang, Y. Wang, S.L. Qiu, Mesoporous metal–organic framework with rare etb topology for hydrogen storage and dye assembly, Angew. Chem. 119 (2007) 6758–6762.

[65] D. Yu, L. Ge, X. Wei, B. Wu, J. Ran, H. Wang, T. Xu, A general route to the synthesis of layer-by-layer structured metal organic framework/graphene oxide hybrid films for high-performance supercapacitor electrodes, J. Mater. Chem. A 5 (2017) 16865–16872.
[66] J.G. Hinman, J.G. Turner, D.M. Hofmann, C.J. Murphy, Layer-by-layer synthesis of conformal metal organic framework shells on gold nanorods, Chem. Mater. 30 (2018) 7255–7261.
[67] M.S. Plesset, A. Prosperetti, Bubble dynamics and cavitation, Annu. Rev. Fluid Mech. 91 (1977) 45–185.
[68] T.G. Leighton, The Acoustic Bubble, Academic Press, 2012.
[69] R. Scheffold, Modern Synthetic Methods, Wiley, New York, 1983.
[70] Y. Koltypin, G. Katabi, X. Cao, R. Prozorov, A. Gedanken, Sonochemical preparation of amorphous nickel, J. Non-Cryst. Solids 201 (1996) 159–162.
[71] T. Hyeon, M. Fang, K.S. Suslick, Nanostructured molybdenum carbide: sonochemical synthesis and catalytic properties, J. Am. Chem. Soc. 118 (1996) 5492–5493.
[72] K.S. Suslick, T. Hyeon, M. Fang, Nanostructured materials generated by high-intensity ultrasound: sonochemical synthesis and catalytic studies, Chem. Mater. 8 (1996) 2172–2179.
[73] L.S. da Silveira Pinto, M.V. de Souza, Sonochemistry as a general procedure for the synthesis of coumarins, including multigram synthesis, Synthesis 49 (2017) 2677–2682.
[74] B. Li, Y. Xie, J. Huang, Y. Qian, Sonochemical synthesis of silver, copper and lead selenides, Ultrason. Sonochem. 6 (1999) 217–220.
[75] T. Kitazume, N. Ishikawa, Ultrasound-promoted selective perfluoroalkylation on the desired position of organic molecules, J. Am. Chem. Soc. 107 (1985) 5186–5191.
[76] J.L. Luche, J.C. Damiano, Ultrasounds in organic syntheses. 1. Effect on the formation of lithium organometallic reagents, J. Am. Chem. Soc. 102 (1980) 7926–7927.
[77] K.S. Suslick, S.B. Choe, A.A. Cichowlas, M.W. Grinstaff, Sonochemical synthesis of amorphous iron, Nature 353 (1991) 414–416.
[78] K.S. Suslick, T. Hyeon, M. Fang, J.T. Ries, A.A. Cichowlas, Sonochemical synthesis of nanophase metals, alloys and carbides, Mater. Sci. Forum 225 (1996) 903–912.
[79] W.J. Son, J. Kim, J. Kim, W.S. Ahn, Sonochemical synthesis of MOF-5, Chem. Commun. 2008 (2008) 6336–6338.
[80] L.G. Qiu, Z.Q. Li, Y. Wu, W. Wang, T. Xu, X. Jiang, Facile synthesis of nanocrystals of a microporous metal–organic framework by an ultrasonic method and selective sensing of organoamines, Chem. Comm. 31 (2008) 3642–3644.
[81] N. Stock, S. Biswas, Synthesis of metal-organic frameworks (MOFs): routes to various MOF topologies, morphologies, and composites, Chem. Rev. 112 (2012) 933–969.
[82] O. Abuzalat, D. Wong, M. Elsayed, S. Park, S. Kim, Sonochemical fabrication of Cu (II) and Zn (II) metal-organic framework films on metal substrates, Ultrason. Sonochem. 45 (2018) 180–188.
[83] H.Y. Cho, J. Kim, S.N. Kim, W.S. Ahn, High yield 1-L scale synthesis of ZIF-8 via a sonochemical route, Microporous Mesoporous Mater. 169 (2013) 180–184.
[84] Z.Q. Li, L.G. Qiu, T. Su, Y. Wu, W. Wang, Z.Y. Wu, X. Jiang, Ultrasonic synthesis of the microporous metal–organic framework $Cu_3(BTC)_2$ at ambient temperature and pressure: an efficient and environmentally friendly method, Mater. Lett. 63 (2009) 78–80.
[85] L. Esrafili, A.A. Tehrani, A. Morsali, Ultrasonic assisted synthesis of two urea functionalized metal organic frameworks for phenol sensing: a comparative study, Ultrason. Sonochem. 39 (2017) 307–312.
[86] A. Tahmasian, A. Morsali, S.W. Joo, Sonochemical syntheses of a one-dimensional Mg (II) metal-organic framework: a new precursor for preparation of MgO one-dimensional nanostructure, J. Nanomater. 2013 (2013) 1–7.

Chapter 8

Synthesis of metal–organic frameworks with ionic liquids

Zaib ul Nisa, Nargis Akhter Ashashi and Haq Nawaz Sheikh
Department of Chemistry, University of Jammu, Jammu and Kashmir, India

8.1 Introduction

Metal–organic frameworks (MOFs), which belong to coordination polymers, represent a novel class of hybrid porous crystalline materials that combine the advantages of both inorganic and organic constituents [1]. This consists of a wide variety of architectures which range from one-dimensional to three-dimensional framework structures. These are generated via cations or minor clusters of metals, i.e., inorganic part, interlinked by coordinating polydentate organic ligands (linkers) possessing O, S, or N atoms in their composition. Because of their distinct structural topologies, well-defined porosity, and customizable functions, MOFs have demonstrated plethoric presentations in the fields of gas adsorption as well as separation, separation of membrane, luminescence and sensing, biomedicine, magnetism, and ionic conductivity [2]. The stupendous performances behind these applications generally rely on the predetermined functional groups that are present within the MOFs architectures. Great success has been achieved to design MOF crystals with ultrahigh porosity, highly selective catalytic activity, high thermal and chemical stability, in addition to other outstanding qualities. The specific design approaches and synthetic routes are vital for achieving the desired structures and characteristics for MOFs [3]. Particularly, the reaction medium implied for the synthesis of MOFs plays avital role in constructing the end products of MOFs. Such as solvent system can provide different reaction conditions such as a chiral environment producing chiral MOFs whereas, a multiphase solvent system favors the formation of hierarchical structured MOF architecture [4]. Solvents effect on MOF fabrication such as dissolving metal-containing precursor with an organic linker, functioning as a ligand in the reaction and a structure-directing mediator is also crucial. Therefore, it is significant to find innovative solvents before constructing diverse solvent systems for creating MOFs with unique structures and characteristics as well as establishing MOF synthesis pathways. Organic liquids including

Synthesis of Metal–Organic Frameworks via Water-Based Routes: A Green and Sustainable Approach.
DOI: https://doi.org/10.1016/B978-0-323-95939-1.00006-X

N, N-dimethylformamide (DMF), acetonitrile (ACN), N, N-dimethylacetamide (DMA), dimethyl sulfoxide (DMSO), N, N-diethyl formamide (DEF), and N, N-diethyl acetamide (DEA) are the most often used solvents for MOF production. Additionally, mixed solvent systems have been employed for the synthesis of MOFs, including water with ethanol, DMF with ethanol, and water with DMF [5]. Solvothermal or hydrothermal processes are the most common methods for MOF development and reaction temperature and pressure are usually organized to be higher. However, the solvency of many reactants in organic solvents and aqueous media is limited. Moreover, environmental expenses are a problem for organic solvents [6]. Furthermore, it is recognized that a wide range of MOFs suffer from loss of crystallinity, phase changes, and decomposition of their structures in water [7]. Therefore, the aqueous medium is completely eluded in the synthesis of water-sensitive MOFs. Additionally, the structure of MOFs produced in ordinary solvents cannot be precisely controlled by the solvent. As a result, it is extremely important to optimize the MOF framework by planning the solvent properties [8].

Ionic liquids (ILs) are regarded as eco-friendly solvents having adjustable properties. The ILs are generally molten salts whose melting points are below some arbitrary temperature usually 100°C [9]. When synthesizing MOFs, ILs can provide practical benefits and access to new materials since they possess a number of features that are significantly different from those of conventional solvents [10]. For instance, large liquid range, low volatility, non-flammability, weakly coordinating characteristics, high polarity, high thermal stability, high chemical stability, high solubility toward both organic and inorganic compounds, and adaptable design ability, and so forth [11]. As in recent years, investigation of the function of ILs in the design and synthesis of novel MOFs has received a great deal of attention. Typically, ILs function as reaction mediums, structure-directional templates, or charge-compensating groups in the reaction systems. Therefore, the coexistence of ionic in addition to organic groups in the ILs along the substantial temperature windows makes them the ultimate reaction medium for metal ions and organic ligands to generate MOFs. More commonly, their cationic parts maneuver as structural templates being embedded in the cavities of skeleton frameworks. Hence, template-framework interactions dominate [12]. The classic hydro/solvothermal reactions, which are frequently used in the exploratory synthesis of inorganic and inorganic–organic hybrid materials, are thought to be less safe and more harmful to the environment than ionothermal synthesis because of these characteristics. This has drawn a lot of interest as a cutting-edge, environmental friendly technique to create new materials with unheard-of topological structures, like zeolite, and zeo-type compounds, besides MOFs [13].

There are several instances where ILs have been successfully used to create innovative MOFs. The main characteristic of ILs is that the synthesis of MOFs takes place in the presence of high ion concentrations, and it is feasible to modify the characteristics of those ions to produce novel materials.

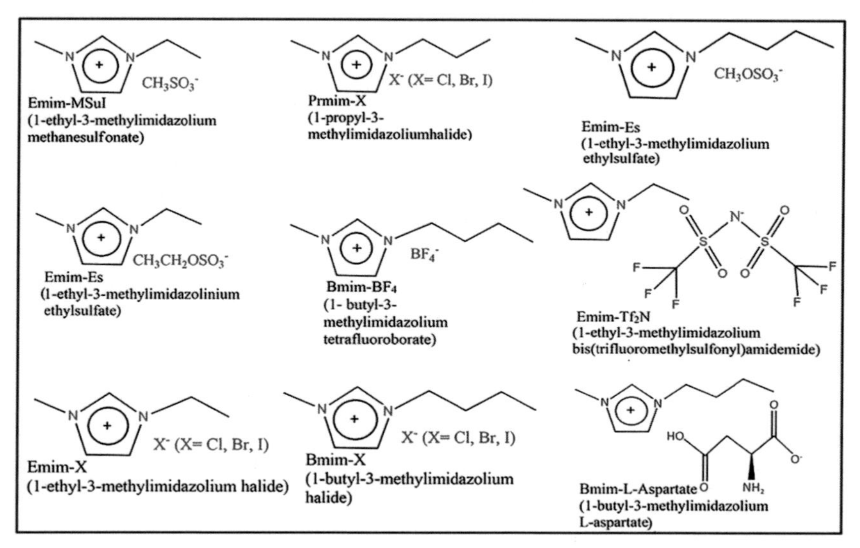

FIGURE 8.1 Ionic liquids used for the synthesis of metal–organic frameworks.

1,3-dialkylimidazolium cations forms the most prevalent root for most of the ILs and inadequately coordinating inorganic anions, for instance, tetrafluoroborate (BF_4^-), bis(trifluoromethanesulfonyl) imide (NTf_2^-), hexafluorophosphate (PF_6^-), and trifluoromethanesulfonate (triflate, OTf^-) (Fig. 8.1).

Numerous 1-methylimidazole-derived 1, 3-dialkylimidazolium cations have methyl groups as one of the alkyl substituents (Cx; a common abbreviation for such cations, where x denotes the length of the alkyl chain and "mim" denotes the methylimidazole part; for example, 1-ethyl-3-methylimidazolium$[C_2mim]^+$ as well as 1-buty anions that are relatively chemically inert and weakly coordinating, such as halide, chloride, or bromide, can also be introduced into the MOF framework (will be discussed later). Since ionic liquids' physical characteristics can be tailored, it makes sense to investigate ionothermal MOF production systems using ILs with systematically different structures. The length of the alkyl group on the imidazolium cation is one example of such variation. The melting point of ILs decreases as the alkyl chain length rises, which causes viscosity to rise. However, this trend is reversible when the alkyl chain has more than seven carbons [14].

8.2 Synthesis of MOFs in ILs

One may essentially categorize the investigations on MOF synthesis in ILs into two categories:

1. Ionothermal synthesis
2. Room-temperature synthesis in ILs using additives

8.2.1 Ionothermal synthesis

Special crystallization environments and structure-directing agents may be produced by ILs due to their distinct physical and solvent properties, which result in networks that are very different from those produced by conventional hydro/solvothermal techniques. Previous research demonstrated that ILs appear to hold promise to develop new MOFs that cannot be otherwise obtained using conventional synthetic pathways.

Synthesis through the ionothermal method was first proposed in 2004 by Cooper et al. Ionic liquids perform a structure-directing role as well as a solvent, and it was a successful procedure for constructing MOFs [15]. Ionothermal synthesis has so far been used to create a large number of MOFs. ILs as a medium for MOF synthesis have various advantages over the traditional methods of solvothermal or hydrothermal synthesis due to their unique characteristics, such as the ability to simultaneously dissolve both inorganic and organic precursors, extraordinary design ability, high boiling point windows, high thermal and chemical stability, as well as extremely low volatility. The conflict between solvent-framework and template-framework structure contacts that exists in solvo/hydrothermal synthesis is ideally eliminated by ionothermal synthesis [16]. Ionothermal conditions, particularly the characteristics of ILs, have an impact on the MOF structures and attributes that follow. For instance, the anions of ILs can join a MOF and are typically coordinated to its metal center or centers. An anionic MOF framework's pores can also accommodate the cation of an IL. These ions have a substantial role in influencing a structure in both scenarios. Another possibility is that an IL will only serve as a solvent during the creation of MOFs, with no actual incorporation of the IL into the finished product. These circumstances will each be covered individually.

8.2.1.1 The idea behind ionothermal synthesis

While developing metal organic structural frameworks, ILs are employed both as a template and a solvent, which makes ionothermal synthesis special. When performing ionothermal synthesis, a salt that melts below the synthesis temperatures, typically between 150°C and 220°C, is referred to as an ionic liquid. Ionic liquids' cation and anion compositions can finally be changed to provide a wide range of reaction settings and template types. The majority of the cations in ionic liquids share chemical similarities with species that are renowned for making excellent templates and are all ideal for dissolving the inorganic components required for synthesis because many of them are fairly polar solvents. The lack of detectable vapor pressure is their significant property which has led to its employment in microwave synthesis and successfully eliminates the safety problems related to high hydrothermal pressures. The breakdown of a tiny fraction of the ionic liquid cations, which could lead to smaller template cations functioning as the structure-directing agent more frequently, makes this utopian situation, however, not always feasible. Recent molecular modeling studies state that the long-range distributions and correlations that characterize the assemblies of ionic liquids reflect the asymmetric structures of the cations. If the process of templating must be under control, the possibility of chemical information being sent from the template cation to the framework may be increased by long-range asymmetric effects.

8.2.2 Room-temperature synthesis in ILs using additives

Obtaining crystalline MOFs in ionic liquid systems at room temperature is of utmost importance. In order to avoid the sluggish nucleation kinetics and high enthalpies of formation that could lead to the disintegration of ILs, the ionothermal synthesis of MOFs typically requires high-temperature widows($>100°C$). It has been discovered that adding additives, such as surfactant and other solvents, to ILs is an efficient way to develop MOFs at room temperature.

Peng et al. investigated the mesoporous Cu (II) MOF assembly led by the surfactant in an IL solution in the ILs 1, 1, 3, and 3-tetramethylguanidine (TMGT) and the surfactant N-ethyl perfluorooctyl sulfonamide ($C_2H_5NHSO_2C_8F_{17}$, N-EtFOSA). At various surfactant concentrations, mesoporous framework nanoplates in circular, square and hexagonal, forms were produced. This approach uses less energy than the more popular solvothermal method since the reaction can take place at a low temperature (30°C). In addition, this process is far more straightforward and environmentally benign than the widely utilized solvothermal method. It demonstrates a dual function for the surfactant creation of the mesoporous MOF nanoplates, controlling both the development of the crystals and the initiation of the mesoporous. Other MOFs with various morphologies and porosities can be synthesized using it with ease. The MOF was also created using only TMGT. When IL has applied alone, no mesoporous framework was created. The surfactant's importance in the creation of mesoporous MOF nanoplates is thus confirmed. The ability to create mesoporous MOF nanoplates at surfactant concentrations far lower than critical micelle concentration (CMC) demonstrates that micelles are not necessary for their production. Due to the benefits of enhanced mass transfer and contact interface, these crystalline materials combine the benefits of mesopores with small particle sizes, and they may find use in controlled medication release, gas separation, and catalysis [17].

8.3 Anion incorporation/structure-directing effects

ILs perform a range of functions in the structure-creation of these materials, according to studies with reference to the ionothermally synthesized MOFs. The IL anion can incorporate into the MOF in a variety of circumstances, most frequently when it binds to the metal center, such as a halide, in the MOF. This section discusses the inclusion of anions and the impacts of anion structure-directing.

In our very first example Lin et al. synthesized [Co(BTC)][C_2mim], [$Co_5(OH)_2(OAc)_8$]·$2H_2O$ and [$Co_3(BTC)_2(OAc)_2$][C_2mim]$_2$,by reacting the $Co(OAc)_2 \cdot 4H_2O$ as well as H_3BTC. The reaction was performed in these separate ILs or IL combinations: neat [C_2mim][Br] and [C_2mim][NTf_2] and an equal mixture of [C_2mim][NTf_2] and [C_2mim][Br]. A fourth MOF, [C_2mim](HBTC)(2,2′-bipy)$_2$, $Co_2(H_2BTC)_3$·[$Co_5(OH)_2(OAc)_8$]·$2H_2O$, was created by the reaction of [C_2mim][NTf_2] with 2,2′-bipyridine (2,2′- bipy), which has none of the planned BTC_3 linking ligand but feature an extensive structural framework with the majority of the bridges among the Co atoms generated through acetate ligands, might be regarded as a MOF for this reason. The 2, 2′-bipy ligands in [C_2mim][$Co_2(H_2BTC)_3$(HBTC)(2,2′-bipy)$_2$] leave four coordination sites on each Co atom open, and the BTC ligand fills these with monodentate carboxylate groups, generating 2D (4, 4)-nets connected by hydrogen bonding. The identity of the IL anion influences the MOF that is generated even if the ionic liquid anion is not integrated into any one of the synthesized MOFs [18].

In one interesting study, Chen et al. obtained ILs@MOF nanocomposites incorporating ionic liquids into porous MOF. The newly synthesized chain of ionic liquids with various counter anions (R-SO_3^-) and their ILs@MOF hybrid materials are known as MSA-EIMS@MIL-101, SA-EIMS@MIL-101, PTSA-EIMS@MIL-101 (SA = sulfate acid, MSA = methane sulfonate acid, EIMS = 1-(1-ethyl-3-imidazolium)propane-3-sulfonate), PTSA = p-toluene sulfonate acid. Even when heated to 150°C in the open air, these hybrid materials demonstrated anhydrous proton conduction with long-term durability. Favorable proton transportation is exhibited along with the long-term retention of ILs. Such hybrid ILs@MOF materials' proton conductivity was noticeably affected by anion volumes of R-SO_3^-. The conductivity of ILs@MOF increases proportionately to the decrease in the van der Waals volume of R-SO_3^-. Their research raises the possibility of the creation of a secure, anhydrous solid-state electrolyte for extraordinary temperature proton-exchange membrane fuel cells by combining a range of integrated ILs with a MOF framework [19].

The loading of ILs into MOF material nanoarchitectures has been reported by Xu et al. A novel form of hybrid ionic conductor made of ionic liquids and MOFs has caused considerable worry. To embed the ionic liquids into the pores of MIL-101, 1-ethyl-3-methylimidazolium thiocyanate ([Emim][SCN]) and 1-ethyl-3-methylimidazolium dicyanamide ([Emim][DCA]) a successful soaking-volatilizing technique was used. The [Emim][SCN] may completely fill the holes of the MIL-101 material, increasing the material's ionic conductivity above that of conventional solid electrolytes. Additionally, this sample's activation energy is comparable to that of other conductive IL@MOF composites. It is noteworthy that the benefits of high conductivity and low activation energy make IL@MOF hybrid composites promising ionic conductors [20].

Additionally, Guo et al. have synthesized two IL@MOF, [Cu_2(EBTC) $(H_2O)_2$ · (2.1[EMIM]Br)] and [Cu_2(EBTC)$(H_2O)_2$ · (1.5[BMIM]Cl)] where

[$EBTC^{4-}$ = 1,1′-ethynebenzene-3,3′,5,5′-tetracarboxylate, BMIM = 1-butyl-3-methylimidazolium]. The promising electrochemical properties of MOFs integrated with imidazolium-based IL with a variety of anion species and alkyl chain sizes have been investigated. The ionic conductivity value of Cl containing MOF is about one order higher than Br containing MOF. The size of the anion plays an important role in the performance of ionic transport in a battery system, even though the IL loading in Br was higher than in Cl [21]. A mixture of 2,4,6-tris(4-pyridyl)-1,3,5-triazine (tpt) ad $Cu(NO_3)_2{\cdot}3H_2O$ and by heating in an IL system of [bmim](BF_4) at 170°C for two days, Danil N. Dybtsevon created crystals of Cu complex having the formula $[Cu_3(tpt)_4](BF_4)_3 \cdot (tpt)_{2/3} \cdot 5H_2O$. The resultant 3D cationic framework contains Cu^+ ions, which are thought to have been produced during the process. The cubic-C_3N_4 topology is formed in the double-interpenetrated cationic framework's channels. 2,4,6-tris(4-pyridyl)-1,3,5-triazine-ligand and BF_4^- anions of the IL function as counter ions. Three Cu^+ ions are connected by each tpt ligand, and four tpt ligands coordinate the Cu^+ ions in a tetrahedral arrangement. The cubic space group's three-fold crystallographic symmetry also links the Cu^+ ions together. Free tpt, H_2O, and disordered BF_4^- anions fill the oversized channels, which have an approximate diameter of 5Å [22].

8.4 IL cation incorporation/templating

An anionic MOF framework can be made through ionothermal synthesis of MOFs in ionic liquids, to balance charge, the cation of the ionic liquid is incorporated into the MOF gallery, acting as a model upon which the structural framework is built. This is the supreme way an ionic liquid controls a MOF's framework during synthesis. This section will include descriptions of several examples.

Dai et al. have fabricated new MOFs with cationic ionic liquids (ILs) and1,3-bis(4-carboxybutyl) imidazolium bromide as organic ligands with nanocrystals modified SiO_2 core-shell microspheres (Table 8.1). Cationic imidazolium-based ionic liquids are introduced to create the MOFs' nanocrystal shells which possess the liquid chromatography's retention mechanism for hydrophilic interactions. Compared to the conventional aminosilica column, the new ILI-01@SiO_2 column demonstrate great separation discrimination in a smaller duration of time. The unauthorized addition of melamine to baby formula was also detected using the novel ILI-01@SiO_2 column. All of the aforementioned findings show that the novel ILI-01@SiO_2 core-shell at the stationary phase has good potential for highly selective polar materials separation [23].

According to Zottnick et al., [EMIM][$B(CN)_4$] ionic liquid was used as the solvent for the manufacture of a [EMIM][$LaNO_3B(CN)_4{\cdot}3(H_2O)$] (EMIM stands for 1-ethyl-3-methylimidazole.) In the $P2_1/n$ space group, it crystallizes in the monoclinic crystal system. It is an anionic coordination polymer with a huge concentration of $[B(CN)_4]^-$ units, and the charge is balanced by the

TABLE 8.1 Some ionic liquids used in synthesis of ionic liquid/metal–organic framework composites.

MOFs/MOF@ILs	ILs	References
$BMIm_2[Zn_3(C_6H_4(COO)_2)_3Cl_2]$	BMIm-Cl (1-butyl-3-methylimidazolium chloride)	[28]
$EMIm_2$ $[Zn_3(C_6H_4(COO)_2)_3Cl_2]$	EMIm-Cl (1-ethyl-3-methylimidazolium chloride)	[28]
[Emim][CdBr(1,4-NDC)] 1,4-$NDCH_2$=1,4-naphthalenedicarboxylic acid)	[Emim]BF_4 [Emim]BF_4 and [Emim]Br (Emim=1-ethyl-3-methylimidazolium)	[29]
[Cu(1,3-bis(4-pyridyl)propane)](BF_4)	[bmim][BF_4] {1-butyl-3-methylimidazolium tetrafluoroborate}	[30]
(EMI)[Cd(BTC)], (BTC = 1,3,5-benzenetricarboxylate)	EMI (1-ethyl-3-methylimidazolium)	[30]
[Cu(bpp)]BF_4 [bpp = 1,3-bis(4-pyridyl)propane]	[bmim][BF_4] {1-butyl-3-methylimidazolium tetrafluoroborate}	[31]
IL@ZIF-8	[BMIM][Tf_2N] 1-butyl-3-methylimidazolium bis(trifluoromethylsulfonyl)imide	[36]
ZIF-8	[BMIM][PF_6] 1-butyl-3-methylimidazolium hexafluorophosphate	[37]
ZIF-8	[BMIM][BF_4] 1-butyl-3-methylimidazolium tetrafluoroborate	[38]
CuBTC copper benzene-1,3,5-tricarboxylate	[BMIM][BF_4] 1-butyl-3-methylimidazolium tetrafluoroborate	[39]
CuBTC copper benzene-1,3,5-tricarboxylate	[BMIM][PF_6] 1-butyl-3-methylimidazolium hexafluorophosphate	[40]

(*continued on next page*)

MIL101(Cr)	[BMIM][Cl] 1-butyl-3-methylimidazolium chloride	[41]
CuBTC copper benzene-1,3,5-tricarboxylate	ABIL-OH Amino-functionalized basic ionic liquid	[42]
MIL-100(Fe)	$[SO_3H\text{-}(CH_2)_3\text{-}HIM][HSO_4]$ 1,4-bis[3-(propyl-3-sulfonate)imidazolium]butane hydrogen sulfate	[43]
UiO-66	[PSMIM][HSO4] 1-methylimidazolium-3-propylsulfonate hydrosulfate	[44]
MIL-100(Fe)	Acidic chloroaluminate IL	[45]
CuBTC copper benzene-1,3,5-tricarboxylate	$[HVIm\text{-}(CH_2)_3SO_3H][HSO_4]$ 1-(propyl-3-sulfonate)vinylimidazolium hydrogen sulfate	[46]
ZIF-8	$[BMIM][Tf_2N]$ 1-ethyl-3-methylimidazolium bis(trifluoromethylsulfonyl)imide	[47]
MIL-100(Fe)@$[SO_3H\text{-}(CH_2)_3\text{-}IM]_2C_4[HSO_4]2$	$[SO_3H\text{-}(CH_2)_3\text{-}IM]_2C_4[HSO_4]_2$ (1,4-bis[3-(propyl-3-sulfonate) imidazolium] butane hydrogen sulfate)	[48]
MOF-5/n-Bu_4NBr	*n*-Bu_4NBr tetrabutylammonium bromide	[49]
ZIF-71	[BMIM][SCN] 1-n-butyl-3-methylimidazolium thiocyanate	[50]
Na-rho-ZMOF	[BMIM][SCN] 1-n-butyl-3-methylimidazolium thiocyanate	[50]

intercalated $[EMIM]^{+}$ cation of the IL. With the help of the compound's three $[B(CN)_4]$ anions, two of which act as terminal ligands and link with nearby La^{3+} ions, a double-stranded one-dimensional coordination polymer is created. This effectively demonstrates the function of the ionic liquid as a structure-directing agent because the synthesis of an anionic coordination polymer is enabled by the ionic character of $[EMIM][B(CN)_4]$ [24].

Xing et al. have synthesized a new compound $[EMIM]^{+}[MIAH]^{+}$ $[Al_6P_7O_{27}(OH)]^{2-}$, using aromatic amine 1-methylimidazole (MIA) and IL 1-methyl-3-ethylimidazolium bromide (EMIMBr) as co-templates. It is shown that a cooperative templating effect occurs when 1-ethyl-3 methylimidazolium (EMIM) bromide is combined with methylimidazolium (MIA), both the aromatic amine MIA and the ionic liquid EMIMBr serve as templates for the production. The framework of MOF is revealed by single X-ray crystallographic structural analysis to be an anionic open framework with protonated $MIAH^{+}$ and $EMIM^{+}$ cations acting as templates to balance the negative charges of the inorganic framework. The intriguing feature of this solid is the proximity of the MIA to one layer and the EMIM to the other, possibly indicating that each cation has a specific function in controlling the direction of the EMIM-"templated" layer in this substance. Additionally, it is highlighted that without MIA in the ionothermal system, MOF could not be manufactured [25].

Xu et al. showed a number of intriguing impacts of the IL cation on the Mn-BTC MOFs as well as an anionic influence although the anion is not part of the MOF structure. A chemical [Cnmim][Mn(BTC)] ($n = 2, 3$) is produced when $Mn(OAc)_2 \cdot 4H_2O$ (OAc = acetate) reacts with H_3BTC in either $[C_2mim][X]$ or $[C_3mim][X]$ (X = Cl, Br, or I). However, with $[C_2mim][X]$ and $[C_3mim][X]$, multiple structural types can be generated. In-depth descriptions reveal that the structures are relatively complex, but they altogether contain Mn_2 units and form a larger (3,6)-connected pyr topological network for topological notations. The anionic Mn(BTC) framework's cavities, which accommodate $[C_2mim]^{+}$ and $[C_3mim]^{+}$ without significant open spaces, act as a catalyst for the creation of other structures [26].

Another example is the formation of $[C_2mim][Br]$ with the reaction of cobalt metal as $Co(NO_3)_2 \cdot 6H_2O$ with the ligand H_2BDC. In this case, Wang et al. showed the additional additives influence on the MOF with the addition of imidazole. The final product, $Co_3(BDC)_3(imidazole)_2$, is a neutrally charged framework made up of Co_3 units with a total of six bridging carboxylate groups and ends that are capped by imidazole ligands rather than bromide ligands. Additionally, the BDC_2 linkers connect these six connected nodes to create a 3D network [27].

Tapalaa ionothermally created two two-dimensional (2D) intercalated zinc benzenedicarboxylates, $EMIm_2[Zn_3(C_6H_4(COO)_2)_3Cl_2]$ likewise $BMIm_2$ $[Zn_3(C_6H_4(COO)_2)_3Cl_2]$, utilizing ionic liquid 1-ethyl-3-methylimidazolium chloride and 1-butyl-3-methylimidazolium chloride as solvents (EMIm-Cl). The constitution of the two compounds is $[Zn_3(C_6H_4(COO)_2)_3Cl_2]^{2-}$ which has an anionic 2D framework and is charge neutralized by the imidazolium cations. The neutral compound that was created has better thermal sturdiness.

Equally the original and the modified assemblies exhibit blue luminescence, which results from ligand-centered $n \rightarrow \pi^*$ and $\pi \rightarrow \pi^*$ shifts. $[Zn_3(BDC)_3Cl_2]$ $[C_2mim]^{2-}$ crystallizes in the *P2₁/n* space group (monoclinic) in 2D networks of the hxl type, $[Zn_3(BDC)_3Cl_2]$ $[C_4mim]^{2-}$ crystalizes in the *P1* triclinic space group system due to a diverse packing of the $[Zn_3(BDC)_3Cl_2]_2$layers due to which ionothermal syntheses leads to such results [28].

Tan et al. described the synthesis of anionic novel MOF ionothermally, viz., [Emim][CdBr(1,4-NDC)] (1,4-$NDCH_2$ = 1,4-naphthalenedicarboxylic acid) using an ionic liquid [Emim]BF_4 or mixed ILs of [Emim]BF_4 and [Emim]Br (Emim = 1-ethyl-3-methylimidazolium) both as solvent and template. Imidazolium cations intercalate into the interlayer gaps, as a charge-compensating agent. The outcome was the first instance of ionothermal MOF synthesis using 1,4-NDC. An anionic 2-dimensional network is featured in the structural frameworks of $[CdBr(1,4\ NDC)]_n^{n-}$sandwiched by organic $Emim^+$ cationic layers [29].

In the ionic liquid Emim-Br, Liao, and colleagues produced a three-dimensional anionic MOF called [Emim][Cd(btc)] in 2006, in which the cations $Emim^+$ became charge-compensating entities in the empty space [30]. Where the cations $Emim^+$ developed into charge-compensating entities in the vacuum space.

8.5 ILs incorporation/combined control of both the cation and anion

In the literature, there are numerous instances of syntheses of MOFs ionothermally where the anion and cation of the IL are absorbed into the MOF and have a certain kind of structure-directing effect. The functionalized MOF materials containing ILs enclosed in nanocages are interesting candidates for electrochemical devices, gas adsorption, and heterogeneous catalysis, among other possible uses. The most well-known way for both IL ions to join the MOF is when the anion binds to the metal center of the resulting anionic MOF structure, although the cation of the ionic liquid serves as a template for the open galleries of the MOF and serves as a counter ion to balance the overall charge of the structure. Such kind of examples are discussed below.

In order to create the CP [Cu(I)(bpp)]BF_4] [bpp=1,3-bis(4-pyridyl)propane], Jin et al. investigated the use of the ionic liquid [bmim][BF_4] (bmim = 1-butyl-3-methylimidozolium) as a thermally stable and inadequately coordinating solvent. This solvent [bmim][BF_4] of this ionic liquid has produced a 2D extended coordination structure. The [bmim][BF_4] is an ionic liquid at room temperature that is neutral, non-volatile, weakly coordinating, and stable in air and moisture. When heated to 400°C, it maintains its integrity thanks to its exceptional thermal stability. Additionally, its anions and cations can both act as synthesis templates or charge-neutralizing groups. This 2D network can be thought of topologically as a brick wall that resembles a wave. The 2D layers' interstices are home to the BF_4^- anions [31].

In Emim-Msul, Long et al. used an ionothermal process to create a heterometallic La^{3+}-Co^{2+} coordination polymer with 2D layers. Both the ions of Emim-Msul, i.e., anions and cations are included in the La^{3+}-Co^{2+} heterometallic structure $[La_2Co(mipt)_2(CH_3COO)_2(CH_3SO_3)_4][Emim]_2$, here the anionic $CH_3SO_3^-$ link the La(III) ions in the layer and the cationic $Emim^+$ serve as the charge-balancing agent or template between the layers [32]. Xu et al. used two groups of ILs as solvents to conduct the reaction between 1,3,5-benzenetricarboxylic acid (H_3BTC) and $Cd(NO_3)_2{\cdot}4H_2O$. The inorganic ligands (ILs) employed were [EMI]X(EMI) 1-ethyl-3-methylimidazolium and [PMI]X(PMI)1-propyl-3-methylimidazolium, where X= Cl, Br, and I. Two kinds of ionic liquids' cation and/or anion effects were investigated. As crystalline phases, three distinct Cd-BTC MOFs, $[EMI][Cd_2(BTC)Cl_2]$, [EMI][Cd(BTC), and [PMI][Cd(BTC), were produced. Reactions in [EMI]Cl created $[EMI][Cd_2(BTC)Cl_2$, whereas the identical reactions with Cl substituted by Br or I produced [EMI][Cd(BTC)]. Regardless of the kind of X, the substitution of PMI for EMI produced [PMI][Cd(BTC)] [33].

Ionic liquid 1-alkyl-3-methylimidazolium bromide is utilized by Xu et al. in ionothermal reactions between H_3BTC and Zn $(NO_3)_2$ and with the alkyl group ranging from ethyl to amyl, H_3BTC, stands for 1,3,5-benzenetricarboxylic acid. $[Zn_3(BTC)_2(H_2O)_2] \cdot 2H_2O$ (H_3BTC = 1,3,5-benzenetricarboxylate acid) came out in two isomeric forms, and four other MOFs namely [EMI][Zn(BTC)] (3), [PMI][Zn(BTC)] (4), $[BMI]_2[Zn_4(BTC)_3(OH)(H_2O)]$ (5), and $[AMI][Zn_2(BTC)(OH)Br]$ (6) (E = ethyl, P = propyl, B = butyl, A = amyl, MI = 3-methylimidazolium), that have been synthesized and structurally analyzed. In isomeric complexes, Zn atoms as well as BTC^{3-}ligands have quite distinct coordination mechanisms. The appropriate ionic liquid cations are integrated into the crystal framework of the other complexes (3-6), which also crystallize. Their crystal structures display a variety of characteristics, such as different Zn^{2+} coordination geometries and different BTC^{3-} ligand bridging modes. The cations that have been added seem to interact strongly with the frameworks [34].

8.6 Ionothermal synthesis when neither the cation nor the anion of the IL are present in the MOF

A number of MOFs were created where they involve ILs but wherein cations and anions are not incorporated in MOF Framework. In our foremost example, ILs influenced the as-formed framework structures.

Xu et al. synthesized a Zinc MOF by ionothermal reaction of $Zn(NO_3)_2.6H_2O$ with the ligand benzene-1,3,5-tricarboxylic acid (H_3BTC) in $[C_2mim][Br]$). The $[Zn_4(4\text{-}O)(H_2O)_2]^{6+}$ SBU of the MOF $Zn_4(BTC)_2$

(μ_4-O)(H_2O)$_2$ is connected to carboxylate groups in a (3, 6) network with a three-dimensional structure, however, the octahedral geometry is unsymmetrical. Due to the coordinated water ligands on two of the Zn^{2+} ions and the asymmetry of the SBU. Zinc MOF crystallizes in an anon-centrosymmetric space group. The $[C_2mim]^+$ cation and the Br^- anion are not included in the framework, although the cation's selection is important [35].

8.7 Exceptional features of ionothermally synthesized MOFs

In the preceding sections, we considered many structures created by ionothermal synthesis and the various angles from which the ionic liquid groups could be seen incorporated into the newly synthesized MOFs. Also, their structural framework may be analyzed for deeper understanding. Finding a useful application for those insights is currently of the utmost importance. Ionothermal synthesis is versatile in that it allows for easy modification of the reaction environment, which, when combined with a change in the end products, directly relates to the variety of structures that may be produced using this process. Ionothermal synthesis thus promises a variety of prospective and real-world applications, while the majority still faces challenges on the path to actual employment.

8.8 Conclusion

In ionic liquids solvents, several MOFs and MOFs-like framework materials have been created. The insertion of the IL cation that acts like a charge-balancing ion inside the MOF, or anion that is often coordinated to a metal center contained by an overall negatively charged framework, or both allows the ionic liquids to frequently exercise structure-directing effects. Various ionic liquids anions lead to the crystallization of different MOF polymorphs or when a chiral ionic liquid anion directs the creation of one enantiomer of a chiral framework, the ionic liquid can have a structure-directing impact even when no portion of the ionic liquid is integrated into the framework. ILs also act as a medium for MOF synthesis with various advantages over the traditional methods of solvothermal or hydrothermal synthesis due to their unique characteristics, such as the ability to simultaneously dissolve both inorganic and organic precursors, extraordinary design ability, high boiling point windows, high thermal and chemical stability, extremely low volatility and also structure directing effect.

References

[1] R.E. Morris, L. Brammer, Coordination change, lability and hemilability in metal-organic frameworks, Chem. Soc. Rev. 46 (2017) 5444–5462.

[2] L. Chen, X. Zhang, X. Cheng, Z. Xie, Q. Kuang, L. Zheng, The function of metal-organic frameworks in the application of MOF-based composites, Nanoscale Adv. 7 (2020) 2628–2647.

[3] X. Liu, L. Zhang, J. Wang, Design strategies for MOF-derived porous functional materials: preserving surfaces and nurturing pores, J. Materiomics 7 (2021) 440–459.
[4] Z. Sharifzadeh, K. Berijani, A. Morsali, Chiral metal-organic frameworks based on asymmetric synthetic strategies and applications, Coord. Chem. Rev. 445 (2021) 214083.
[5] R.A. Dodson, A.P. Kalenak, A.J. Matzger, Solvent choice in metal-organic framework linker exchange permits microstructural control, J. Am. Chem. Soc. 142 (2020) 20806–20813.
[6] S.H. Feng, G.H. Li, Hydrothermal and solvothermal syntheses (4th Chapter), Modern inorganic synthetic chemistry (Second Edition), Elsevier (2017) 73–104.
[7] C.A. Grande, R. Blom, A. Spjelkavik, V. Moreau, J. Payet, Life-cycle assessment as a tool for eco-design of metal-organic frameworks (MOFs), Sustain. Mater. Technol. 14 (2017) 11–18.
[8] F.P. Kinik, U. Alper, S. Keskin, Ionic liquid/metal–organic framework composites: from synthesis to applications, ChemSusChem 10 (2017) 2842–2863.
[9] R.D. Rogers, K.R. Seddon, Ionic liquids-solvents of the future, Science 302 (2003) 792–793.
[10] S. Wan, O. Xu, X. Zhu, Synthesis of ionic liquid modified metal-organic framework composites and its application in solid-phase extraction: a review, Ionics 27 (2021) 445–456.
[11] T.P. Vaid, S.P. Kelleya, R.D. Rogers, Structure-directing effects of ionic liquids in the ionothermal synthesis of metal–organic frameworks, IUCrJ 4 (2017) 380–392.
[12] Y. Yoshida, H. Kitagawa, Ionic conduction in metal-organic frameworks with incorporated ionic liquids, ACS Sustain. Chem. Eng. 7 (2018) 70–81.
[13] G.E.M. Schukraft, S. Ayala, B.L. Dick, S.M. Cohen, Isoreticular expansion of poly MOFs achieves high surface area materials, Chem. Commun. 53 (2017) 10684–10687.
[14] H.C. Oh, S. Jung, I.J. Ko, E.Y. Choi, Ionothermal synthesis of metal-organic framework, in Recent Advancements in the Metallurgical Engineering and Electrodeposition, IntechOpen (2018) 83–106.
[15] E.R. Cooper, C.D. Andrews, P.S. Wheatley, P.B. Webb, P. Wormald, R.E. Morris, Ionic liquids and eutectic mixtures as solvent and template in synthesis of zeolite analogues, Nature 430 (2004) 1012–1016.
[16] E.R. Parnham, R.E. Morris, Ionothermal synthesis of zeolites, metal-organic frameworks, and inorganic–organic hybrids, Acc. Chem. Res. 40 (2007) 1005–1013.
[17] L. Peng, J. Zhang, J. Li, B. Han, Z. Xue, G. Yang, Surfactant-directed assembly of mesoporous metal–organic framework nanoplates in ionic liquids, Chem. Commun. 48 (2012) 8688–8690.
[18] Z. Lin, D.S. Wragg, J.E. Warren, R.E. Morris, Anion control in the ionothermal synthesis of coordination polymers, J. Am. Chem. Soc. 129 (2007) 10334–10335.
[19] H. Chen, S. Han, R. Liu, T. Chen, K. Bi, J. Liang, et al., High conductive, long-term durable, anhydrous proton conductive solid-state electrolyte based on a metal-organic framework impregnated with binary ionic liquids: synthesis, characteristic and effect of anion, J. Power Sources 376 (2018) 168–176.
[20] Q. Xu, X. Zhang, S. Zeng, L. Bai, S. Zhang, Ionic liquid incorporated metal organic framework for high ionic conductivity over extended temperature range, ACS Sustain. Chem. Eng. 7 (2019) 7892–7899.
[21] P. Guo, M. Song, Y. Wang, Promising application of MOF as composite solid electrolytes via clathrates of ionic liquid, Inorganica Chim. Acta 491 (2019) 128–131.
[22] D.N. Dybtsev, H. Chun, K. Kim, Three-dimensional metal–organic framework with (3, 4)-connected net, synthesized from an ionic liquid medium, Chem. Commun. 14 (2004) 1594–1595.
[23] Q. Dai, J. Ma, S. Ma, S. Wang, L. Li, X. Zhu, et al., Cationic ionic liquids organic ligands based metal–organic frameworks for fabrication of core–shell microspheres for hydrophilic interaction liquid chromatography, ACS Appl. Mater. Interfaces 8 (2016) 21632–21639.

[24] S.H. Zottnick, M. Finze, K.M. Buschbaum, Transformation of the ionic liquid [EMIM][B(CN)$_4$] into anionic and neutral lanthanum tetracyanoborate coordination polymers by ionothermal reactions, Chem. Commun. 53 (2017) 5193–5195.
[25] H. Xing, J. Li, W. Yan, P. Chen, Z. Jin, J. Yu, et al., Cotemplating ionothermal synthesis of a new open-framework aluminophosphate with unique Al/P ratio of 6/7, Chem. Mater. 20 (2008) 4179–4181.
[26] L. Xu, Y.U. Kwon, B. Castro, L.C. Silva, Novel Mn (II)-based metal–organic frameworks isolated in ionic liquids, Cryst. Growth Des. 13 (2013) 1260–1266.
[27] X.R. Wang, J. Du, Z. Huang, K. Liu, Y.Y. Liu, J.Z. Huo, L.L. Chen, B. Ding, et al., Anion directing self-assembly of 2D and 3D water-stable silver (I) cation metal organic frameworks and their applications in real-time discriminating cysteine and DNA detection, J. Mater. Chem. B 6 (2018) 4569–4574.
[28] W. Tapala, T.J. Prior, A. Rujiwatra, Two-dimensional anionic zinc benzenedicarboxylates: ionothermal syntheses, structures, properties and structural transformation, Polyhedron 68 (2014) 241–248.
[29] B. Tan, Z.L. Xie, X.Y. Huang, X.R. Xiao, Ionothermal synthesis, crystal structure, and properties of an anionic two-dimensional cadmium metal organic framework based on paddle wheel-like cluster, Inorg. Chem. Commun. 14 (2011) 1001–1003.
[30] J.H. Liao, P.C. Wu, W.C. Huang, Ionic liquid as solvent for the synthesis and crystallization of a coordination polymer:(EMI)[Cd (BTC)](EMI= 1-Ethyl-3-methylimidazolium, BTC= 1, 3, 5-benzenetricarboxylate), Cryst. Growth Des. 6 (2006) 1062–1063.
[31] K. Jin, X. Huang, L. Pang, J. Li, A. Appel, S. Wherland, [Cu (i)(bpp)] BF 4: the first extended coordination network prepared solvothermally in an ionic liquid solvent, Chem. Commun. 23 (2002) 2872–2873.
[32] W.X. Chen, Y.P. Ren, L.S. Long, R.B. Huang, L.S. Zheng, Ionothermal synthesis of 3d–4f and 4f layered anionic metal–organic frameworks, CrystEngComm 11 (2009) 1522–1525.
[33] L. Xu, E.Y. Choi, Y.U. Kwon, Combination effects of cation and anion of ionic liquids on the cadmium metal–organic frameworks in ionothermal systems, Inorg. Chem. 47 (2008) 1907–1909.
[34] L. Xu, E.Y. Choi, Y.U. Kwon, Ionothermal syntheses of six three-dimensional zinc metal–organic frameworks with 1-alkyl-3-methylimidazolium bromide ionic liquids as solvents, Inorg. Chem. 46 (2007) 10670–10680.
[35] L. Xu, E.Y. Choi, Y.U. Kwon, Ionothermal synthesis of a 3D Zn–BTC metal-organic framework with distorted tetranuclear [Zn4 (μ4-O)] subunits, Inorg. Chem. Commun. 11 (2008) 1190–1193.
[36] Y. Ban, Z. Li, Y. Li, Y. Peng, H. Jin, W. Jiao, A. Guo, P. Wang, Q. Yang, C. Zhong, Confinement of ionic liquids in nanocages: tailoring the molecular sieving properties of ZIF-8 for membrane-based CO_2 capture, Angew. Chem. 54 (2015) 15483–15487.
[37] F.P. Kinik, C. Altintas, V. Balci, B. Koyuturk, A. Uzun, S. Keskin, [BMIM][PF_6] incorporation doubles CO_2 selectivity of ZIF-8: elucidation of interactions and their consequences on performance, ACS Appl. Mater. Inter. 8 (2016) 30992–31005.
[38] B. Koyuturk, C. Altintas, F.P. Kinik, S. Keskin, A. Uzun, Improving gas separation performance of ZIF-8 by [BMIM][BF4] incorporation: interactions and their consequences on performance, J. Phys. Chem. C. 121 (2017) 10370–10381.
[39] K.B. Sezginel, S. Keskin, A. Uzun, Tuning the gas separation performance of CuBTC by ionic liquid incorporation, Langmuir 32 (2016) 1139–1147.

[40] F.W.M. Da Silva, G.M. Magalhães, E.O. Jardim, J. Silvestre-Albero, A. Sepúlveda-Escribano, D.C.S. de Azevedo, S.M.P. de Lucena, CO_2 adsorption on ionic liquid—modified Cu-BTC: experimental and simulation study, Adsorp. Sci. Technol. 33 (2015) 223–242.
[41] N.A. Khan, Z. Hasan, S.H. Jhung, Ionic liquids supported on metal-organic frameworks: remarkable adsorbents for adsorptive desulfurization, Chem. Eur. J. 20 (2014) 376–380.
[42] Q.X. Luo, X.D. Song, M. Ji, S.E. Park, C. Hao, Y.Q. Li, Molecular size-and shape-selective Knoevenagel condensation over microporous Cu_3 $(BTC)_2$ immobilized amino-functionalized basic ionic liquid catalyst, Appl. Catal. A 478 (2014) 81–90.
[43] H. Wan, C. Chen, Z. Wu, Y. Que, Y. Feng, W. Wang, L. Wang, G. Guan, X. Liu, Encapsulation of heteropolyanion-based ionic liquid within the metal–organic framework MIL-100 (Fe) for biodiesel production, ChemCatChem 7 (2015) 441–449.
[44] J. Wu, Y. Gao, W. Zhang, Y. Tan, A. Tang, Y. Men, B. Tang, Deep desulfurization by oxidation using an active ionic liquid-supported Zr metal–organic framework as catalyst, Appl. Organomet. Chem. 29 (2015) 96–100.
[45] A. Nasrollahpour, S. Moradi, A simple vortex-assisted magnetic dispersive solid phase microextraction system for preconcentration and separation of triazine herbicides from environmental water and vegetable samples using Fe_3O_4@ MIL-100 (Fe) sorbent, Microporous Mesoporous Mater. 243 (2017) 47–55.
[46] C. Chen, Z. Wu, Y. Que, B. Li, Q. Guo, Z. Li, L. Wang, H. Wan, G. Guan, Immobilization of a thiol-functionalized ionic liquid onto HKUST-1 through thiol compounds as the chemical bridge, RSC Adv. 6 (2016) 54119–54128.
[47] H. Li, L. Tuo, K. Yang, H.K. Jeong, Y. Dai, G. He, W. Zhao, Simultaneous enhancement of mechanical properties and CO_2 selectivity of ZIF-8 mixed matrix membranes: Interfacial toughening effect of ionic liquid, J. Membrane Sci. 511 (2016) 130–142.
[48] M. Han, Z. Gu, C. Chen, Z. Wu, Y. Que, Q. Wang, H. Wan, G. Guan, Efficient confinement of ionic liquids in MIL-100 (Fe) frameworks by the "impregnation-reaction-encapsulation" strategy for biodiesel production, RSC Adv. 6 (2016) 37110–37117.
[49] J. Song, Z. Zhang, S. Hu, T. Wu, T. Jiang, B. Han, MOF-5/n-Bu 4 NBr: an efficient catalyst system for the synthesis of cyclic carbonates from epoxides and CO_2 under mild conditions, Green Chem. 11 (2009) 1031–1036.
[50] K.M. Gupta, Y. Chen, J. Jiang, Ionic Liquid membranes supported by hydrophobic and hydrophilic metal—organic frameworks for CO_2 capture, J. Phys. Chem. C 117 (2013) 5792–5799.

Chapter 9

Solubility and thermodynamic stability of metal–organic frameworks

Mohd Khalid, Samrah Kamal and Shaikh Arfa Akmal
Functional Inorganic Materials Lab (FIML), Department of Chemistry, Aligarh Muslim University, Aligarh, Uttar Pradesh, India

9.1 Introduction

Metal–organic frameworks are a fascinating class of crystalline porous materials, made up of inorganic metal nodes and organic linkers and joined by strong coordination bonds [1]. Their topologies, physicochemical properties, and potential functionalities can be easily tuned by employing a reticular chemistry approach and choosing suitable constituents for the MOF assembly process [2]. As a result, a large number of MOF materials have been successfully synthesized and exhibit potential applications such as gas storage and separation, catalysis and energy storage, drug delivery, and sensing [3]. Specifically for drug delivery applications, the poor water solubility of MOFs is one of the major drawbacks because it results in low oral bioavailability [4]. The MOF solubility in water depends on its crystallinity, and particle surface area and amorphous solids show greater dissolution rates [5]. It is reported that the utilization of MOF as a drug carrier could improve the solubility of drugs in water [6]. The adsorption of drugs into the pores of MOFs greatly enhances the drug surface area and prevents its reaggregation. CD-MOF (γ-cyclodextrin MOF) increases the water solubility of azilsartan drug due to the formation of its clusters in the pores [6]. It was also reported that the encapsulation of drug molecules into the MOF structure which is soluble in water exhibits fast drug release due to the collapse of the MOF framework [7]. Therefore, in the recent past MOF solubility has gathered much attention in the pharmaceutical field for drug delivery applications. On the other hand, thermodynamically driven water-stable MOFs still remain a major task for several applications such as adsorption and separation, proton conduction, catalysis, and sensing. The water stability of adsorbents (MOF) is the main issue for the adsorption and detection of pollutants in industrial effluents and nuclear waste containing radioactive metal ions, oxyanions, and hazardous metal cations. In such applications, the integrity of the frameworks should be preserved to

Synthesis of Metal–Organic Frameworks via Water-Based Routes: A Green and Sustainable Approach.
DOI: https://doi.org/10.1016/B978-0-323-95939-1.00004-6

uphold their features and functionalities [8]. The poor water stability of MOFs is the main obstruction towards adsorption-based gas separation applications due to the presence of water vapor in various industrial spills, for instance, biogas, flue gas, and natural gas [9]. Thus, for these applications, the robustness and stability of the adsorbents over a wide pH range are necessary for their successful employment. Walton et al. proposed criteria for the water stability of MOFs, accordingly, a "thermodynamic stable" MOF shows greater stability on long-term exposure to aqueous solution, acidic/basic medium, or boiling conditions. Nevertheless, a high "kinetically stable" MOF exhibits decomposition on short exposure to an aqueous solution or harsh conditions [9], while low kinetically stable MOFs are stable in low humidity conditions whereas, unstable MOFs rapidly collapse after exposure to any moisture [9]. Famous MOFs, such as M-MOF-74 where M = Mg, Ni, Co, Zn are attributed to low kinetic stabile, while Bio MOF 11-12, MOF-5, and MOF-177 are assigned to be unstable MOFs [9]. Sometimes, post-synthetic modifications are employed to enhance the stability of MOFs and a few strategies have also been adopted. Generally, increasing the steric hindrance around the metal center plays an important role in enhancing the water stability of MOFs [10]. For the scientific community, the design and synthesis of hydrophobic MOFs still remain a great challenge. Recently, few thermodynamic-stable MOFs which exhibit a wide range of pH stability are reported [11]. At different temperatures, the molar heat capacities of MOFs are the primary data in engineering and chemistry, from which many thermodynamic properties can be evaluated, such as entropy and enthalpy. The determination of heat capacities has attracted the attention of many researchers [12]. TMDSC (Temperature modulated differential scanning calorimetry) is one of the simpler and most accurate methods for calculating heat capacity and it helps in determining the heat capacities of several materials both non-isothermally and isothermally [13]. Jiang et al. calculated the molar heat capacity of Li-MOF using the TMDSC method and also determines its thermodynamic parameters [14]. Moreover, Navrotsky et al. reported a high-temperature drop combustion calorimetry method to study the thermodynamic stability of water-sensitive MOFs. This method is solvent-free and employs heat of combustion at elevated temperatures, a drop-in combustion calorimetry permits the inspection of the thermodynamics parameters of MOFs [15]. Sahoo et al. synthesized thermodynamically stable Ni-based MOF which can withstand drastic conditions, for example, a wide range of pH, and boiling water and it shows the great capability to separate C2/C1 selectively [16]. In this chapter, solubility and thermal stability of MOFs are discussed in detail which will promote basic understanding and helps researchers in the rational design and construction of MOFs for applications in various domains.

9.2 Fundamentals of MOFs

The MOFs are a novel class of infinite crystalline structures composed of metal ions/clusters bonded with organic ligands (Scheme 9.1) [17]. MOFs possess

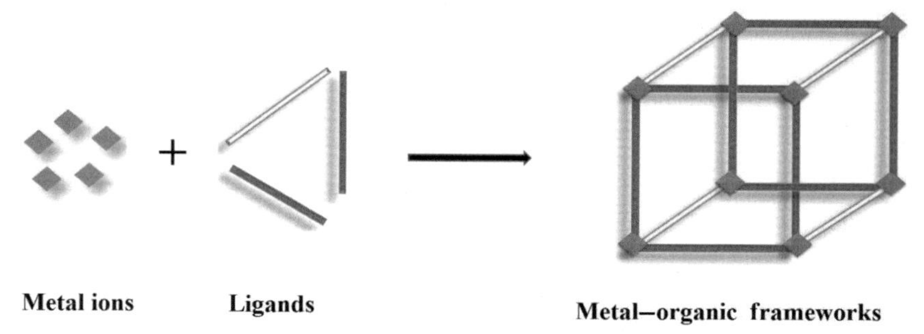

SCHEME 9.1 Scheme for the preparation of a metal–organic framework. (*Reproduced from [17] with permission from the Elsevier*).

characteristic properties, such as they are mostly flexible and porous owing to their chemical and structural tunability. Due to these feasible properties, MOFs have employed various applications. Water-soluble MOFs show promising applications in ion separations [18] and drug delivery [6]. While water-stable MOFs are used as adsorbents for the liquid phase and gas-phase separation and also exhibit good adsorption towards organics, H_2, and CO_2[19]. In the recent past, MOF shows good catalytic properties specifically photocatalysis due to its optic characteristics which are easily modified to improve its light harvesting and could be employed for photosensitive molecules [20]. The stability of MOF in water can be enhanced by strengthening the coordination between metal nodes and organic moieties. Thermodynamically stable MOF is a new class of breed and great advancements have been taken place to construct them. This class of MOFs can immerse in water for longer durations and withstand wide ranges of pH. The modified, defective, and optimized MOF can increase the performance of MOFs under harsh conditions. The optimized MOFs possess unique properties and are considered as 1-D (one-dimensional) linear molecules, 2-D (two-dimensional) polymeric sheets, and 3-D (three-dimensional) metal–organic systems have attracted much attention [21]. MOF-based composite materials have also been synthesized to improve thermodynamic stability, surface areas, and straightforward separations [21]. We hope that new possibilities in this area highlighting the structure-activity relationship on the basis of structural changes should be elaborated by the coordination chemist to improve the material importance (MOF).

9.3 Solubility of MOF

It is reported in literature that stable MOFs possess chemical, thermal, and mechanical strength and exhibit various applications in various fields. MOFs chemical stability, in particular, refers to their resistance to the effects of various chemicals in their environment, for instance, moisture, solvent, acids, bases, and aqueous solution [22]. Herein, we discuss the dissolution of MOFs and their uses

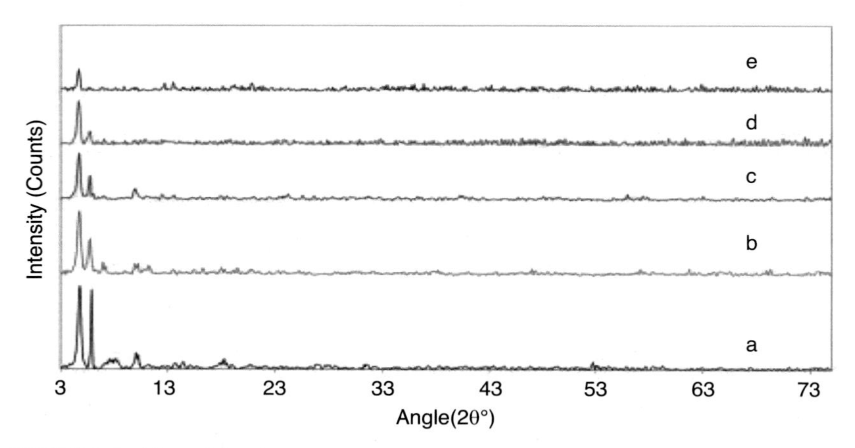

FIGURE 9.1 X-ray diffraction analysis patterns of MOF-177 after different durations of exposure to ambient air: (A) 1 week, (B) 2 weeks, (C) 3 weeks, (D) 4 weeks, and (E) 5 weeks. *(Reproduced from [23] with permission from the American Chemical Society).*

which makes everyday life easier in some aspects. There are several reports on MOFs that become soluble in different solvents and humid conditions. MOF-177 [framework consist of a $[Zn_4O_6]^{6+}$ cluster and linker 1,3,5-benzene-tri benzoate (BTB) ligands], when it is exposed to ambient air with a relative humidity of 16% and 25°C for 1–5 weeks, it decomposes gradually changing from hexagonal to orthogonal, and finally to monoclinic phase in five weeks at ambient conditions, shows poor stability of MOF-177 (Fig. 9.1) [23]. Crystallographic data in Table 9.1, describes the changes that occur during the humid conditions with time.

MOF-177 samples were dried at 150°C under vacuum for 12 hours after being submerged in water for 12 hours and produced the X-ray diffraction analysis (XRD) data displayed in Fig. 9.2. Which confirms that the MOF-177 gets crumbled in water since it lacked distinguish MOF-177 peak. Further, this dissolution is explained by Brunauer–Emmett–Teller (BET), which specifies that the surface area is considerably less than that of the fresh MOF-177 by 1 m^2/g [24]. This further corroborates that molecules of water can destroy crystal structure.

There are a few explanations in the literature regarding the degradation of zinc-based MOFs in the presence of water. Greathouse et al. conducted a molecular dynamics simulation demonstrating that the bonds between zinc ions and the oxygen atom are attacked by water molecules and break down the MOF structure [25]. Yu et al. suggested that Zn_4O clusters in MOF material hydrolysis form zinc ions and the benzendicarboxylic acid (corresponding organic linker acid) of MOF-5 and MOF-177, respectively [5]. Greathouse et al. proposed that the hydrolysis reaction occurs favorably in the presence of acid [26]. Li and Yang also proposed the same hydrolysis process [27]. In another concept, the structural

TABLE 9.1 Summary of chronological structural property changes of a MOF-177 sample. Reproduced from [23] with permission from the American Chemical Society.

Crystallographic properties	Fresh sample	After one week	After two weeks	After three weeks	After four weeks	After five weeks
Cell type	Hexagonal	Orthogonal	monoclinic	monoclinic	monoclinic	monoclinic
Space group	P63(173)	Pbam(55)	P2/c (13)	P2/c(13)	P21(4)	P2/m(10)
Lattice parameter (*A*)	$a = 20.905$	$a = 19.161$	$a = 18.700$	$a = 18.878$	$a = 18.641$	$a = 19.874$
	$b = 20.905$	$b = 23.691$	$b = 17.870$	$b = 15.307$	$b = 16.223$	$b = 13.066$
	$c = 22.718$	$c = 17.527$	$c = 21.540$	$c = 18.947$	$c = 23.724$	$c = 24.785$
Lattice angle	$R = 90$	$R = 90$	$R = 90$	$R = 90$	$R = 90$	$R = 90$
	$\beta = 90$	$\beta = 90$	$\beta = 123.2$	$\beta = 97.7$	$\beta = 138.7$	$\beta = 114$
	$\gamma = 120$	$\gamma = 120$	$\gamma = 90$	$\gamma = 90$	$\gamma = 90$	$\gamma = 90$
Cell volume ($\AA_3$)	8598.4	7956.5	6025.2	5425.2	4739.1	5879.9

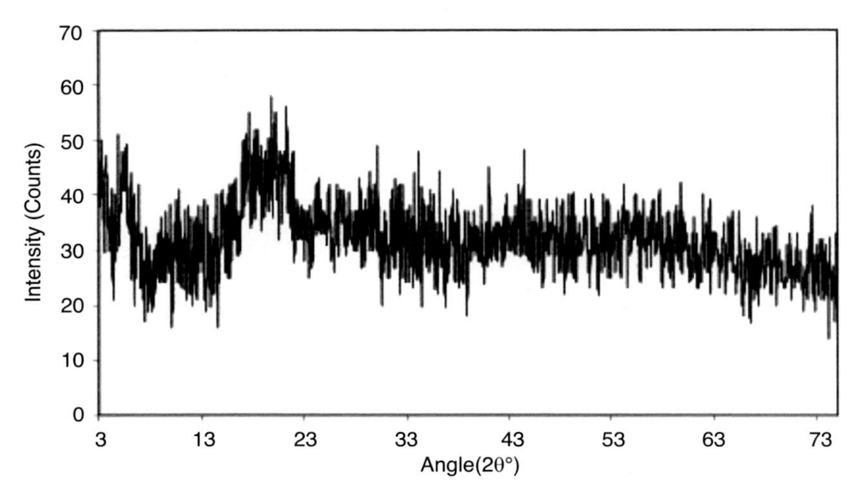

FIGURE 9.2 X-ray diffraction analysis pattern of MOF-177 after exposure to direct water. (*Reproduced from [24] with permission from the Elsevier*).

differences around the metal ions, cause changes in the solubility of MOFs. Thus, surrounding ligands also play an important role in the degradation of the framework. Kobayashi D et al. synthesized two novel bioactive MOFs containing TBZ and Zn^{2+} ions by a solvothermal reaction [28]. TBZ-MOF-1 was prepared from a solution of TBZ and $ZnCl_2$ in DMF. TBZ-MOF-2 was synthesized from a solution containing TBZ, 1,4-benzene dicarboxylic acid (H_2BDC), and $ZnCl_2$ in DMF. The differences in water solubility between the two TBZ-MOFs due to the structural designs allow for the controlled release of the desired bioactive component. Accordingly, TBZ-MOFs with Zn–N bond frameworks are expected to exhibit improved biocidal activity, thermostability, water solubility, and resistance to degradation because of the synergistic effects of bioactive Zn^{2+} ions and TBZ ligands [29]. Coordination polymers composed of TBZ have been reported but no studies have systematically investigated their properties and biocidal activities [30]. We evaluated water solubility to explore the physical properties of TBZ-MOF-1 and TBZ-MOF-2. TBZ-MOF-1 and TBZ-MOF-2 were tested for water solubility. TBZ-MOFs produced free TBZ during the water solubility test through framework breakdown. The concentration of free TBZ served as a determinant of the TBZ-MOFs' solubility. When the TBZ-MOF/water mixture reached equilibrium after 168 hours of supersaturation, the solubilities of TBZ-MOF-1 (0.9 ppm) and TBZ-MOF-2 (13.7 ppm) were lower than those of TBZ alone (20.4 ppm). As a result, TBZ-MOF-1 was 15 and 23 times less soluble in water than TBZ-MOF-2 and TBZ alone, respectively, and indicates low water solubility. After immersing TBZ-MOFs in water for 24 hours, no changes in their XRD patterns were observed. Additionally, TBZ-MOFs' solubility did not alter even after the immersion period was increased, suggesting that they have high water stability because they share ZIFs' strong Zn-N framework, which is a desirable structure for materials that resist degradation. Zhang et al. synthesized soluble polymeric MOFs (poly MOFs) with versatile linkers obtained by Williamson ether polycondensation [31]. The poly-ligands lead to crystalline and porous MOFs with high solubility in N-methyl pyrrolidone for membrane (NMP) fabrication via a casting method and explain the solubility limit and stability of MOF via an equation that is

$$X\,(\text{solubility limit}) = (M - m) \tag{9.1}$$

The poly-MOFs' solubility limit was determined by dissolving 300 mg (M) of PM-Cx powder in 1 mL of NMP and sonicated for a specified time at two different temperatures, i.e., 25°C and 80°C, then it was centrifuged, dried and weighs to recover the undy's solved poly-MOFs (m). After that, the poly-MOFs' structural integrity was assessed again for XRD analysis.

9.3.1 Uses of soluble MOF

Soluble MOFs could use for drug administration purposes due to their easy synthesis, high compliance, cost-effectiveness, and oral delivery. If a medicine dissolves quickly in the digestive tract, relatively quick absorption and the commencement of therapeutic activity will take place [32]. Several approaches for improving solubility have been pursued, such as amorphous solid dispersions using appropriate polymeric excipients [33], co-crystallization [34], and solubilization in co-solvents [35]. However, there is still a need for improved drug solubility methodologies, with the holy grail being a general platform for amorphous drug stabilization during storage and administration. A few general approaches to improve drug solubility is explained such as, by incorporating drugs into a MOF, the MOF acts as a drug carrier that inhibits crystallization of the drug by confining it within its nanoscale pores while still undergoing rapid hydrolytic decomposition in simulated gastric (SG) and phosphate buffer saline (PBS) media, which results in immediate release of an amorphous drug (Fig. 9.3). High supersaturation is produced by the undeveloped delivery system which increased the drug-free energy as compared to the crystalline form [36].

MOFs are excellent candidates for micro/nanomedicine-extended or controlled cargo delivery. Recently, MOFs were employed in drug delivery centers using the controlled release strategy. This strategy competes with polymers used as controlled-release medication delivery vehicles. [37]. Contrarily, in this case, drug dissolution and solubility are improved by using the quick release of an amorphous drug in aqueous media from a drug@MOF composite via MOF hydrolytic degradation; this strategy serves as a generic platform for delivering poorly soluble compounds.

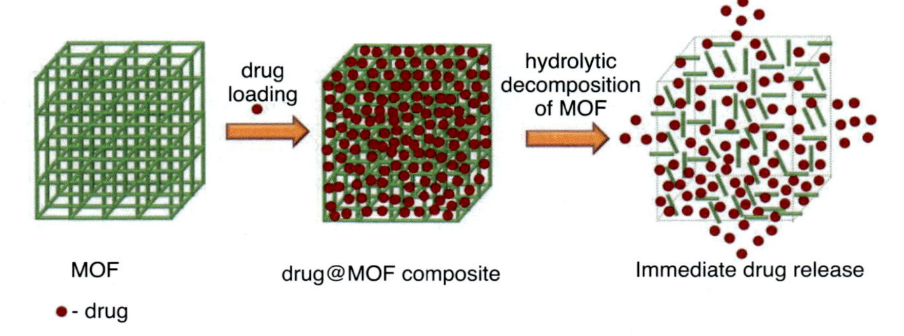

FIGURE 9.3 Schematic representation of a drug encapsulated into metal–organic framework followed by the immediate release of drug from drug@MOF composite via metal–organic framework hydrolytic decomposition. (*Reproduced from [36] with permission from the Wiley*).

9.4 Stability of MOF

9.4.1 Water-stable MOF

From the application point of view stability of MOF in the aqueous phase is an important parameter, if employing it for wastewater treatment [38]. There are various famous reports of MOFs that show excellent water stability such as ZIF-8, MIL-53 (Cr, Al), MIL-101 (Cr, Fe, Al), MOF-74, and MIL-100 (Cr, Fe) [39]. The weak coordination between metal ions and organic linkers leads to the instability of the MOF framework. The unique properties of MOFs have been recorded in varied environments, for instance, water vapor, water, and acid/base medium. Low et al. designed a series of MOFs employed in the high-throughput steam treatment and demonstrated that the stability of MOF can be altered by oxidation states of metals ions, enhancing the strength between metal and ligands, and study unveils that on water treatment MIL-100(Fe), and ZIF-8, MIL-101 exhibits stability [40]. Schoenecker et al. performed water relative stabilities and water vapor adsorption of MOFs such as Mg MOF-74, HKUST-1, DMOF-1, DMOF-1-NH_2, UMCM-1, UiO-66(–NH_2), in the 90% relative humidity [41]. It was observed that the MOF Mg MOF-74, HKUST-1, and UiO-66(–NH_2) were stable in water but MOFs that possess Zn-COOH (UMCM-1, DMOF-1, and DMOF-1-NH_2) completely lost their crystalline nature [41]. Accordingly, Zr-based MOFs like UiO-66(–NH_2) and UiO-66 are stable in water. Moreover, Zn-based MOFs, for instance, MOF-177 and MOF-5 were found very sensitive to moisture and due to the hydrolysis of the MOF framework, the surface area of MOF-5 was dramatically reduced [42]. Therefore, at 90% relative humidity decomposition of HKUST-1 takes place. The Cu-BTC MOF framework decomposed easily at 90% relative humidity at 460°C [43]. Cychosz et al. employed the PXRD technique to explain the framework stability of different MOFs having carboxylate groups such as MOF-177, MOF-5, MOF-505, MIL-100 (Cr),

HKUST-1, ZIF-8, and UMCM-150 in water-containing solutions. Among them, MIL-100 (Cr) and ZIF-8, due to the inertness of chromium metal, the MOFs are found to be stable in water for several months [44]. Furthermore, MOFs chemical stability were also checked in the acidic/alkaline or buffer solution, and a few Zr-based MOFs, i.e., PCN 222 and 224 are very stable under harsh alkaline conditions [45,46]. In comparison to other MOFs Zr-based MOF also exhibits great stability under acidic or basic conditions. The Al-MOFs show less chemical stability and dissolve in an acidic phase owing to the protonation of organic ligands [47]. In literature, Park et al. research group reported that ZIFs maintain their chemical stability for a week in boiling water, NaOH, methanol, and benzene, at varying temperatures. They state that ZIF-8 retains its chemical stability in boiling solvent for 7 days and in 0.1 M and 8 M NaOH for a day at 100°C [48], stability of MOF is also estimated in buffer medium. Cunha et al. synthesized carboxylate-based MOFs and investigate their stability at pH = 7.4 and 37°C in a phosphate buffer solution which shows stability as follows, Fe-MIL-100/-127 > Fe-MIL-53, UiO-66- NH_2 > Fe-MIL-53-Br UiO-66 > UiO-66-Br [49]. Understanding the MOF decomposition mechanism of MOF in water is essential for the designing of water-insensitive MOFs. There are several reports on theoretical studies which explained interactions between MOF and water provide various potential mechanisms along with some possible reaction pathways related to these interactions [50]. Some researchers reported that the famous MOF-5 can show instability in water and is easily protonated because the oxygen present in water attacks directly to the ZnO_4 tetrahedron cluster and releases the linker present in the framework. In another report, rapid displacement of linkers takes place along with the cluster formation at the ZnO_4 site. It leads to the coordination of water to the framework and modifies the framework into a ligand-free state. In the literature, several detailed review reports on the mechanism of decomposition of MOF framework in the water have been published [51].

9.4.2 Thermodynamic stability of MOF

When we talk about stability, it introduces the binary comparison between reactant and product energies; in the same consequence, the term "thermodynamic stability" describes how the free energy of a reaction changes from reactant to product when the system comes to an equilibrium state. The opposite of thermodynamic stability is "kinetic stability," which describes a system's reactivity rather than that system's equilibrium state [52] MOF stability is determined by several factors, including thermodynamic and kinetic aspects. The lability of metal clusters and the strength of the metal-ligand coordination bond play crucial roles in regulating the chemical stability of MOFs [53]. Consequently, the MOFs' thermodynamic stability is caused by how strong their coordinate bonds are. Thus, a metal-ligand coordination bond helps in distinguishing MOFs from chemically inert porous materials like zeolites and activated carbons, and, the strength of this bond can be a strong indicator of the MOFs hydrolytic stability [54]. Therefore, the stronger the coordinate linkages are the more stable framework is anticipated. The binding strength can be evaluated using Pearson's hard and soft acids and bases (HSAB) principle. [55] The HSAB principle states that stable MOFs can be created by combining either soft commands (such as imidazolate, pyrazolate, triazolam, and tetrazolate ligands) with soft divalent metal ions (such as Zn(II), Cu(II), Mn(II), etc.) or complex bases (such as carboxylate ligands) with high-valent metal ions (such as Ti(IV), and Zr(IV). The charges of the metal cations have a positive correlation with the metal-ligand bond strengths for a particular ligand, while the ionic radius has a negative correlation. Even though the enthalpy of formation of a MOF from its component parts can be used to predict the thermodynamic stability of the evacuated or solvated structure [56] this property will be of little use if the component parts do not correspond to the MOFs product state following the hydrolysis reaction. This emphasizes the significance of conducting additional experimental and computational research into MOF breakdown mechanisms.

9.4.3 Factors affecting the thermodynamic stability of MOF

Thermodynamics stability can be recognized by various variables such as entropy, enthalpy, molar volume, temperature, etc. An inert metal cluster in MOF makes it difficult for an irreversible hydrolysis reaction, this is an essential structural characteristic shown by thermodynamically stable MOFs. While the metal coordination centers in MOFs are electrophilic, oxygen in water is a nucleophile. Water can coordinate with the metal cluster and modify or destroy the MOF's crystal structure if the metal core is not completely inert. A structure's thermodynamic stability in the presence of water can be readily calculated by looking at the free energy of the hydrolysis reaction using the relationship shown below.

$$\Delta G_{hydrolysis} = \Delta G_{prod\,(MOF+nH_2O)} - \Delta G_{react\,(MOF+nH_2O)} \tag{9.2}$$

where, G_{prod} is the free energy of the coordination complex that results from the hydrolysis reaction between the MOF and the water molecules, and G_{react} is the free energy of the MOF and the water molecules before the hydrolysis reaction. This stability can be determined experimentally by immersing the MOF in an aqueous phase for long-time and comparing the sample structural characteristics before and after water exposure using PXRD and BET surface area analysis. Both enthalpic and entropic parameters play significant roles to control the thermodynamic stability of MOFs. For instance, high-temperature reactions typically result in denser structures with higher levels of framework connection and lower levels of solvation, and polymorph formation often follows thermodynamic stability [9]. Anthony et al. show how thermodynamic factors are responsible for the crystallization of MOFs. They describe enthalpic and entropic variables that played significant roles for the dense MOFs crystallization,

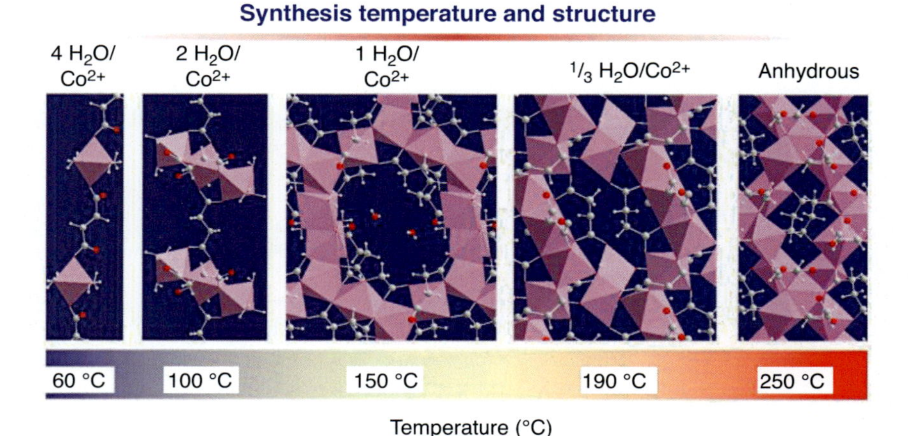

FIGURE 9.4 Progression of the five phases of cobalt succinate, from low temperature (far left) to high temperature (far right) in the same reaction mixture. *(Reproduced from [58] with permission from the Royal Society of Chemistry).*

which was firmly under thermodynamic control. As an illustration, reactions at elevated temperatures mostly result in denser structures with higher framework connectivity levels, decreased solvation rates of framework connectivity, and reduced solvation rate. Polymorphs typically form with thermodynamic stability using calorimetry had been seen in metal tartrates [57]. But it has long been known that some phases crystallize under kinetic control, mainly when the production of the thermodynamic product may require a change in ligand shape or coordination around a metal center. Researchers explain how this can result in time-dependent crystallization processes that progress in stages by the Ostwald rule and can be seen in real-time usage of in-situ techniques. We consider the crystallization of porous MOFs due to solvation effects, which poses additional difficulties.

The relation between reaction temperature and phase behavior was successfully established by synthesizing five different cobalt succinates at various temperatures. (Fig. 9.4). A distinct pattern of trends was observed, as reaction temperature climbed from 60°C to 250°C, the product density, dimensionality, and the number of coordinated Co^{2+} ions per succinate ligand increased while their hydration declined [58]. The primary driving force behind such developments was the entropy, releasing of water molecules from the solid-state confinement to the relative aqueous form. As a result, either hydroxide ions or carboxylate ligands occupied the vacant coordination sites on Co^{2+}, increases the connection between the metal core, and forms a denser framework simultaneously. This result shows thermodynamic modulation throughout MOF synthesis and offered guidance on how temperature affects the reaction condition. Besides, experimental thermodynamics of MOFs can be done by mechanochemistry, near-room

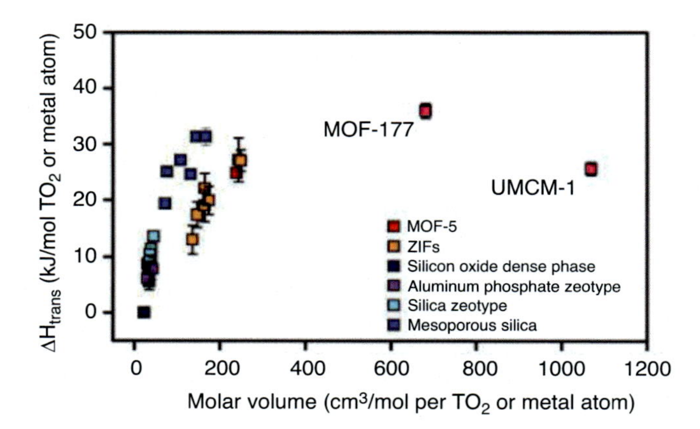

FIGURE 9.5 Energy landscape of inorganic and hybrid porous materials concerning their corresponding dense assemblages. "T" represents a T-atom in the zeolite frameworks or ZIFs tetrahedrally coordinated (Color online). *(Reproduced from [59] with permission from the Elsevier)*.

temperature, solution calorimetry, low-temperature heat capacity measurements, the energetic landscape, entropy trends, and Gibbs free energy. Earlier research and reviews have examined and discussed the energy stability of a few typical MOFs, such as MOF-5, HKUST-1, and zinc zeolite imidazolate frameworks (ZIFs) [59]. Akimbekov et al. measured the formation enthalpies of two MOFs with ultrahigh porosity, i.e., MOF-177 (681.7 cm^3 per mole of Zn) and UMCM-1 (1068.3 cm^3 per mole of Zn), in 2015 to investigate the potential limit of thermodynamic stability of MOFs as a function of molar volume [60]. The molar volumes of these ultra-porous MOFs are more than ten times greater than those of the comparable dense phases. The calorimetric data show that the ultra-porous MOF-177 and UMCM-1 formation enthalpies, which vary from 7 kJ to 36 kJ per mole of metal, are comparable to other MOFs with substantially less porosity, despite the significantly higher molar volume. This formation enthalpy implies that MOFs have a low energetic cost for being ultrahigh porous (Fig. 9.5). In other words, they are acquiring MOF materials with extraordinarily large surface areas. This is terrific news for chemists and materials scientists who want to design and creates such materials.

9.4.4 Methods improving the stability of MOF

To increase the stability of MOFs, two main strategies have been used: (Eq. 9.1) de novo synthesis of MOFs and (Eq. 9.2) enhancing the stability of already-existing MOFs.

As shown in Fig. 9.6, exclusion of pepsin from the MOF framework and exposure of insulin and insulin@NU-1000 to stomach acid. Free insulin denatures stomach acid and is digested by pepsin. The insulin release of NU-1000 happens when insulin@NU-1000 is exposed to a PBS solution [11]. The HSAB

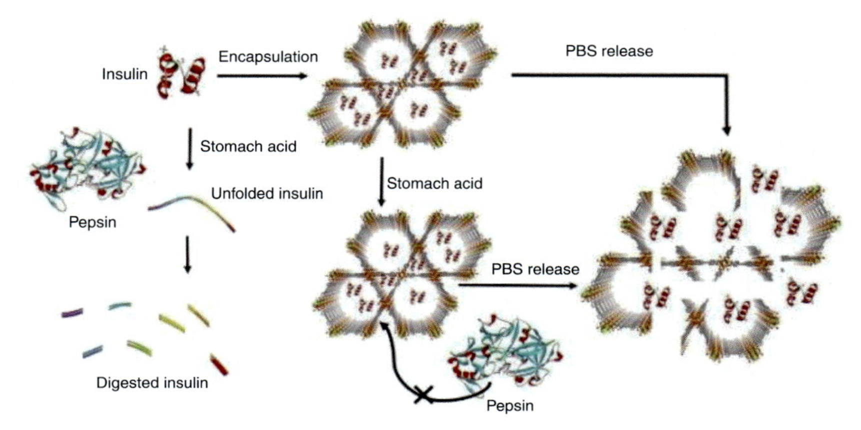

FIGURE 9.6 Schematic representation of exposure of free insulin and insulin@-NU-1000 to stomach acid. Schematic illustration of insulin encapsulation in the mesopores (32 A) of NU-1000. (*Reproduced from [62] with permission from the Elsevier*).

principle is the basis of de novo synthesis, creating complex compounds out of simpler ones like sugars or amino acids rather than recycling them after partial destruction [61]. Chemically stable MOFs are mostly formed using carboxylate ligands and high-valent metals. Phosphonate- and phenolate-based ligands are likewise expected to create stable frameworks using high-valent metals, even though they have not yet been extensively applied in medical fields. Suppose soft ligands such as ligands containing N are coupled with soft low-valent metal ions. In that case, they can form stable frameworks in addition to high-valent metal ions (imidazolates, triazolates, and tetrazolates). Due to the material toxicity, fewer components are available to synthesize novel MOFs for medical purposes. Because of this, MOFs with iron and zinc ions, as well as minimally toxic ligands, are mostly utilized. In addition to the coordinate bond strength, which controls thermodynamic stability, kinetic considerations such as framework rigidity should also be considered. Structures that are stiff and dense are usually more stable. For instance, it was found that as ligand length increased, the stability of the isoreticular MOFs UiO-66 and UiO-67 reduced. On the other side, kinetically tuning the dimensions enhances the stability of already-existing MOFs.

The rationalizing method to synthesize a MOF that already constitutes inorganic building blocks and accounts for the kinetic and thermodynamic aspects of the MOF development process (Fig. 9.7) [62]. Single crystals of ultrastable MOFs can be developed using the kinetically tuned dimensional augmentation (KTDA) method, which involves metal clusters and monotopic carboxylates as equilibrium-shifting agents. To create ultra-stable and incredibly crystalline MOFs, post-synthesis metathesis and oxidation (PSMO) of metal ions, quick ligand exchange rate at low oxidation state, and the kinetic inertness of the

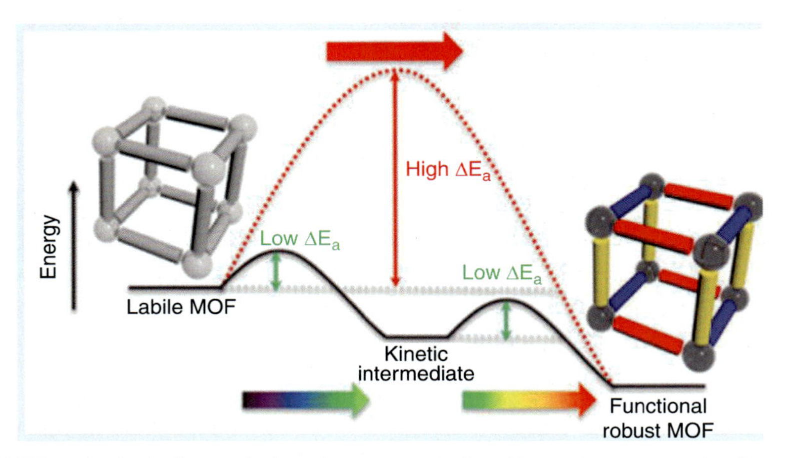

FIGURE 9.7 Kinetically tuned dimensional augmentation (KTDA) is an approach using preformed metal clusters as starting materials and monotopic carboxylates as equilibrium shifting agents to make single crystals of ultra-stable metal–organic frameworks. (*Reproduced from [62] with permission from the American Chemical Society*).

same metal at a high oxidation state. Fe-based MOFs have been successfully metathesized to Cr^{3+} using various comparable techniques. Additionally, several Ti-MOFs with excellent crystallinity have been made. Utilizing a stable MOF with naturally occurring coordinatively unsaturated sites as the matrix, kinetically controlled linker installation, and cluster metalation procedures post-synthesize the linkers or grow clusters on the matrix. A strong MOF with strategically arranged functions can be constructed, this technique has a variety of uses, particularly when specific functional groups or metals with synergistic effects are sought in proximity.

9.4.5 Applications of thermodynamically stable MOF

An emerging family of porous materials called MOFs has the potential to be used for chemical sensing, separation, gas storage, and catalysis. Despite having several benefits, the stability of many MOFs under adverse environments ultimately limits their applicability. The most recent developments in the study of stable MOFs are discussed here, including the fundamental mechanisms behind MOF stability and the design and synthesis of regular MOF architectures. Stable MOFs have numerous prospective and uses in biological systems, such as medication delivery and enzyme immobilization. Due to their biocompatibility and high level of stability in the natural environment, nontoxic metal-based MOFs, including Fe-MOFs, are frequently used as drug delivery systems [63]. A more recent study has demonstrated that MOFs flexibility and multivariate nature enable researchers to better control drug release in a highly controllable and programmable way. For instance, a recent MTV-MIL101(Fe) report suggests

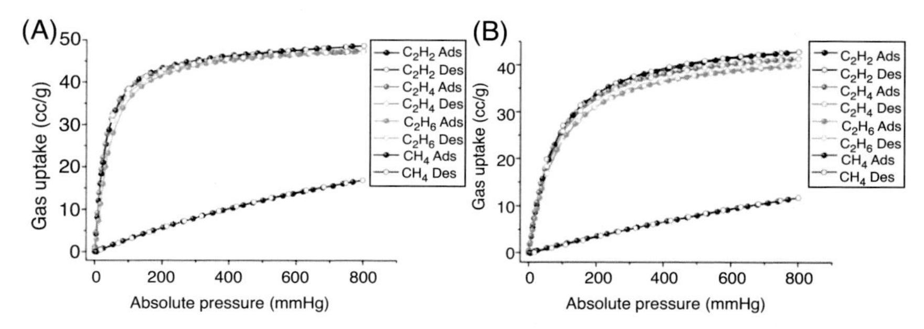

FIGURE 9.8 Adsorption and desorption isotherms for C_2H_2, C_2H_4, C_2H_6, and CH_4 of IITKGP-20a (A) at 273 K and (B) at 295 K (adsorption and desorption isotherms are represented by filled and hollow circles, respectively). *(Reproduced from [64] with permission from the Wiley).*

that controllable drug release could be achieved by fine-tuning interactions between drugs. Moreover, a novel thermodynamically stable Ni-MOF, [Ni(L)(1,4-NDC)(H2O)2]n, IITKGP-20, with moderate porosity and micropores in the direction has a BET surface area of 218 m^2/g. The development of appropriate pore diameters and persistent porosity in conjunction with this MOF is "thermodynamically stable," Sahoo et al. decided to investigate its potential as an adsorbent for commercially significant C2s/C1 separations (Fig. 9.8) [64]. Based on two disparate temperatures, Single-component gas sorption tests have been carried out at 273 K and 295 K. The distinct uptake capacities for the C2 series at 273 K/1 bar pressure. Over C1 methane (17 cc/g), hydrocarbons like C_2H_2 (49 cc/g), C_2H_6 (47 cc/g), and C_2H_4 (48 cc/g), but the uptake capabilities are substantially decreased as expected at 295K and 1 bar pressure (C_2H_2 (43 cc/g), C_2H_6 (40 cc/g), C_2H_4 (41 cc/g) over CH_4 (12 cc/g). These findings demonstrated that IITKGP-20 has a limited absorption ability for all C2 hydrocarbons compared to C1, making it a promising sorbent for the separation of C2s from C1. IAST (Ideal Adsorbed Solution Theory), a well-known method for calculating binary gas mixture selectivity, was used to calculate the binary gas mixture selectivity of C2s/C1 (50:50) to assess IITKGP-20's ability to separate C2 hydrocarbons. Pressures up to 100 kPa were used in this calculation [65].

9.5 Concluding remarks and future prospective

In recent years, MOFs have received much attention from material scientists due to their outstanding characteristic properties including, crystallinity, large surface area, high porosity, and tunable pore size to implement it for practical applications. In the pharmaceutical field, soluble MOF has been evolving rapidly for drug delivery functions. This chapter describes some methods to improve the water solubility of MOFs. The present studies will help to develop new solubilization technology and elucidate mechanisms for drug solubilization through MOF. However, the thermodynamic stability of MOF is another important criterion for various applications such as adsorption, proton conduction, separation, sensing, and catalysis. The instability of MOFs in water is the main barrier to such applications. Therefore, it is crucial to develop thermodynamically stable MOFs to successfully employ them in water as well as in harsh acidic or basic conditions without significant changes in the porosity and integrity of the MOF framework. This chapter discussed in detail, the MOFs fundamentals, synthesis, characterization, solubility, water stability, thermodynamic stability, and approaches to improve the solubility and thermodynamic stability of MOFs for several applications. Thus, this chapter will be of special interest for scientists and researchers to explore the potential of soluble and thermodynamic stable MOF for a better sustainable environment.

References

[1] J.R. Long, O.M. Yaghi, The pervasive chemistry of metal–organic frameworks, Chem. Soc. Rev. 38 (2009) 1213–1214.

[2] C.E. Wilmer, M. Leaf, C.Y. Lee, O.K. Farha, B.G. Hauser, J.T. Hupp, R.Q. Snurr, Large-scale screening of hypothetical metal–organic frameworks, Nat. Chem. 4 (2012) 83–89.

[3] S. Kamal, M. Khalid, M.S. Khan, M. Shahid, M. Ashafaq, I. Mantasha, et al., Synthesis, characterization and DFT studies of water stable Cd(II) metal–organic clusters with better adsorption property towards the organic pollutant in waste water, Inorganica Chim. Acta 512 (2020) 119872.

[4] G.L. Amidon, H. Lennernas, V.P. Shah, J.R. Crison, A theoretical basis for a biopharmaceutic drug classification: the correlation of in vitro drug product dissolution and in vivo bioavailability, Pharm. Res. 12 (1995) 413–420.

[5] S. Yu, H. Pang, S. Huang, H. Tang, S. Wang, S.M. Qiu, X. Wang, et al., Recent advances in metal-organic framework membranes for water treatment: A review, Sci. Total Environ. 800 (2021) 149662.

[6] Y. He, W. Zhang, T. Guo, G. Zhang, W. Qin, L. Zhang, C. Wang, et al., Drug nanoclusters formed in confined nano-cages of CD-MOF: dramatic enhancement of solubility and bioavailability of azilsartan, Acta Pharm. Sin. B 9 (2019) 97.

[7] S. Rojas, A. Arenas Vivo, P. Horcajada, Metal–organic frameworks: a novel platform for combined advanced therapies, Coord. Chem. Rev. 388 (2019) 202–226.

[8] S. Kamal, M. Khalid, M.S. Khan, M. Shahid, M. Ahmad, A Zinc(II) MOF for recognition of nitroaromatic explosive and Cr(III) ion, J. Solid State Chem. 315 (2002) 123482.

[9] N.C. Burtch, H. Jasuja, K.S. Walton, Water stability and adsorption in metal–organic frameworks, Chem. Rev. 114 (2014) 10575–10612.

[10] K.A. Cychosz, A.J. Matzger, Water stability of microporous coordination polymers and the adsorption of pharmaceuticals from water, Langmuir 26 (2010) 17198–17202.

[11] M. Ding, X. Cai, H.L. Jiang, Improving MOF stability: approaches and applications, Chem. Sci. 10 (2019) 10209–10230.

[12] D. Sedmidubsky, J. Leitner, P. Svoboda, Z. Sofer, J. Machacek, Heat capacity and phonon spectra of $A^{III}N$: experiment and calculation, J. Therm. Anal. Calorim. 95 (2009) 403–407.

[13] J. Zhang, J.L. Zeng, Y.Y. Liu, L.X. Sun, F. Xu, W.S. You, Y. Sawada, Thermal decomposition kinetics of the synthetic complex Pb(1,4-BDC)·(DMF)(H_2O), J. Therm. Anal. Calorim. 91 (2008) 189–193.

[14] C.H. Jiang, L.F. Song, C.L. Jiao, J. Zhang, L.X. Sun, F. Xu, Y. Du, Z. Cao, Exceptional thermal stability and thermodynamic properties of lithium based metal–organic framework, J. Therm. Anal. Calorim. 103 (2011) 373.

[15] A.A. Voskanyan, V.G. Goncharov, N. Novendra, X. Guo, A. Navrotsky, Thermodynamics drives the stability of the MOF-74 family in water, ACS Omega 5 (2020) 13158.

[16] R. Sahoo, S. Chand, M. Mondal, A. Pal, S.C. Pal, M.K. Rana, M.C. Das, et al., A thermodynamically stable 2D nickel metal–organic framework over a wide pH range with scalable preparation for efficient C2s over C1 hydrocarbon separations, Compos. B: Eng. 26 (2020) 12624–12631.

[17] S. Kamal, M. Khalid, M.S. Khan, M. Shahid, Metal organic frameworks and their composites as effective tools for sensing environmental hazards: an up to date tale of mechanism, current trends and future prospects, Coord. Chem. Rev. 474 (2022) 214859.

[18] X. Wang, B. Wu, N. Afsar, Y. Zhu, T. Xu, Z. Zhao, et al., Soluble polymeric metal–organic frameworks toward crystalline membranes for efficient cation separation, J. Membr. Sci. 639 (2021) 119757.

[19] J.R. Li, J. Sculley, H.C. Zhou, Metal–organic frameworks for separations, Chem. Rev. 112 (2012) 869–932.

[20] J.E. Mondloch, M.J. Katz, W.C. Isley, P. Ghosh, P. Liao, W. Bury, Destruction of chemical warfare agents using metal-organic frameworks, Nat. Mater. 14 (2015) 512.

[21] M.S. Khan, M. Khalid, M. Shahid, What triggers dye adsorption by metal organic frameworks? The current perspectives, Mater. Adv. 1 (2020) 11575–11601.

[22] M. Ding, X. Cai, H.L. Jiang, Improving MOF stability: approaches and applications, Chem. Sci. 44 (2019) 10209–10230.

[23] D. Saha, S. Deng, Structural stability of metal organic framework MOF-177, J. Phys. Chem. Lett. 1 (2010) 73–78.

[24] D. Saha, Z. Wei, S. Deng, Equilibrium, kinetics and enthalpy of hydrogen adsorption in MOF-177, Int. J. Hydrogen Energy 33 (2008) 7479–7488.

[25] J.A. Greathouse, M.D. Allendorf, The interaction of water with MOF-5 simulated by molecular dynamics, J. Am. Chem. Soc. 128 (2006) 10678.

[26] M.V. Parkes, C.L. Staiger, J.J. Perry IV, M.D. Allendorf, J.A. Greathouse, Screening metal–organic frameworks for selective noble gas adsorption in air: effect of pore size and framework topology, Phys. Chem. Chem. Phys. 15 (2013) 9093–9106.

[27] Y. Li, R.T. Yang, Gas adsorption and storage in metal—organic framework MOF-177, Langmuir 23 (2007) 12937–12944.

[28] D. Kobayashi, A. Hamakawa, Y. Yamaguchi, T. Takahashi, M. Yanagita, S. Arai, M. Minoura, Broadened bioactivity and enhanced durability of two structurally distinct metal–organic frameworks containing Zn^{2+} ions and thiabendazole, Dalton Trans. 50 (2021) 7176–7180.

[29] B. Liu, K. Vikrant, K.H. Kim, V. Kumar, Kailasa, Critical role of water stability in metal–organic frameworks and advanced modification strategies for the extension of their applicability, Environ. Sci. 7 (2020) 1319–1347.

[30] S.Q. Wei, C.W. Lin, X.H. Yin, Y.J. Huang, P.Q. Luo, Hydrothermal synthesis, crystal structure of four novel complexes based on thiabendazole ligand, Bull. Korean Chem. Soc. 33 (2012) 2917–2924.

[31] L. Zhang, W.X. Fang, C. Wang, H. Dong, S.H. Ma, Y.H. Luo, Porous frameworks for effective water adsorption: from 3D bulk to 2D nanosheets, Inorg. Chem. Front. 8 (2021) 898–913.

[32] D.D. Sun, P.I. Lee, Evolution of supersaturation of amorphous pharmaceuticals: the effect of rate of supersaturation generation, Mol. Pharm. 10 (2013) 4330–4346.

[33] T. Vasconcelos, B. Sarmento, P. Costa, Solid dispersions as strategy to improve oral bioavailability of poor water soluble drugs, Drug Discov. Today 12 (2007) 1068–1075.

[34] C.B. Aakeröy, S. Forbes, J. Desper, Using cocrystals to systematically modulate aqueous solubility and melting behavior of an anticancer drug, J. Am. Chem. Soc. 131 (2009) 17048–17049.

[35] N. Seedher, M. Kanojia, Co-solvent solubilization of some poorly-soluble antidiabetic drugs, Pharm. Dev. Technol. 14 (2009) 185–192.

[36] M.X. Wu, Y.W. Yang, Metal–organic framework (MOF)-based drug/cargo delivery and cancer therapy, Adv. Mater. 29 (2017) 1606134.

[37] K. Suresh, A.J. Matzger, Enhanced drug delivery by dissolution of amorphous drug encapsulated in a water unstable metal–organic framework (MOF), Angew. Chem. Int. Ed. Engl. 58 (2019) 16790–16794.

[38] M.W. Tibbitt, J.E. Dahlman, R. Langer, Emerging frontiers in drug delivery, J. Am. Chem. Soc. 138 (2016) 704–717.

[39] S. Kamal, M. Khalid, M.S. Khan, M. Shahid, M. Ahmad, Amine-and imine-functionalized Mn-based MOF as an unusual turn-on and turn-off sensor for d10 heavy metal ions and an efficient adsorbent to capture iodine, Cryst. Growth Des. 22 (2022) 3277–3294.

[40] J.J. Low, A.I. Benin, P. Jakubczak, J.F. Abrahamian, S.A. Faheem, R.R. Willis, et al., Virtual high throughput screening confirmed experimentally: porous coordination polymer hydration, J. Am. Chem. Soc. 131 (2009) 15834.

[41] P.M. Schoenecker, C.G. Carson, H. Jasuja, C.J. Flemming, K.S. Walton, Evaluation of MOFs for air purification and air quality control applications: ammonia removal from air, Chem. Eng. Sci. 51 (2012) 6513–6519.

[42] L. Huang, H. Wang, J. Chen, Z. Wang, J. Sun, D. Zhao, Y. Yan, Synthesis, morphology control, and properties of porous metal–organic coordination polymers, Microporous Mesoporous Mater. 58 (2003) 105–114.

[43] J.B. DeCoste, G.W. Peterson, B.J. Schindler, K.L. Killops, M.A. Browe, J.J. Mahle, The effect of water adsorption on the structure of the carboxylate containing metal–organic frameworks Cu-BTC, Mg-MOF-74, and UiO-66, J. Mater. Chem. 1932 (2013) 11922–11932.

[44] I.J. Kang, N.A. Khan, E. Haque, S.H. Jhung, Chemical and thermal stability of isotypic metal–organic frameworks: effect of metal ions, Chemistry 17 (2011) 6437–6442.

[45] D. Feng, Z.Y. Gu, J.R. Li, H.L. Jiang, Z. Wei, H.C. Zhou, Zirconium-metalloporphyrin PCN-222: mesoporous metal–organic frameworks with ultrahigh stability as biomimetic catalysts, Angew. Chem. 124 (2012) 10453–10456.

[46] D. Feng, W.C. Chung, Z. Wei, Z.Y. Gu, H.L. Jiang, Y.P. Chen, D.J. Darensbourg, H.C. Zhou, Construction of ultrastable porphyrin Zr metal–organic frameworks through linker elimination, J. Am. Chem. Soc. 135 (2013) 17105–17110.

[47] C. Volkringer, S.M. Cohen, Generating reactive MILs: isocyanate-and isothiocyanate-bearing MILs through postsynthetic modification, Angew. Chem. Int. Ed. 49 (2010) 4644–4648.

[48] K.S. Park, Z. Ni, A.P. Côté, J.Y. Choi, R. Huang, F.J. Uribe-Romo, H.K. Chae, M.O. Keeffe, O.M. Yaghi, Exceptional chemical and thermal stability of zeolitic imidazolate frameworks, Proc. Natl. Acad. Sci. U. S. A. 103 (2006) 10186–10191.

[49] D. Cunha, M. Ben Yahia, S. Hall, S.R. Miller, H. Chevreau, E. Elkaïm, G. Maurin, P. Horcajada, C. Serre, Rationale of drug encapsulation and release from biocompatible porous metal–organic frameworks, Chem. Mater. 25 (2013) 2767–2776.

[50] S.S. Han, S.H. Choi, A.C. Duin, Molecular dynamics simulations of stability of metal–organic frameworks against H_2O using the ReaxFF reactive force field, Chem. 46 (2010) 5713–5715.

[51] J. Canivet, A. Fateeva, Y. Guo, B. Coasne, D. Farrusseng, Water adsorption in MOFs: fundamentals and applications, Chem. Soc. Rev. 43 (2014) 5594–5617.
[52] S. Muthaiah, A. Bhatia, M. Kannan, Stability of metal complexes, Stability and Applications of Coordination Compounds, IntechOpen, 2020.
[53] S. Yuan, L. Feng, K. Wang, J. Pang, M. Bosch, C. Lollar, et al., Stable metal–organic frameworks: stable metal–organic frameworks: design, Synthesis, and Applications, Adv. Mater. 30 (2018) 1870277.
[54] J.J. Low, A.I. Benin, P. Jakubczak, J.F. Abrahamian, S.A. Faheem, R.R.J. Willis, Virtual high throughput screening confirmed experimentally: porous coordination polymer hydration, J. Am. Chem. Soc. 131 (2009) 15834–15842.
[55] R.G. Pearson, H.R. Sobel, J. Songstad, Nucleophilic reactivity constants toward methyl iodide and trans-dichlorodi(pyridine) platinum(II), J. Am. Chem. Soc. 90 (1968) 319–326.
[56] J.T. Hughes, A. Navrotsky, MOF-5: enthalpy of formation and energy landscape of porous materials, J. Am. Chem. Soc. 133 (2011) 9184–9187.
[57] R.I. Walton, F. Millange, In situ studies of the crystallization of metal–organic frameworks, The Chemistry of Metal–Organic Frameworks: Synthesis, Characterization, and Applications, vol. 2, Wiley, 2016, pp. 729–764.
[58] P.M. Forster, A.R. Burbank, C. Livage, G. Ferey, A.K. Cheetham, The role of temperature in the synthesis of hybrid inorganic–organic materials: the example of cobalt succinates, Chem. Commun. 4 (2004) 368–369.
[59] D. Wu, A. Navrotsky, Thermodynamics of metal–organic frameworks, J. Solid State Chem. 223 (2015) 53–58.
[60] Z. Akimbekov, D. Wu, C.K. Brozek, M. Dincă, A. Navrotsky, Thermodynamics of solvent interaction with the metal–organic framework MOF-5, Phys. Chem. Chem. Phys. 18 (2016) 1158–1162.
[61] R.G. Pearson, Hard and soft acids and bases, J. Am. Chem. Soc. 85 (1963) 3533–3539.
[62] M. Bosch, S. Yuan, W. Rutledge, H.C. Zhou, Stepwise synthesis of metal–organic frameworks, Acc. Chem. Res. 18 (2017) 857–865.
[63] D. Feng, K. Wang, Z. Wei, Y.P. Chen, C.M. Simon, R.K. Arvapally, R.L. Martin, T.F. Liu, M. Bosch, S. Fordham, D. Yuan, Kinetically tuned dimensional augmentation as a versatile synthetic route towards robust metal–organic frameworks, Nat. Commun. 4 (2014) 5723.
[64] R. Sahoo, S. Chand, M. Mondal, A. Pal, S.C. Pal, M.K. Rana, M.C. Das, A "thermodynamically stable" 2D nickel metal–organic framework over a wide pH range with scalable preparation for efficient C2s over C1 hydrocarbon separations, Eur J. Chem 26 (2020) 12624–12631.
[65] A.L. Myers, J.M. Prausnitz, Thermodynamics of mixed-gas adsorption, AIChE J. 11 (1965) 121–127.

Chapter 10

Preparation and applications of water-based zeolitic imidazolate frameworks

Farhat Vakil and M. Shahid
Functional Inorganic Materials Lab (FIML), Department of Chemistry, Aligarh Muslim University, Aligarh, Uttar Pradesh, India

10.1 Introduction

The topologies of ZIFs, which are classified as a subclass of MOFs and have a lot in common with zeolites, are quite comparable [1]. Their structure is composed of metal–imidazolate–metal (metals like Zn or Co) that is made by the process of self-assembly. The ZIF structures are very diverse and are somehow alike to conventional aluminosilicate zeolites, structures of zeolites are made of tetrahedral units of $Si(Al)O_4$, which is bonded covalently through bridging atoms of oxygen to generate more than 150 non-identical types of frameworks [2]. This coordination exposes distinct ZIF structures as well as various crystal structures with molecular images, as seen in Fig. 10.1 [3]. A time-honored dare to us is to subsume organic linkers and d-block metal ions inside pores and more admirably, to make it done is a bit essential to the ZIFs. Due to the presence of pores that would be lined with a high concentration of ordered transition metal positions, whose characteristics, including steric and electronic, could be altered through the functionalization of organic links, this potential is expected to be useful in a variety of applications, including catalysis. However, there is still a cloud over the prospect of obtaining these zeolites that fulfill the above properties. Therefore, we outlined a general synthesis method of structures containing ZIFs topologies in which transition metals are tetrahedral atoms, and the bridging species are imidazolate units.

Following the loss of one proton, imidazole transforms into imidazolate (IM). The dense phases of $Co(IM)_2$ and Zn(IM) when examined, its structures are built from networks of connected CoN_4 or ZnN_4 tetrahedra [4,5], we found that IM bridges create an M-IM-M angle, 1, roughly 145 degrees, that is concurrent with the Si–O–Si angle, 2, that is similar to better and frequently found in

Synthesis of Metal–Organic Frameworks via Water-Based Routes: A Green and Sustainable Approach.
DOI: https://doi.org/10.1016/B978-0-323-95939-1.00003-4

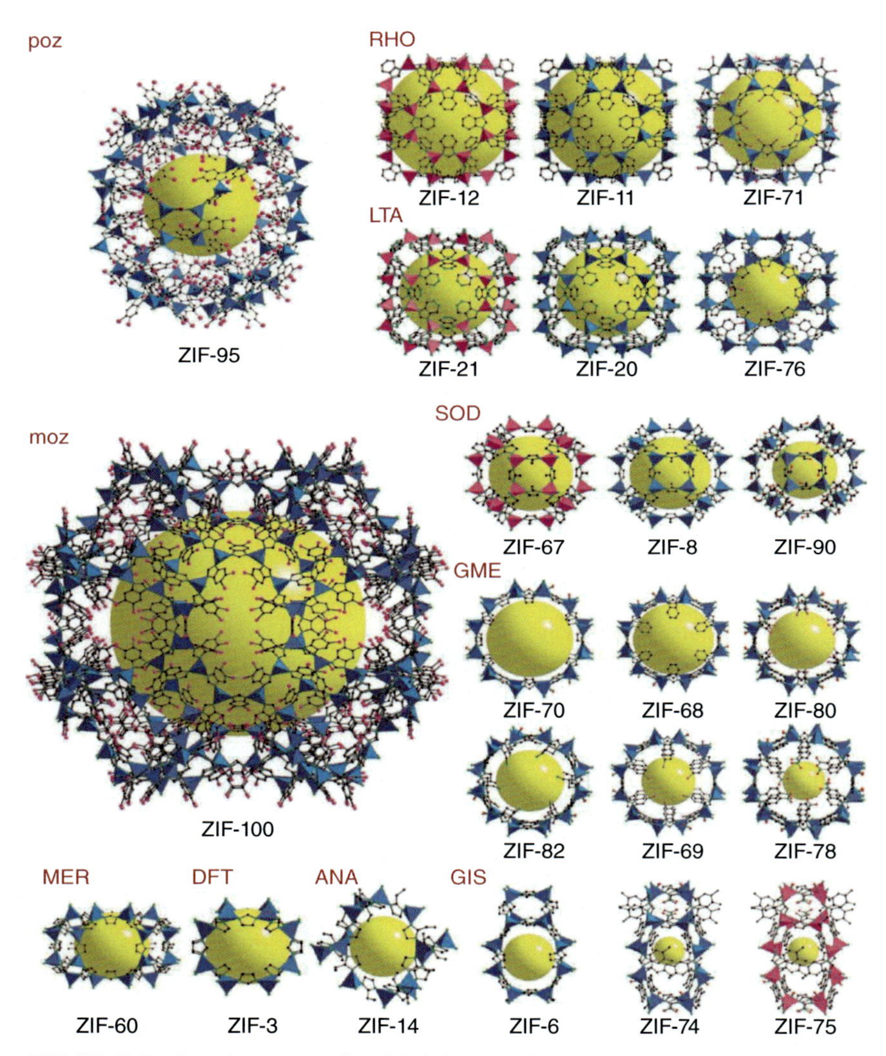

FIGURE 10.1 Crystal structures of zeolitic imidazolate frameworks (ZIF) are categorized based on their topology (three-letter symbol). The links are illustrated as a ball-and-stick diagram, and each largest cage of ZIF is shown with ZnN_4 in blue and CoN_4 in pink polyhedra. The yellow ball in the cage denotes a free area. For clarity, H atoms are not included (C, *black*; N, *green*; O, *red*; Cl, *pink*). (*Reproduced from [3] with permission from American Chemical Society, 2010*).

a number of zeolites (Scheme 10.1). In our opinion, it should be possible to create metal IMs that embrace open-framework zeolite structures under the right conditions. Undoubtedly, structures of a sufficient number of relatively recent IM compounds such as Fe(II) [6], Co(II) [7,8], Cu(II) [9], and Zn(II) [10] tetrahedral nets-based resembling zeolites. Nevertheless, these ZIF materials have

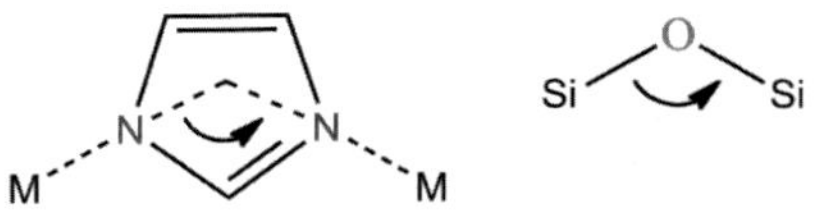

1. M-IM-M 2. Si-O-Si

SCHEME 10.1 Bridging angles: (A) metal imidazolates and (B) zeolites. (*Reproduced from [17]*).

low-symmetry structures or are quite thick (nonporous). ZIFs, which resemble Zn(II) IMs and have features like symmetrical porous structures akin to zeolites, have just been announced [11]. More importantly, we confirmed the porosity of ZIFs and found that they differ from other organometallic compounds in that they have very high chemical stability in water and aqueous alkaline solution. This is a previously unreported finding. The aforementioned effects on the potential uses and rich variety of architectures of this until now the underutilized class of porous materials. In this chapter, we will shed more light on water-based ZIFs their chemical and thermal stability, porosity, and applications.

ZIFs commonly exhibit properties that overlap those of zeolites and MOFs, including exceptionally high surface areas, unimodal micropores, crystallinities, a wide range of functions, and outstanding chemical and thermal stabilities [12–14]. In light of this, ZIFs show significant promise in a number of application areas, including catalysis, gas separation, evaporation, and electronic devices. As demonstrated in Table 10.1, the unusual features of ZIFs distinguish them from conventional zeolites in a variety of ways.

10.2 Porosity and stability of water-based ZIF-8 and microwave-assisted ZIF-11

To understand the thermal, architectural, porosity, and chemical stability of ZIFs, we are concentrating on two archetypical structures, that is to say, ZIF-8 and ZIF-11 which were being synthesized at the scale of grams to permit investigation in detail of the above-given properties. The two ZIFs mentioned have large pores in their structures, with diameters of 11.6Å for ZIF-8 and 14.6Å for ZIF-11, and they are connected by tiny gaps, with 3.4Å for ZIF-8 and 3.0Å for ZIF-11, respectively. Due to the longer IM connecting units (look up), the size of their pores is somehow two times larger than that of their zeolite equivalents; yet molecular sieves have smaller hole sizes due to the presence of side chains or linker rings (Table 10.2).

To learn about the thermal stabilities of water-based ZIF-8 and microwave-assisted ZIF-11, the extraordinary stability of ZIF-8 and ZIF-11 was discovered by thermal gravimetric analysis (TGA).

The TGA trace for ZIF-8 demonstrated a gradual loss of weight step of 28.3% (25–450°C), which was correlated with the fragmented loss of guest molecules (N,N-dimethylformamide (DMF) and $3H_2O$, which was estimated at 35.9%),

TABLE 10.1 The zeolites and zeolitic imidazolate frameworks are contrasted.

Comparison content	Zeolites	ZIFs
Framework	Inorganic type	Organic–inorganic
Composition	Si; Al; O	Co; C; Zn; N; H; etc.
Secondary building units	[SiO_4] and [AlO_4]	M(Im)2
Topology	~200	Over 100 were discovered; the number may rise exponentially
Stability	Thermodynamic and chemical stability are often very high, depending on the Si/Al ratio.	High chemical stability in organic and aqueous media and thermal stability at 500°C
Compatibility	Ineffective contact with polymers	Better closeness with organic polymers
Functionality	Si/Al ratio tunable but typically challenging to functionalize	Enhanced chemical capabilities using organic linkers
Development	Over 50 years	~10 years
Application prospects	Large-scale and affordable for industrial applications	Costly; possibility of industry uses

TABLE 10.2 ZIF-8 and ZIF-11 structural properties as determined by single crystal X-ray diffraction analysis.

	Pore aperture diameter, Å			Pore	Surface area	Pore volume
ZIF-*n*	8-ring	6-ring	4-ring	diameter, Å	m^2/gm	cm^3/gm
ZIF-8	—	3.4	*	11.6	1,947	0.663
ZIF-11	3.0	3.0	*	14.6	1,676	0.582

The free volume procedure of the CERIUS2 software and the single crystal X-ray structures of ZIF-8 and ZIF-11 with guests-free and averaged the disorder effects served as the foundation for all calculations (Version 4.2; MatSci; Accelrys, Inc., San Diego; probe radius 1.4, medium grid).
**The aperture sizes of 4-ring in ZIF-8 and -11 are insignificant.*

followed by the plateau (450–550°C). More imposingly, the TGA trace for ZIF-11 disclosed a keen loss of weight step of 22.8% (25–250°C), correlating with the loss of all solvent N,N-diethylformamide (DEF) molecules confined in the pores (0.9DEF; calculated 23.3%), in spite of the fact that DEF is literally bigger than the crevice of ZIF-11 in size. The TGA trace of ZIF-11 likewise showed a lengthy plateau in the temperature range of 250–550°C, indicating that in the absence of guest species, its thermal stability is extremely high. When heated to and maintained at 500°C and 300°C, respectively, the PXRD patterns of both ZIF-8 and ZIF-11 samples coincide in a nitrogen atmosphere served as proof that the guest molecules in ZIF-8 and ZIF-11 were acquired without altering the frameworks. Only a very small number of MOFs with similarly intensive structures can match the remarkable thermal stability of ZIFs (up to 550°C in N_2) with the permanently porous cubic structure of MOF-5, which decomposes at 450°C in a nitrogen atmosphere [15,16].

By switching solvents, the amide guest molecules integrated with ZIF-8 and ZIF-11 as they were created can be eliminated more quickly. The thermogravimetric behavior of ZIF-8 and ZIF-11 was significantly clarified after being dissolved in organic solvents like methanol. The as-produced samples of ZIF were submerged in methanol (CH_3OH) at ambient temperature for a prolonged period of 48 hours, moved out at ambient temperature for 5 hours, then heated to a higher temperature (300°C for ZIF-8; 180°C for ZIF-11) for 2 hours in order to separate the guest molecules from the frameworks and make the move out ZIF-8 and ZIF-11 forms for gas-sorption examination. The samples of ZIF so acquired were pushed out as efficiently as possible, as seen by their well-maintained patterns of powdered X-ray diffraction and the extended plateau (25–550°C) in their thermogravimetric analysis track downs [17].

Gas-sorption testing clearly demonstrated the move out structural rigidity of ZIF-8 and ZIF-11 and, as a result, their long-term porosity. ZIF-8 exhibited type I N_2 sorption isotherm behavior (Fig. 10.2A), revealing that it is microporous. By using data points on the adsorption subclass in the range of *P_P0*_0.01–0.10, evident surface areas of 1810 m^2/g (Langmuir model) and 1630 m^2/g (Brunauer–Emmett–Teller [BET] model) for ZIF-8 were obtained, and a micropore volume of 0.636 cm^3/g for ZIF-8 was acquired based on a single data point at *P_P0*_0.10.

Figures of ZIF-8 micropore volume and surface area match the augury well because of its single-crystal structure (Table 10.2). The reported greatest values for ordered mesoporous materials and zeolites are lower than these surface areas [18–21].

Due to its 3.0 pore size, which was considerably smaller than the kinetic diameter of nitrogen (3.6), ZIF-11 was not porous to N_2, yet it was nevertheless permitted to consume H_2. ZIF-8 and ZIF-11 both displayed reversible hydrogen sorption (Fig. 10.2B). Curiously, the initial hydrogen grip of ZIF-11 was substantially higher than that of water-based ZIF-8 due to its distinctive intramural cage, which is made up of protruding benzene side rings of the Ph IM linkages and can be used to form favorable H_2 sorption sites.

ZIF-8 and ZIF-11, however, were equal in terms of H_2 absorption when the pressure of adsorbate reached one atmosphere (145 cm^3/g at STP or 12.9 mg/g for ZIF-8 and 154 cm^3/g STP or 13.7 mg/g for ZIF-11). Given the larger surface area and increased pore capacity of ZIF-8, this outcome is expected (Table 10.2). A significant amount of removed ZIF-8 (0.724 g) was tested for hydrogen sorption at 77 K under high pressure (up to 80 bar), which demonstrated 350 cm^3/g STP (31 mg/g) at 55 atm. We recently proposed a linear relationship between the hydrogen grip of ZIF-8 and its Langmuir surface area (1810 m^2/g)

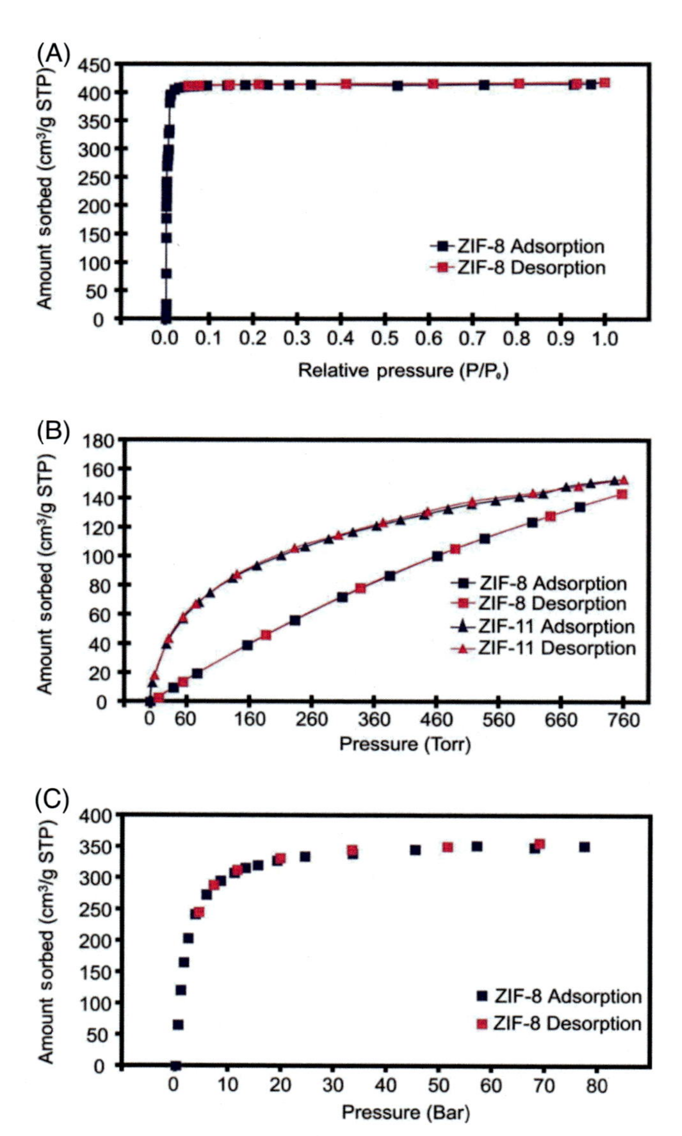

FIGURE 10.2 The prototype gas-sorption isotherms of zeolitic imidazolate frameworks (ZIF) (A) Nitrogen isotherm for ZIF-8 sod at 77 K. (B) Hydrogen isotherms for ZIF-11 rho and sod of ZIF-8 at 77 K. (C) For sod of ZIF-8, high-pressure isotherm of hydrogen at 77 K. (*Reproduced from [17]*).

using measurements of high-pressure sorption of hydrogen on a variety of MOFs with large surface areas. ZIFs have been shown to outdistance conventional crystalline microporous materials like zeolites and ordered mesoporous silicas and to have pore volumes and surface areas that are comparable to porous MOF compounds. We hypothesize that this presentation is brought on by the fully exposed borders and faces of the organic linker, which have been cited as being essential for developing extraordinarily high surface areas [22].

In order to test the chemical stability of ZIFs, samples of ZIF-8 and ZIF-11 were submerged in boiling solutions of C_6H_6, MeOH, H_2O, and aqueous NaOH (Fig. 10.3). These settings reflect the ideal operating conditions of traditional industrial chemical procedures. Samples of ZIF were submerged in the necessary solvent at 50°C for 1 to 7 days, and the boiling point was moderate. ZIF samples were sporadically observed during this process and discovered to be insoluble under each of the settings indicated under an optical microscope.

PXRD (powder X-ray diffraction) patterns obtained at various time intervals revealed that solid samples of ZIF-8 and ZIF-11 crystallinity remained intact and were obviously resistant to the boiling organic solvents for many days up to seven. ZIF-8 and ZIF-11 structures can both be kept stable in water at 50°C for up to 7 days. However, what we saw was that only the structure of ZIF-8 could be preserved in boiling water for 7 days, while ZIF-11 was converted to a different crystalline substance after 3 days.

ZIF-8 was then investigated further and discovered to have remained unchanged at 100°C for 1 day in 0.1 M aq. and 8 M aq. sodium hydroxide. We found that the stability of ZIF-8 is much greater than that of the original mesoporous silica mobile composition of matter (MCM) and santa barbara amorphous (SBA) types of sequenced mesoporous silica on hydrothermal bases [20], even rivaling the extremely stable offspring of such materials [23].

Among organometallic solids, this unusual hydrolysis-repellent property is exceptional. While the bonding between Zn(II), Co(II), and IM is among the least unstable for nitrogen donor ligands, the water-repellent pore, and structure of ZIFs are expected to repel molecules of water, preventing the assault of units of ZnN_4 and the termination of the framework. These are two credible explanations that can be considered state-of-the-art (on metal–complex formation constants scale) [24]. The extraordinary stability of ZIFs on hydrothermal bases was also communicated by these two ZIF properties, which were more comparable to covalent solids. In the work presented here, we have discovered a typical route to strong porous materials with very high flexibility and chemical stability in changing their inorganic and organic components. We believe that these new discoveries will open a number of formerly untapped uses for oxide-based porous materials.

10.3 Synthesis

Zeolitic imidazolate frameworks have been mainly synthesized by hydrothermal or solvothermal procedures. Crystals grow unhurriedly from a metal salt heated solution that was hydrated, an ImH (acidic proton of imidazole), base and a solvent [17]. Functionalized ImH organic linkers permits to maintain the structure of ZIFs [25].

For producing monocrystalline materials for doing single crystal X-ray diffraction (XRD) the given procedure is cool [26,27]. A broad scale of bases,

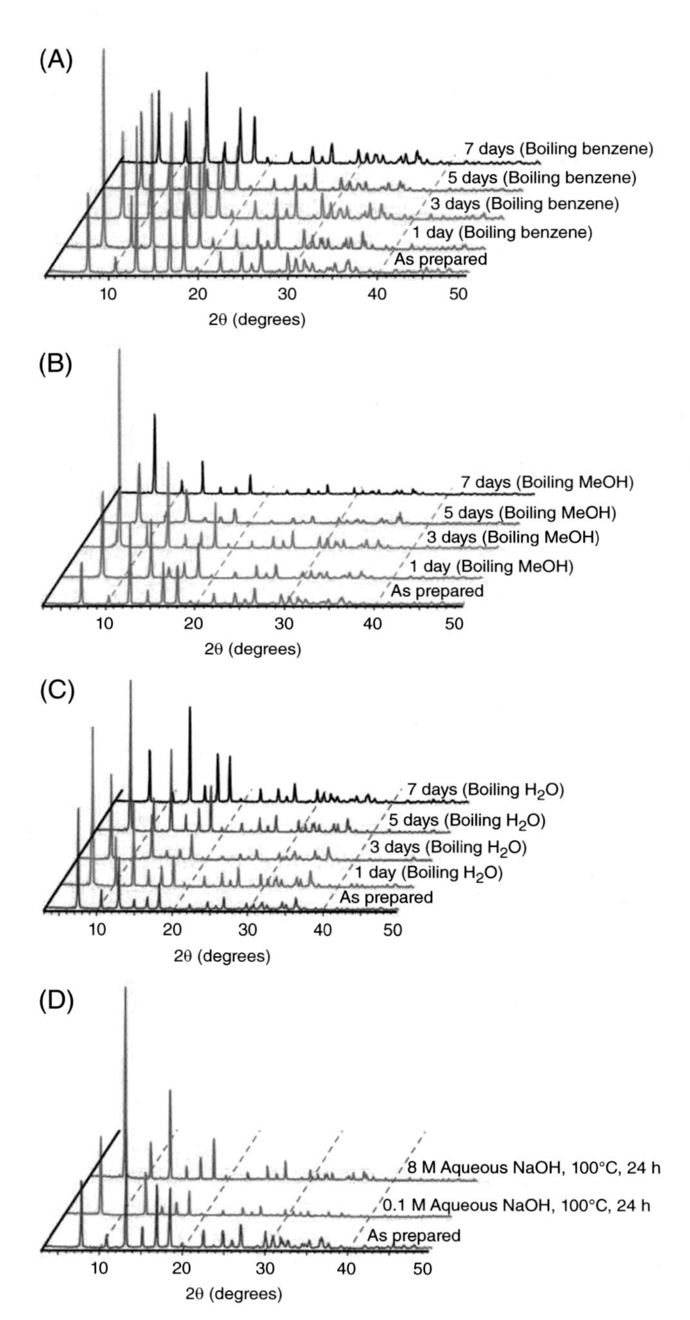

FIGURE 10.3 Chemical stability experiments were used to assess the PXRD patterns of ZIF-8 sample. (A) By benzene being refluxed up to 7 days at 80°C. (B) At 65°C for up to 7 days, refluxing CH_3OH. (C) In water that is refluxing for up to 7 days at 100°C. (D) In an aq. sodium hydroxide solution that is refluxing for 24 hours. *(Reproduced from [17])*.

solvents and conditions have been surveyed, with a vision towards enhancing functions of crystal, cytology, and dispersity [28]. Characteristically, the use of DMF which is amide solvent. The heat given decays the solvent DMF to produce amines, turning to produce the imidazolate from imidazole species. CH_3OH [29,30], C_2H_5OH [31], isoC_3H_7OH [32], and H_2O [33–35] have also been surveyed as substitute solvents for making ZIFs but need some bases like pyridine [36], tri ethyl amine [37], sodium formate [38], and sodium hydroxide [39]. Some polymers like poly(ethylene oxide)–poly(propylene oxide)–poly(ethylene oxide) [40], polyvinylpyrrolidone [41], and poly-(diallyldimethylammonium chloride) [42] have been identified to act as crystal dissipates, recounting control of morphology and particle size.

Due to ZIFs encouraging material properties, notable interest reclines in inexpensive large-scale generation procedures. Such as sonochemical method of synthesis, that allows rainmaking reactions to begin quickly through audile production of limited pressure and heat, has been surveyed as a way of quick synthesis of ZIFs [43,44]. ZIF synthesis with microwave assistance has also been used to create ZIFs more quickly [45,46] whereas it has been demonstrated that sonochemical and microwave-assisted approaches can shorten reaction durations from a few days to a few hours, and even hours to minutes. It has also been discovered that solvent-less techniques, such as ball milling or chemical vapor deposition, can produce high-quality ZIF-8 [47,48]. Because of its high level of uniform structure and feature ratio control and flexibility to be incorporated into conventional blocking procedures for useful thin films, chemical vapor deposition has a special ability (e.g., microelectronics). Supercritical carbon dioxide (scCO_2)-based low-cost synthesis has also been suggested as a feasible strategy for producing water-based ZIF-8 on an industrial scale [49]. ZIF-8 can be produced under stoichiometric conditions in around 10 hours without the need for extra ligands, organic solvents, additives, or cleaning techniques (Fig. 10.4).

10.3.1 Synthesis of some water-based ZIFs

Despite the fact that early ZIF research has been driven and dominated by synthesis methods for ZIFs such as solvothermal. Organic solvents are unquestionably expensive, combustible, and non-biodegradable. Many efforts have been made recently to produce ZIFs in environmentally friendly and straightforward ways by using organic solvents sparingly or eventually refraining from using them. Pan et al. perceived the making of zeolitic imidazolate framework-8 hydrothermally at room temperature for the first time by a simple process [50]: the zinc nitrate solution merged with 2-methylimidazole (MIm) solution, and the results were collected after stirring by centrifugation for around 6 minutes. Nevertheless, it is understandable that too much quantity of Mim (molar ratio of Zn^{2+}:Mim 1:70) was frittered away in this process because the stoichiometric molar ratio of Zn^{2+} and Mim in ZIF-8 is Zn^{2+}:Mim 14 1:2. As a result, numerous attempts to build ZIFs from a stoichiometric molar ratio of Im and metal ions by-products in an

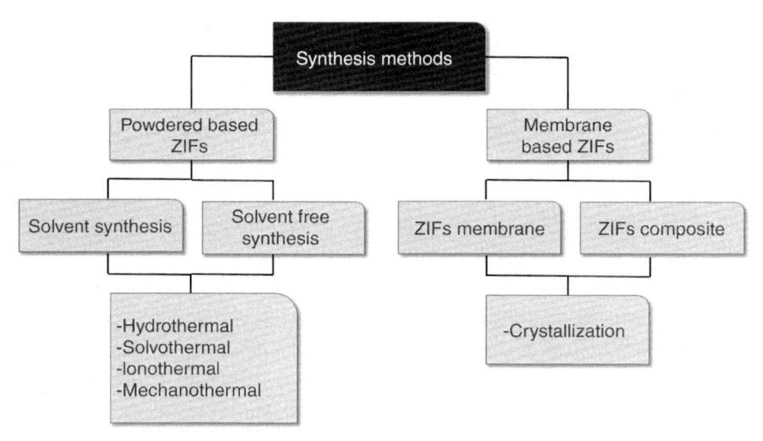

FIGURE 10.4 Summarized synthesis processes of zeolitic imidazolate framework materials/composites.

aqueous medium have failed. In distinct, Tanaka et al. [51] has done considerable work in this field. Pure ZIF-8 crystals could be prepared successfully in an aqueous media at room temperature by utilizing the molar ratio Zn^{2+}:MIm 1/4 1:20 [52]. Additionally, Qian's group demonstrated that the molar ratio of Co^{2+}:MIm:H_2O 1/4 1:58:1100 in aqueous mediums at ambient temperature can result in nanocrystals of ZIF-67 [53].

Synthesis of water-based ZIFs is also altered by using add-ons. Such as tri ethanol amine [54] and NH_4OH [55] were put in as deprotonation agents to lessen the ligand MIm use and started making ZIFs. ZIF-8 and ZIF-67 were created by Gross et al. in an aqueous solution with a few drops of triethyl amine added at room temperature without producing any byproducts [54], where the metal ion: ion molar ratio may be lessened to 1:4. Yao et al. additionally discover that zeolitic imidazolate framework-8 can also be produced with the molar ratio of Zn^{2+}:MIm:NH^{4+}:H_2O 1/4 1:4:16:547 when ammonium hydroxide is present [55].

Strangely, ZIFs can also be created by a stoichiometric ratio of metal ions and MIm when additional components, such as tri-block copolymers, are present. (PEO-PPO-PEO) Poly (ethylene oxide, poly (propylene oxide), and poly (ethylene oxide) [55] and polyvinyl pyrrolidone (PVP) [56] in an aqueous medium, i.e., to create ZIF-67 and ZIF-8, stoichiometric metal ions and MIm were combined with an aq. ammonia medium and a surfactant (triblock copolymer) that contained PEO groups [55], and it was hypothesized that this surfactant can speed up the synthesis of porous ZIF-67 and ZIF-8 due to the metal ions electrostatic attraction to it. Additionally, Shieh et al. propose that ZIF-90 micron-sized crystals might form in an aq. solution when triblock copolymer PVP was present (having 40,000 molecular weight) [56]. Copolymer PVP was thought to influence crystal cytology and avert the formation of crystal seeds.

We and other organizations have both developed a thriving understanding of the ZIF manufacturing from stoichiometric forerunners by aqueous ammonia

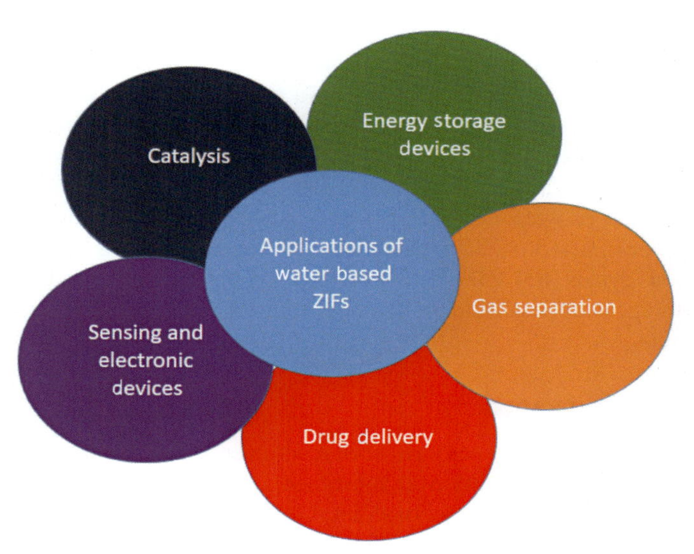

FIGURE 10.5 Applications of water-based zeolitic imidazolate frameworks.

refashioning in recent years [57,58]. In concentrated NH_4OH solutions at room temperature, He's group revealed the manufacture of water-based ZIF-8 from stoichiometric precursors (Zn^{2+}:MIm 1/4 1:2) [58]. The emergence of the ZIF-8 crystal and its development through deprotonation and coordination processes were designed to depend on the mixing of an adequate amount of NH_4OH. Additionally, it is discovered that ZIF-8 can be easily made at room temperature from stoichiometric precursors in solution of aqueous ammonia without the use of any additional components [57]. The ZIF-8 material particle size, shapes, and textural characteristics can be easily adjusted in the synthesis mixture by carefully regulating the aqueous ammonia concentration.

Unquestionably, the large-scale manufacturing of ZIFs for potential applications will be strongly encouraged by the master plans of these environmentally beneficial synthesis.

10.4 Applications

On the bases of significant properties such as flexibility toward substituents, tunable porosity, various compositions, and thermal and chemical stabilities of water based ZIFs and ZIF materials have wider applications. Out of which some applications of great importance have been discussed in this chapter (Fig. 10.5).

10.4.1 Energy storage devices

High demand of devices that stores energy like batteries and super capacitors have come forth in recent years. Two types of batteries are known rechargeable and non-rechargeable in which batteries that is rechargeable are mainly classified into lithium sulfur batteries, lithium-ion batteries, and supercapacitors [59–62]. An electrode with extemporized electrochemical behavior is needed to succeed in preconditions of a traditional battery. To take control of this matter, ZIF-obtained materials that is transition metal based can be practically employed for energy storage devices as an electrode material. Zhang et al. evolved a new avenue to develop ZnO@ZnO QDs/C NRAs on cloth of carbon substrate by use of material of water based ZIF-8 and put in as an anode for lithium-ion batteries. The material of ZIF expands high certain capacity and superb stability over various cycles [63]. ZIF-derived materials could be used for other types of energy storage batteries due to the extemporized electrochemical presentation of the materials.

10.4.2 Gas separation

ZIF composites or membranes have been recommended for separation of gases, their properties like flexibility over substituents, porosity, numerous structures, and having many chemical functionalities. Gases like carbon dioxide separation and adsorption in an enlarging area, in which they adsorbed only pure carbon dioxide gas, on the other hand the separation of carbon dioxide gas it is related with some other gas systems namely CO_2/NH_2, CO_2/N_2, and CO_2/H_2. The crystals present within the material of ZIFs have the ability of adsorbing carbon dioxide, and where ZIFs can come up with Langmuir sites. The adsorbed carbon dioxide molecule has restored those locations. Song et al. designed water based ZIF-8 nanoparticle of membrane based distributed between the polymers. This membrane of ZIF-8 can hugely expand the penetrability of carbon dioxide in the tests of gas absorption, that could intelligibly appear that the stuffing of ZIFs notably impacts the penetrability of carbon dioxide [1].

10.4.3 Drug delivery

Due to their sensitivity to pH-like properties, materials of ZIFs are thought to be one of the encouraging choices for the drug delivery party line. ZIF-8 made of water has been used as a cancer-fighting substance and modified to provide heat generation for other medications. Water based zeolitic imidazolate framework-8 was produced by Sun et al. and used with the cancer treatment 5-fluorouracil (5-FU) (around 660 mg/5 g).

The ZIF-8 substance appears to have exceptional anticancer potential and could be employed in the cancer treatment [64]. Adhikari et al. discovered that after successfully enclosing the anticancer medicine DOX in ZIF-8 and ZIF-7, when they link with lipid membranes and micelles, they can also function as a remarkable drug-liberating characteristics [65].

10.4.4 Catalysis

It has been discovered that water-based ZIF-8 works well as a catalyst for the trans-esterification of vegetable oils, the Friedel–Craft acylation reaction of benzoyl chloride (C_7H_5ClO) and anisole ($C_6H_5NH_2$), and for forming carbonates. Water based ZIFs hold a great potential to act as heterogeneous catalysts and are used in number of catalytic reactions. It is possible to accelerate the staging in the Knoevenagel condensation reaction between benzaldehyde (C_6H_5CHO) and malononitrile by using nanoparticles of ZIFs-8 ($C_3H_2N_2$) [66].

A more condensed list of ZIFs that can serve as catalysts for various organic processes can be found in the table below [67].

ZIF material	Additional materials	Reaction(s) catalyzed
ZIF-8	Nanoparticles of gold	Oxidation of both ketone and CHO groups
ZIF-8	Silver and gold nanoparticles with a core shell	4-nitrophenol reduction
ZIF-8	Nanoparticles of platinum, gold, and silver	Hydrogenation of n-hexene and oxidation of CO
ZIF-8	Nanoparticles of platinum	Alkene hydrogenation
ZIF-8	Dioxide nanotubes of platinum and titanium	Phenol degradation
ZIF-8	Nanoparticles of palladium	Aminocarbonylation
ZIF-8	Nanoparticles of iridium	Cyclohexene and phenylacetene hydrogenation
ZIF-8	Nanoparticles of ruthenium	Acetophenone's asymmetric hydrogenation
ZIF-8	Microspheres of iron oxide	Knoevenagel condensation
ZIF-8	Nanorods of Zn_2GeO_4	Conversion of Carbon dioxide
ZIF-65	Oxide of molybdenum	Breakdown of methyl orange and orange II dyes

10.4.5 Sensing and electronic devices

Because of their integrated adsorbance qualities, they are a strong option for chemical sensors. ZIFs exposed to mixture of C_2H_5OH and water vapor ZIFs exhibit sensitivity, and this behavior is reliant on the C_2H_5OH mixture concentration [68]. ZIFs are also intriguing building blocks for biosensors that function similarly to electrochemical biosensors for in vivo electrochemical quantifications. ZIFs may potentially be used in exploration since they are luminous and can be used to detect metal ions and tiny molecules. Zeolitic imidazolate framework-8 luminescence based on water is highly delicate to ions and acetone. Fluorescently marked single stranded DNA can also be detected in zeolitic imidazolate framework (ZIF) nanoparticles [68].

10.5 Comparison of zeolitic imidazolate frameworks with other compounds

10.5.1 Zeolitic imidazolate frameworks versus metal–organic frameworks

Despite being a subclass of hybrids of organic–inorganic MOFs that combine organic linkers and frameworks of metal to produce hybrid micro pores and structures with high crystallinity, ZIFs are highly crowded structurally. Similar to MOFs, the majority of ZIFs attributes depend very little on the characteristics of ligands, metal clusters, and the synthetic settings in which they are created [69].

However, due to control in designing production and method, the majority of ZIFs modifications up to now, have involved switching the linkers—bridging oxygens—anions and ligands that is based on imidazolate ions [70]—or merging two kinds of organic linkers to change pore size or bond angles [71]. A large section of switching linkers incorporated functional groups addition with numerous symmetries and imidazolate ligand polarities to change the ZIFs adsorption of CO_2 capability without interrupting the transition metals [72]. Comparing this thing to MOFs, which possess a very much larger level of diversity on the bases of their building unit types. Zeolitic imidazolate frameworks offer specific qualities that set them out as particularly relevant to carbon capture processes, despite their similarity to other MOFs. Because ZIFs resemble the crystalline structure of zeolites, they have substantially higher chemical and thermal stability than other MOFs, allowing them to operate over a larger temperature range and being suitable for chemical operations [69]. It is known that ZIFs, characteristics like water repellent properties and stability in water are the most noticeable things. A main matter with zeolites and those of MOFs, to a definite stretch, was their H_2O adsorption along with carbon dioxide. Water vapor is frequently discover in carbon-rich tire out gases, and MOFs would absorb H_2O, lowering the amount of carbon dioxide needed to make it to saturation [69]. As a result of metal–oxygen bonds in situations where hydrolysis is taking place that are humid and oxygen-rich, MOFs are also not much more stable. However, ZIFs operate virtually equally well in dry and humid environments and have a significantly higher CO_2 alertness than H_2O, enabling the adsorbent to store additional carbon before reaching saturation [71].

10.5.2 Zeolite imidazolate frameworks versus commercially available products

Zeolite imidazolate frameworks being hydrophobic have the most unique property in comparison to other materials. In some conditions like dry condition ZIFs when compared, activated carbon was almost similar with its grasp capacity [72]. Nevertheless, the activated carbon uptake by ZIFs reduced to half in wet condition. When this saturation and reanimation tests were slide at the given circumstances, ZIFs have also shown very little to no structural humiliation, re-usability of adsorbent is a good indication [71]. However, ZIFs are expensive to synthesize. MOFs need long reaction periods to synthesize, under high pressure and temperatures, which is not found to be very easy to scale up [69]. Compared to commercially available non-ZIF MOFs, ZIFs have a tendency to be more affordable. Research determined that membranes of hybrid polymer-ZIFs sorbent no prolonged supported the chained of the Robeson plot of selectivity as a function of permeability for membrane gas separation when polymer-sorbent materials combined [70].

10.6 Conclusion and future outlook

Research and development on ZIFs, including their synthesis, characterization, and applications, have proliferated recently. ZIFs can be successfully made utilizing a number of synthetic approaches due to their intrinsic porosity and structures that resemble those of conventional aluminosilicate zeolites. Although water-based ZIFs synthesis and applications have undergone tremendous advancement, there will undoubtedly be more cutting-edge synthesis methods and application fields in the future.

In this chapter, the origin of ZIFs, a synthesis procedure which is usually available to synthesize pure water based ZIFs, has briefly been discussed. In addition, the exceptional chemical and thermal stability of water based, and non-water based ZIFs and the application of ZIFs and their materials in many fields with their superiority in terms of properties like drug delivery, gas separation, catalysis, sensing, and electronic devices have also been discussed.

Due to the diversity of organic linkers and link–link interactions, which enables the development of rich chemistry, there are significantly more opportunities to build ZIFs than there are with typical zeolites. To develop new synthesis pathways and look into the prospective applications of ZIFs, it is essential to draw on the expertise and knowledge acquired from other research areas, such as zeolites. Despite having a highly distinct evolutionary history, ZIFs continue to receive more attention and have shown significant growth over the past five years. Given the intense interest that various research groups have in ZIFs, we anticipate that new ideas and methods for ZIFs synthesis as well as uncharted uses for the rapidly evolving distinctive materials, will keep developing in the future.

References

[1] Q. Song, S.K. Nataraj, M.V. Roussenova, J.C. Tan, D.J. Hughes, W. Li, P. Bourgoin, M.A. Alam, A.K. Cheetham, S.A. Al-Muhtaseb, et al., Zeolitic imidazolate framework (ZIF-8) based polymer nanocomposite membranes for gas separation, Energy Environ. Sci. 5 (8) (2012) 8359–8369.

[2] C. Baerlocher, W.M. Meier, D.H. Olson, Atlas of Zeolite Framework Types, fifth ed., Elsevier, Amsterdam, 2001.

[3] A. Phan, A. Doonan, F.J. Uribe-Romo, C.B. Knobler, M. O'Keeffe, O.M. Yaghi, Synthesis, structure, and carbon dioxide capture properties of zeolitic imidazolate frameworks, Acc. Chem. Res. 43 (1) (2010) 58–67.

[4] M. Sturm, F. Brandl, D. Engel, W. Hoppe, Die Kristallstruktur von Diimidazolylkobalt, Acta Crystallogr. B 31 (1975) 2369–2378.

[5] R. Lehnert, F.Z. Seel, Topology analysis of metal-organic frameworks based on metal-organic polyhedra as secondary or tertiary building units, Z. Anorg. Allg. Chem. 464 (1980) 187–194.

[6] S.J. Rettig, A. Storr, D.A. Summers, R.C. Thompson, J. Trotter, Transition metal azolates from metallocenes. 2. Synthesis, X-ray structure, and magnetic properties of a three-dimensional polymetallic iron(II) imidazolate complex, a low-temperature weak ferromagnet, Can. J. Chem. 77 (1999) 425–433.

[7] Y.Q. Tian, C.X. Cai, J. Ji, X.Z. You, S.M. Peng, G.H. Lee, Reproducible synthesis and high porosity of mer-Zn(Im)2 (ZIF-10): exploitation of an apparent double-eight ring template, Angew. Chem. Int. Ed. 41 (2002) 1384–1386.
[8] Y.Q. Tian, C.X. Cai, X.M. Ren, C.Y. Duan, Y. Xu, S. Gao, X.Z. You, Zeolite CAN and AFI-type zeolitic imidazolate frameworks with large 12-membered ring pore openings synthesized using bulky amides as structure-directing agents, Chem. Eur. J. 9 (2003) 5673–5685.
[9] N. Masciocchi, S. Bruni, E. Cariati, F. Cariati, Extended polymorphism in copper(II) imidazolate polymers: a spectroscopic and XRPD structural study, Inorg. Chem. 40 (2001) 5897–5905.
[10] X.C. Huang, J.P. Zhang, X.M. Chen, Design and synthesis of zeolitic tetrazolate-imidazolate frameworks, Chin. Sci. Bull. 48 (2003) 1531–1534.
[11] X.C. Huang, Y.Y. Lin, J.P. Zhang, X.M. Chen, Ligand-directed strategy for zeolite-type metal-organic frameworks: zinc(II) imidazolates with unusual zeolitic topologies, Angew. Chem. Int. Ed. 45 (2006) 1557–1559.
[12] S.A. Moggach, T.D. Bennett, A.K. Cheetham, The effect of pressure on ZIF-8: increasing pore size with pressure and the formation of a high-pressure phase at 1.47 GPa, Angew. Chem. 121 (2009) 7221–7223.
[13] D. Fairen-Jimenez, S.A. Moggach, M.T. Wharmby, P.A. Wright, S. Parsons, T. Düren, Opening the gate: framework flexibility in ZIF-8 explored by experiments and simulations, J. Am. Chem. Soc. 133 (2011) 8900–8902.
[14] F. Wang, Y.X. Tan, H. Yang, H.X. Zhang, Y. Kang, J. Zhang, A new approach towards tetrahedral imidazolate frameworks for high and selective CO_2 uptake, Chem. 47 (2011) 5828–5830.
[15] S.Y. Yang, L.S. Long, Y.B. Jiang, R.B. Huang, L.S. Zheng, Exceptional chemical and thermal stability of zeolitic imidazolate frameworks, Chem. Mater. 14 (2002) 3229–3231.
[16] N. Masclocchi, G.A. Ardizzoia, G. LaMonica, A. Maspero, A. Sironi, Exceptional chemical and thermal stability of zeolitic imidazolate frameworks, Eur. J. Inorg. Chem. (2000) 2507–2515.
[17] K. Park, Z. Ni, A. Côté, J. Choi, R. Huang, F. Uribe-Romo, H. Chae, M. O'Keeffe, O. Yaghi, Exceptional chemical and thermal stability of zeolitic imidazolate frameworks, Proc. Natl. Acad. Sci. U.S.A. 103 (2006) 10186–10191. https://doi.org/10.1073/pnas.0602439103.
[18] D.W. Breck, Zeolite molecular sieves: structure, chemistry, and use, Wiley, New York, 1974, pp. 593–724.
[19] M. Kruk, M. Jaroniec, A. Sayari, Adsorption study of surface and structural properties of MCM-41 materials of different pore sizes, J. Phys. Chem. B 101 (1997) 583–589.
[20] D. Zhao, Q. Huo, J. Feng, B.F. Chmelka, G.D. Stucky, Nonionic triblock and star diblock copolymer and oligomeric surfactant syntheses of highly ordered, hydrothermally stable, mesoporous silica structures, J. Am. Chem. Soc. 120 (1998) 6024–6036.
[21] A.G. Wong-Foy, A.J. Matzger, O.M. Yaghi, Exceptional H_2 saturation uptake in microporous metal-organic frameworks, J. Am. Chem. Soc. 128 (2006) 3494–3495.
[22] H.K. Chae, D.Y. Siberio-Perez, J. Kim, Y.B. Go, M. Eddaoudi, A.J. Matzger, M. O'Keeffe, O.M. Yaghi, A route to high surface area, porosity and inclusion of large molecules in crystals, Nature 427 (2004) 523–527.
[23] Z. Zhang, Y. Han, L. Zhu, R. Wang, Y. Yu, S. Qiu, D. Zhao, F.S. Xiao, Mesoporous zeolitic materials (MZMs) derived from zeolite Y using a microwave method for catalysis, Angew. Chem. Int. Ed. 40 (2001) 1258–1262.
[24] R.J. Sundberg, R.B. Martin, Interactions of histidine and other imidazole derivatives with transition metal ions in chemical and biological systems, Chem. Rev. 74 (1974) 471–517.

[25] H. Hayashi, A.P. Côté, H. Furukawa, et al., Zeolite A imidazolate frameworks, Nat. Mater. 6 (7) (2007) 501–506.

[26] R. Banerjee, A. Phan, Bo. Wang, et al., High-throughput synthesis of zeolitic imidazolate frameworks and application to CO_2 capture, Science 319 (5865) (2008) 939–943.

[27] Bo Wang, A.P. Côté, H. Furukawa, et al., Colossal cages in zeolitic imidazolate frameworks as selective carbon dioxide reservoirs, Nature 453 (7192) (2008) 207–211. https://doi.org/10.1038/nature06900.

[28] D. Madhav et al., Synthesis of nanoparticles of zeolitic imidazolate framework ZIF-94 using inorganic deprotonators. New J. Chem. 44 2020, 20449–20457.

[29] X.C. Huang, Y.Y. Lin, J.P. Zhang, X.M. Chen, Ligand-directed strategy for zeolite-type metal-organic frameworks: zinc(II) imidazolates with unusual zeolitic topologies, Angew. Chem. Int. Ed Engl. 45 (10) (2006) 1557–1559. https://doi.org/10.1002/anie.200503778.

[30] J. Cravillon, S. Münzer, S.-J. Lohmeier, et al., Rapid room temperature synthesis and characterization of nanocrystals of a prototypical zeolitic imidazolate framework, Chem. Mater. 21 (8) (2009) 1410–1412.

[31] M. He, J. Yao, L. Li, et al., Synthesis of zeolitic imidazolate framework-7 in a water/ethanol mixture and its ethanol-induced reversible phase transition, Chempluschem 78 (10) (2013) 1222–1225.

[32] T.D. Bennett, P.J. Saines, D.A. Keen, et al., Ball-milling induced amorphization of zeolitic imidazolate frameworks (ZIFs) for the irreversible trapping of iodine, Chemistry 19 (22) (2013) 7049–7055.

[33] Y. Pan, Y. Liu, G. Zeng, et al., Rapid synthesis of zeolitic imidazolate framework-8 (ZIF-8) nanocrystals in an aqueous system, Chem. Commun. 47 (7) (2011) 2071–2073.

[34] S. Tanaka, K. Kida, M. Okita, et al., Size-controlled synthesis of zeolitic imidazolate framework-8 (ZIF-8) crystals in an aqueous system at room temperature, EPA Newsl. 41 (10) (2012) 1337–1339. https://doi.org/10.1246/cl.2012.1337.

[35] K. Kida, M. Okita, K. Fujita, et al., Formation of high crystalline ZIF-8 in an aqueous solution, CrystEngComm 15 (9) (2013) 1794.

[36] T. Yang, T.-S. Chung, Room-temperature synthesis of ZIF-90 nanocrystals and the derived nano-composite membranes for hydrogen separation, J. Mater. Chem. A 1 (19) (2013) 6081. https://zenodo.org/record/998611.

[37] Y. Ban, Y. Li, X. Liu, Y. Peng, W. Yang, Solvothermal synthesis of mixed-ligand metal–organic framework ZIF-78 with controllable size and morphology, Microporous Mesoporous Mater. 173 (29-36) (2013) 1387–1811.

[38] J. Cravillon, C.A. Schröder, H. Bux, et al., Formate modulated solvothermal synthesis of ZIF-8 investigated using time-resolved in situ X-ray diffraction and scanning electron microscopy, CrystEngComm 14 (2) (2011) 492–498. http://www.repo.uni-hannover.de/handle/123456789/2191.

[39] D. Peralta, G. Chaplais, A. Simon-Masseron, K. Barthelet, G.D. Pirngruber, Synthesis and adsorption properties of ZIF-76 isomorphs, Microporous Mesoporous Mater. 153 (2012) 1–7. https://hal.archives-ouvertes.fr/hal-01589660/file/article.pdf.

[40] J. Yao, M. He, K. Wang, et al., High-yield synthesis of zeolitic imidazolate frameworks from stoichiometric metal and ligand precursor aqueous solutions at room temperature, CrystEngComm 15 (18) (2013) 3601.

[41] F.-K. Shieh, S.-C. Wang, S.-Y. Leo, K.C.-W. Wu, Water based synthesis of zeolitic imidazolate framework-90 (ZIF-90) with a controllable particle size, Chemistry 19 (34) (2013) 11139–11142.

[42] S.K. Nune, P.K. Thallapally, A. Dohnalkova, et al., Synthesis and properties of nano zeolitic imidazolate frameworks, Chem. Commun. 46 (27) (2010) 4878–4880. https://zenodo.org/record/1230000.

[43] B. Seoane, J.M. Zamaro, C. Tellez, J. Coronas, Sonocrystallization of zeolitic imidazolate frameworks (ZIF-7, ZIF-8, ZIF-11 and ZIF-20), CrystEngComm 14 (9) (2012) 3103.

[44] H.-Y. Cho, J. Kim, Se-Na Kim, W.-S. Ahn, High yield 1-L scale synthesis of ZIF-8 via a sonochemical route, Microporous Mesoporous Mater. 169 (2013) 180–184.

[45] H. Bux, F. Liang, Y. Li, et al., Zeolitic imidazolate framework membrane with molecular sieving properties by microwave-assisted solvothermal synthesis, J. Am. Chem. Soc. 131 (44) (2009) 16000–16001.

[46] F. Hillman, J.M. Zimmerman, P. Seung-Min, et al., Rapid microwave-assisted synthesis of hybrid zeolitic–imidazolate frameworks with mixed metals and mixed linkers, J. Mater. Chem. A 5 (13) (2017) 6090–6099.

[47] T.D. Bennett, S. Cao, J.C. Tan, et al., Facile mechanosynthesis of amorphous zeolitic imidazolate frameworks, J. Am. Chem. Soc. 133 (37) (2011) 14546–14549.

[48] I. Stassen, M. Styles, G. Grenci, et al., Chemical vapour deposition of zeolitic imidazolate framework thin films, Nat. Mater. 15 (3) (2016) 304–310. https://lirias.kuleuven.be/handle/123456789/551545.

[49] P. López-Domínguez, A.M. López-Periago, F.J. Fernández-Porras, et al., Supercritical CO_2 for the synthesis of nanometric ZIF-8 and loading with hyperbranched aminopolymers. Applications in CO_2 capture, J. CO_2 Util. 18 (2017) 147–155.

[50] Y. Pan, Y. Liu, G. Zeng, L. Zhao, Z. Lai, Rapid synthesis of zeolitic imidazolate framework-8 (ZIF-8) nanocrystals in an aqueous system, Chem. 47 (2011) 2071–2073.

[51] S. Tanaka, K. Kida, M. Okita, Y. Ito, Y. Miyake, Size-controlled synthesis of zeolitic imidazolate framework-8 (ZIF-8) crystals in an aqueous system at room temperature, Chem. Lett. 41 (2012) 1337–1339.

[52] K. Kida, M. Okita, K. Fujita, S. Tanaka, Y. Miyake, Formation of high crystalline ZIF-8 in an aqueous solution, CrystEngComm 15 (2013) 1794–1801.

[53] J. Qian, F. Sun, L. Qin, Hydrothermal synthesis of zeolitic imidazolate framework-67 (ZIF-67) nanocrystals, Mater. Lett. 82 (2012) 220–223.

[54] A.F. Gross, E. Sherman, J.J. Vajo, Aqueous room temperature synthesis of cobalt and zinc sodalite zeolitic imidizolate frameworks, Dalton Trans. 41 (2012) 5458–5460.

[55] J. Yao, M. He, K. Wang, R. Chen, Z. Zhong, H. Wang, High-yield synthesis of zeolitic imidazolate frameworks from stoichiometric metal and ligand precursor aqueous solutions at room temperature, CrystEngComm 15 (2013) 3601–3606.

[56] F.-K. Shieh, S.-C. Wang, S.-Y. Leo, K.C.W. Wu, Economical, environmental friendly synthesis, characterization for the production of zeolitic imidazolate framework-8 (ZIF-8) nanoparticles with enhanced CO_2 adsorption, Chemistry 19 (2013) 11139–11142.

[57] B. Chen, F. Bai, Y. Zhu, Y. Xia, A cost-effective method for the synthesis of zeolitic imidazolate framework-8 materials from stoichiometric precursors via aqueous ammonia modulation at room temperature, Microporous Mesoporous Mater. 193 (2014) 7–14.

[58] M. He, J. Yao, Q. Liu, K. Wang, F. Chen, H. Wang, Facile synthesis of zeolitic imidazolate framework-8 from a concentrated aqueous solution, Microporous Mesoporous Mater. 184 (2014) 55–60.

[59] M. Pramanik, Y. Tsujimoto, V. Malgras, S.X. Dou, J.H. Kim, Y. Yamauchi, Mesoporous iron phosphonate electrodes with crystalline frameworks for lithium-ion batteries, Chem. Mater. 27 (3) (2015) 1082–1089.

[60] S.M. Hwang, Y.G. Lim, J.G. Kim, Y.U. Heo, J.H. Lim, Y. Yamauchi, M.S. Park, Y.J. Kim, S.X. Dou, J.H. Kim, A case study on fibrous porous SnO_2 anode for robust, high-capacity lithium ion batteries, Nano Energy 10 (2014) 53–62.

[61] K.N. Jung, J. Kim, Y. Yamauchi, M.S. Park, J.W. Lee, J.H. Kim, Rechargeable lithium-air batteries: a perspective on the development of oxygen electrodes, J. Mater. Chem. 4 (37) (2016) 14050–14068.

[62] J. Lee, J. Moon, S.A. Han, J. Kim, V. Malgras, Y.U. Heo, H. Kim, S.M. Lee, H.K. Liu, S.X. Dou, et al., Everlasting living and breathing gyroid 3D network in Si@SiOx/C nanoarchitecture for lithium ion battery, ACS Nano 13 (8) (2019) 9607–9619.

[63] G. Zhang, S. Hou, H. Zhang, W. Zeng, F. Yan, C.C. Li, H. Duan, High-performance and ultra-stable lithium-ion batteries based on MOF-derived ZnO@ZnO quantum dots/C core-shell nanorod arrays on a carbon cloth anode, Adv. Mater. 27 (14) (2015) 2400–2405.

[64] C.Y. Sun, C. Qin, X.L. Wang, G.S. Yang, K.Z. Shao, Y.Q. Lan, Z.M. Su, P. Huang, C.G. Wang, E.B. Wang, Zeolitic imidazolate framework-8 as efficient PH-sensitive drug delivery vehicle, Dalt. Trans. 41 (23) (2012) 6906–6909.

[65] C. Adhikari, A. Das, A. Chakraborty, Zeolitic imidazole framework (ZIF) nanospheres for easy encapsulation and controlled release of an anticancer drug doxorubicin under different external stimuli: a way toward smart drug delivery system, Mol. Pharmaceutics 12 (9) (2015) 3158–3166.

[66] Y. Guan, J. Shi, M. Xia, et al., Monodispersed ZIF-8 particles with enhanced performance for CO_2 adsorption and heterogeneous catalysis, Appl. Surf. Sci. 423 (2017) 349–353.

[67] A. Phan, C.J. Doonan, F.J. Uribe-Romo, et al., Synthesis, structure, and carbon dioxide capture properties of zeolitic imidazolate frameworks, Acc. Chem. Res. 43 (1) (2010) 58–67.

[68] B. Chen, Z. Yang, Y. Zhu, Y. Xia, Zeolitic imidazolate framework materials: recent progress in synthesis and applications, J. Mater. Chem. A 2 (40) (2014) 16811–16831.

[69] S.A. Basnayake, J. Su, X. Zou, K.J. Balkus. "Carbonate based zeolitic imidazolate frame for highly selective CO_2 capture". Inorganic Chem. 2015, 54 (4): 1816–1821.

[70] B. Smit, J.A. Reimer, C.M. Oldenburg, I.C. Bourg, Introduction to Carbon Capture and Sequestration, first ed., Imperial College Press, Hackensack, NJ, 2014.

[71] N.T.T. Nguyen, T.N.H. Lo, J. Kim, Mixed-metal zeolitic imidazolate frameworks and their selective capture of wet carbon dioxide over methane, Inorg. Chem. 55 (12) (2016) pp. 6201–6207. http://globalscience.berkeley.edu/sites/default/files/mm-zif-2016.pdf.

[72] S. Wang, X. Wang, Imidazolium ionic liquids, imidazolylidene heterocyclic carbenes, and zeolitic imidazolate frameworks for CO_2 capture and photochemical reduction, Angew. Chem. 55 (7) (2015) 2308–2320.

Chapter 11

Preparation and applications of water-based isoreticular metal–organic frameworks

Sami-Ullah Rather
Department of Chemical and Materials Engineering, King Abdulaziz University, Jeddah, Saudi Arabia

11.1 Introduction

Metal–organic frameworks (MOFs) are a combination of organic–inorganic blend crystalline highly porous type of materials that are made of a regular arrangement of positively charged metal ions surrounded by different types of organic linker molecules [1–3]. The existence of inorganic and organic constituents in the MOFs makes them very interesting for various kinds of applications particularly in storage, separation, catalyst, biomedical imaging, and sensors [4–12]. The different types of metal ions construct nodes that bind the branches of the linkers together to build repeating, cage-like structures as presented in Fig. 11.1. The large surface area extending beyond 7000 m^2/g and high pore volume is due to the existence of vast hollow structure [13]. Two different types of components namely aromatic materials and organic linkers also called building blocks to construct these types of materials. These two types of simple blocks build a very sophisticated complex material such as MOFs. The size and uniformity of cages, cavities, and windows developed inside MOFs can be controlled in a programmable way. The amalgam of two important constituents of a MOF including organic linker and metal ions or clusters brings forth unlimited opportunities [1–3].

The preparation of MOFs has attracted tremendous attention from researchers all over the world by their elevated surface area, easily moldable pore framework and accommodating inner surface characteristics, and numerous types of applications. Based on extending topology, there is another class of MOFs called isoreticular metal–organic frameworks (IRMOFs). Yaghi and co-workers designed and synthesized successfully a series of IRMOFs [14,15]. IRMOFs is a combination of Zn_4O tetrahedral with oxygen in the

Synthesis of Metal–Organic Frameworks via Water-Based Routes: A Green and Sustainable Approach.
DOI: https://doi.org/10.1016/B978-0-323-95939-1.00005-8

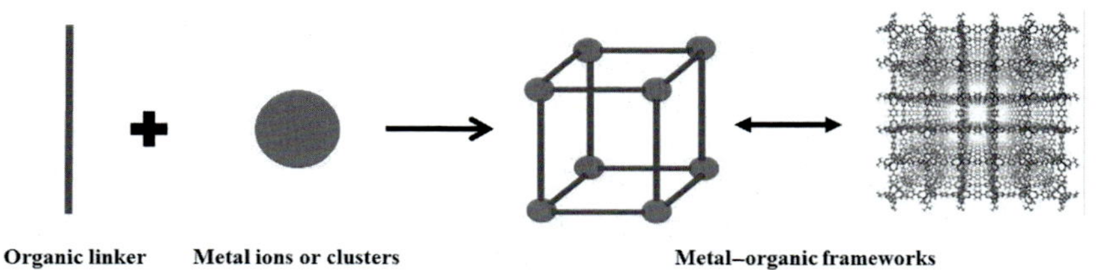

FIGURE 11.1 Schematic formation of metal–organic frameworks from organic linker and metal ions or clusters.

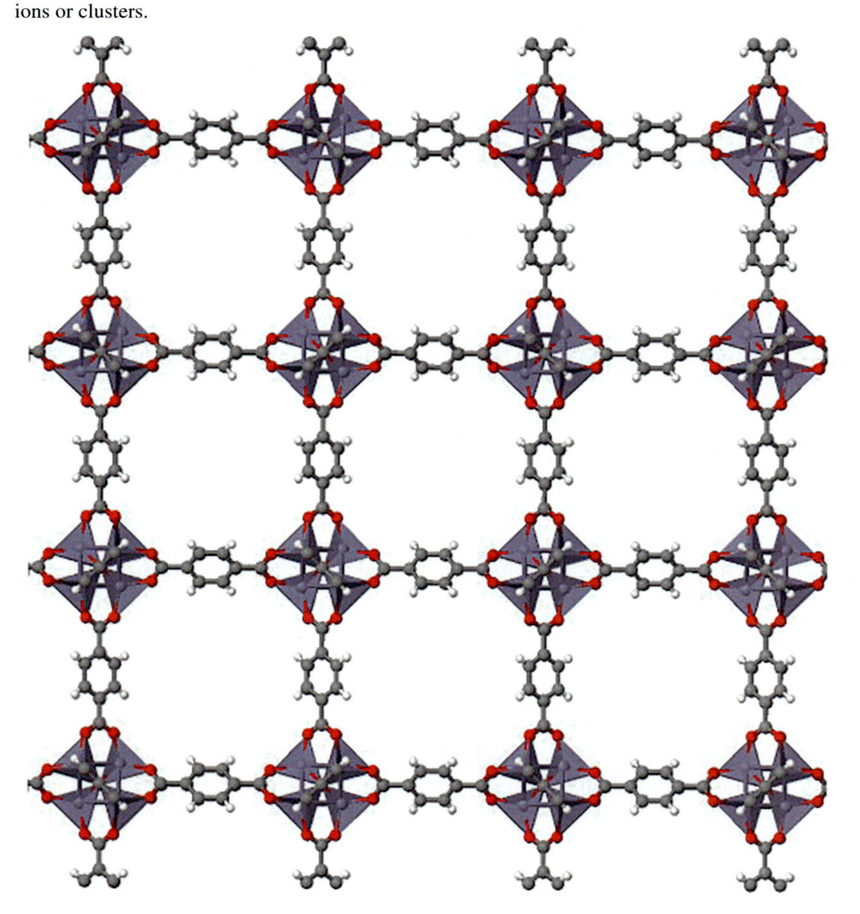

FIGURE 11.2 Isoreticular metal–organic framework constructed from Zn_4O nodes and 1,4-benzodicarboxylic acid ligands between nodes.

center attached with different types of dicarboxylate linkers forming a cubic-like three-dimensional (3D) highly porous framework as shown in Fig. 11.2 [16–18]. There are a huge number of IRMOFs based on the MOFs series. The IR stands for isoreticular, which means it is an array of MOFs with similar topology but

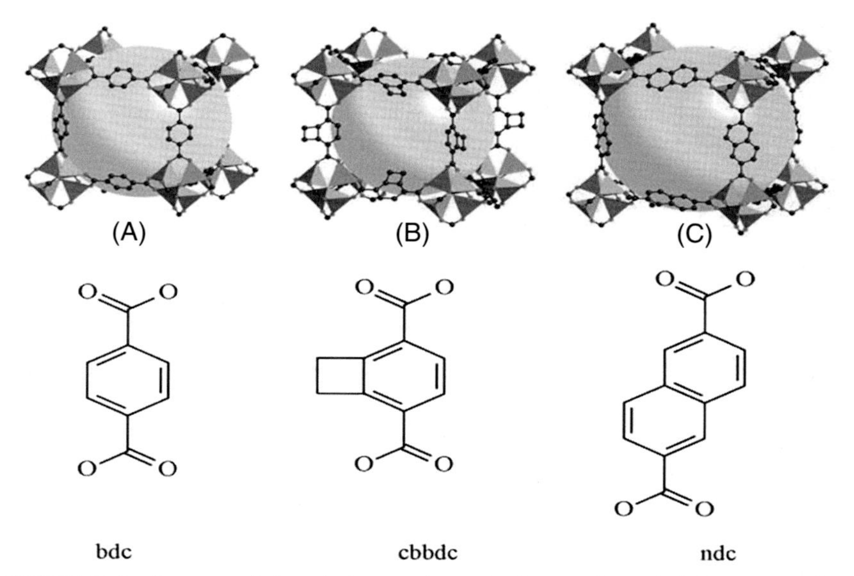

FIGURE 11.3 Structure of isoreticular metal–organic frameworks and their corresponding ligands: A = IRMOF-1, B = IRMOF-6, C = IRMOF-8 [17,18].

distinct types of pore diameters. The MOF-5 also called the IRMOF-1 composed of Zn_4O nodes and 1,4-benzodicarboxylic acid as a ligand is an early example of isoreticular MOFs as shown in Fig. 11.3 [17,18].

11.2 Preparation

The topology development of the IRMOFs and their significant characteristics such as ultrahigh BET surface area and tunable pore diameters to build different types of applications including storage, separation, sensors, biomedicine, and catalysts were the only consideration until recently [4–12]. However, both academic and industrial research are being utilized to develop cheap, environmentally friendly, and high-production preparation methods for IRMOFs. The topology, high surface area, and pore volume of IRMOFs possess numerous advantages over other kinds of crystalline materials [13]. The expensive precursors and organic solvents utilized in the synthesis method increase the production cost and toxicity. However, the utilization of water in the form of solvent as compared to organic solvent in the facile and green preparation method can curtail cost, energy, and environmental hazards which can subsequently reduce production cost [19,20]. For the synthesis of toxic or hazardous-free IRMOFs, different types of synthesis methods were recommended to replace organic-based solvents with water-based solvents. Various routes such as microwave irradiation, ultrasound-assisted, microwave-assisted solvothermal, and in situ-growth methods not only eliminate hazardous solvents in the preparation method but also boost porosity tenability, stability, and exceptional production rate [21–24]. However, these preparation methods involve complications, little generality, and expensive solvents. Therefore, a facile, efficient, and cheap synthesis method for IRMOFs without utilizing an organic solvent is mandatory.

Water as a solvent for the synthesis of IRMOFs is considered the perfect choice owing to its low cost and environmental friendliness. The solubility of IRMOFs precursor in water as a solvent is free from its toxic nature, economical, stable and easy to obtain, and disposable as compared to organic solvent. The organic solvent utilized for the synthesis of IRMOFs is difficult to eliminate from channels and pores while water is easy to dispose of from pores and channels. Moreover, the easy elimination of water from pores and channels of IRMOFs enhances overall material characteristics. The water-based medium solvent utilized for the synthesis of most of the IRMOFs is stable. The improvement of structure and characteristics of IRMOFs are also related to synthesis when water is used as a solvent. The production of cheap IRMOFs from pristine and renewable water mediums is possible on an industrial scale. Furthermore, the development of IRMOFs with unique structure and properties are also possible by utilizing the academicians to open new research areas. Sometimes both water and organic solvents are employed for the synthesis of aqueous base IRMOFs, however, the amount of the latter is very small. Based on the recognition of water as an excellent solvent owing to its various unique characteristics mentioned above, in the following section, various routes utilized for the synthesis of some IRMOFs are discussed.

The two main differences between IRMOFs and MOFs are (1) the decoration of pores by functional groups and (2) the measurement of pore structure frameworks. In the water-based synthesis of IRMOF-2 to IRMOF-7, benzene dicarboxylic acid interconnection with cyclobutyl, amino n-propoxy, bromo, and assimilated benzene groups blend into an adaptable framework where functional groups point inside the voids are displayed in Fig. 11.4 [17,18]. Moreover, different types of carboxylate bonds diversified in their prominent functional groups while this characteristic largely remains absent in other crystalline solids and porous materials. The same series of IRMOFs presented in Fig. 11.4 also illustrates the pore enlargement within the outlook of chemistry as explained by the presented framework structures of IRMOF-8 to IRMOF-12 in which longer links have been successfully utilized. A series of renewable aqueous synthesized IRMOFs including MOF-5 also called IRMOF-1, MOF-74, MOF-177, and HKUST-1 known as Cu-BTC or MOF-199 has been extensively examined [5,15,17,18,23,25]. The pure metal–oxygen coordination of HKUST-1 or CU-BTC or MOF-199 shown in Fig. 11.5 is vulnerable to water molecules, which makes it unstable upon exposure to water content. However, HKUST-1 synthesis is still possible in an aqueous medium. Huo et al. reported easy room temperature (RT) route synthesis of MOF-199 via dynamic mingling of a metal salt such as cupric acetate anhydrous [$Cu(OAC)_2$] with overload benzene-1,3,5-tricarboxylic acid [H_3BTC] employing water as a solvent [25]. Modulation of copper origin

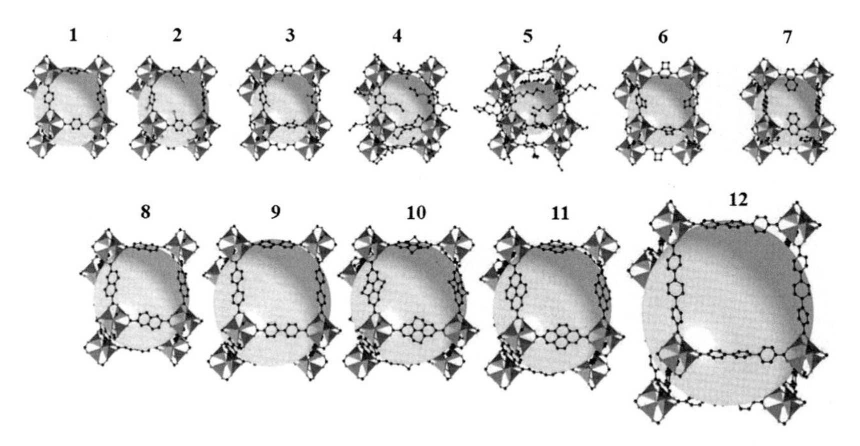

FIGURE 11.4 The single crystal X-ray framework of IRMOF-n (n = 1–12). Zn (*blue polyhedra*), O (*red spheres*), C (*black spheres*), Br (*green spheres in 2*), amino-groups (*blue spheres in 3*) presents color scheme. The large yellow spheres acts as the largest van der Waals spheres that would meet in the chamber without distressing the stable frameworks. The hydrogen atoms of all isoreticular metal–organic frameworks have been discontinued and only one direction of dislocated atoms is displayed for clarity [17,18].

and reaction direction tuned the crystallize size and porosity characteristics of HKUST-1. Furthermore, the preparation can be easily scaled-up so that the synthesized HKUST-1 acquires efficient space-time yield (STY) attributing to the pristine aqueous medium such as water and compressing crystallization duration. However, in the process of purification step of as produced HKUST-1, unreacted H3BTC is eliminated utilizing ethanol; therefore, the use of hazardous organic-related solvents has not been effectively eliminated from the entire procedure. Siew et al. revealed that during the preparation of HKUST-1, a little amount of methanol was exchanged with the existing ethanol during the solvent activation process [26]. Furthermore, Gerardo and Javier utilized an ethanolic-based medium in the form of solvent to synthesize MOF-199 [27]. Recently according to the literature, HKUST-1 was prepared by using water as a solvent and a surplus amount of methylamine as a supplement. To obtain a yield of 89%, the preparation time was curtailed only to 5 minutes. Even though HKUST-1 activation was accomplished in distilled water, methylamine used in the preparation method was not helpful to eliminate expensive and adverse organic solvents. Therefore, it indicates that water stability and IRMOFs synthesized in the water atmosphere are not interconnected. Stephane and co-workers reported another kind of IRMOF such as MOF-74 preparation in a water-based medium [28]. The synthesis involves combining a nickel (Ni) starting material solution with a 2,5-dihydroxyterephthalic acid (H_4dhtp) solution at high temperatures. Furthermore, a similar kind of reaction method was utilized by Garzón-Tovar et al. in an optimized condition where a list of MOF-74 including Mg, Ni, Co, and Zn in water at RT was obtained [29]. The as-produced MOF-74-M acquires a

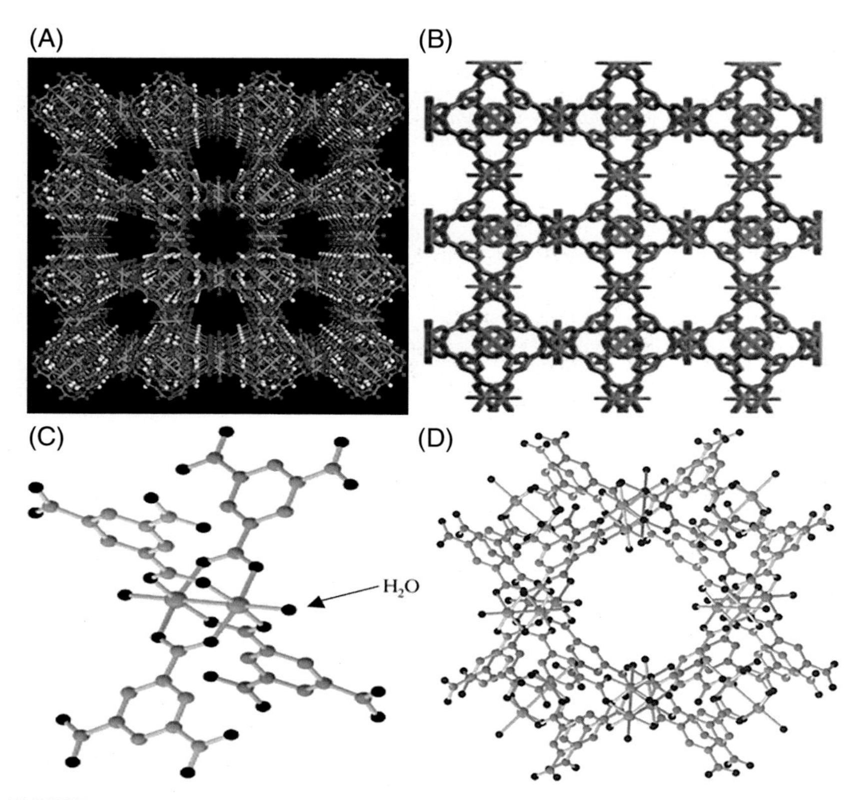

FIGURE 11.5 (A and B) structure of HKUST-1 (Cu-BTC) (C) structure of the Cu-paddle wheel in $Cu_3(BTC)_2(H_2O)_3$, and (D) view inside the pore of $Cu_3(BTC)_2(H_2O)_3$.

high surface area (>1200 m^2/g) and remarkable STYs (18,720 kg/m^3 per day). Didriksen et al. advanced a simple continuous-flow preparation process, wherein MOF-74-Ni inside a tubular reaction setup using water as a solvent was performed [30]. At moderate circumstances, transformation above 90% was achieved in approximately 20 minutes with high crystallinity and surface area. A mechanical approach method such as high-energy ball milling for the preparation of MOF-74-Zn was developed in which a blend of zinc oxide, ligand dihydroxyterephthalic acid, and little amount of water was used [31]. Another zeolitic imidazolate framework (ZIF-93) based IRMOF was prepared using a stoichiometric combination of metal-ligand in the water-based medium at RT with a yield of 80% of the pristine product. The same product prepared by a solvothermal method using DMF instead of water confirms the same material with similar properties [8]. However, the water-based synthesis method is cheap and environmentally friendly. Another series of water medium IRMOFs based on aluminum blend-linker MOFs were prepared by balancing the linker amount. In IRMOF $[Al(OH)(X)_a(Y)_{1-a}]$, where X stands for IPA (isophthalate) and

Y stands for FDC (2,5-furandicarboxylate) [24]. The CAU-10-H and MIL-160 based on the Al-linker series are prepared by utilizing the different ratios of linkers as shown in Fig. 11.6. The enhancing portion of IPA in MIL-160 generates the display of characteristics bands for IPA e.g. vibrations at 745/cm and 723/cm. The BET-specific surface area of CAU-10-H and MIL-160 was found to be 1153 m^2/g and 668 m^2/g. Moreover, it was reported that the BET surface area of mixed linker IRMOF was higher than pristine CAU-10-H and MIL-160. This type of synthesis process can be outlined for the green and continuous preparation of IRMOFs and have the can industrial production.

Another series of IRMOFs based on lanthanide elements was developed where Eu, Gd, and Tb elements were used [32]. Lanthanide IRMOFs directly developed on the cotton continuous fibers at RT utilizing a water-based precipitation method as shown in Fig. 11.7. In this preparation method equimolar amount of aqueous Ln^{3+} salts and 1,3,5-benzenetricarboxylic acid which facilitates active specific crystallization of high quantity of Ln-IRMOF on cotton continuous fibers. One-dimensional (ID) slim, continuous, and dense wire-like framework achieved. Eu-, Gd-, and Tb-based IRMOFs formed were strongly inclined to the cotton continuous fibers by keeping their crystal structure intact. Under the influence of ultraviolet-C (UVC) exposure, intensified emissions of red, blue, and green were obtained for Eu-, Gd-, and Tb-IRMOFs as shown in Fig. 11.8. In the application perspective, these types of Ln-IRMOFs can be used in protecting cloth textile-based sensors and smart tagging. Tao et al. reported another series of water-based IRMOFs synthesis based on cobalt-adeninate bio-MOFs such as bio-MOF-11, bio-MOF-12, bio-MOF-13, and bio-MOF-14. The bio-MOF 11-14 is embedded with valerate, butyrate, acetate, and propionate respectively [6]. In the bio-MOF series, it was observed that as the aliphatic chain size elevates, the surface area decreases from 1148 m^2/g to 17 m^2/g. The water stability of bio-MOFs enhanced upon increasing the aliphatic chain size. Furthermore, no effect of porosity or crystallinity upon exposure to water for several days in bio-MOF-14 was observed.

11.3 Applications

The highly porous water-based IRMOFs are utilized for different types of main applications including adsorption and separation, catalysts, sensors, and biomedical. In the following section, the above-mentioned applications are discussed in detail.

11.3.1 Adsorption and separation

Several advantages are possible by utilizing different types of water-based IRMOFs in adsorption and separation:

- Several IRMOFs are chemically, thermally, and physically well stable to utilize as an adsorbent in difficult environmental circumstances.

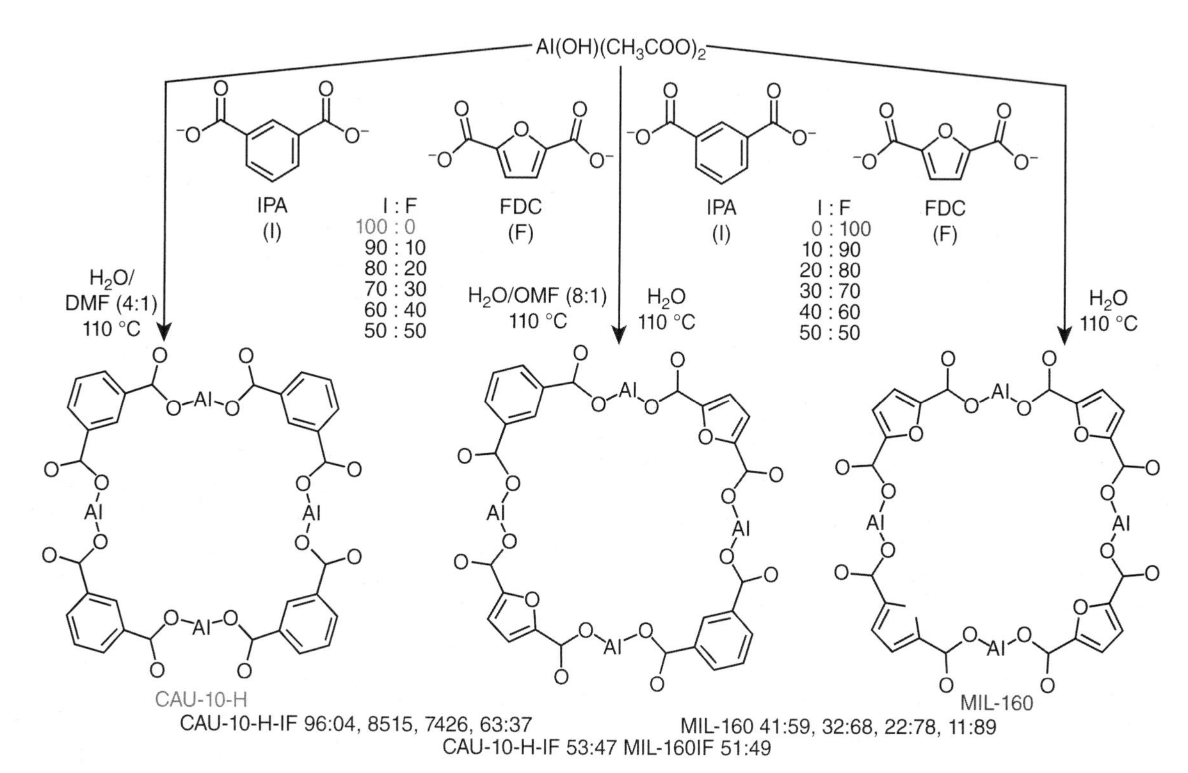

FIGURE 11.6 Schematic pattern of mixed-linker isoreticular metal–organic frameworks by alternating the linker proportions [24].

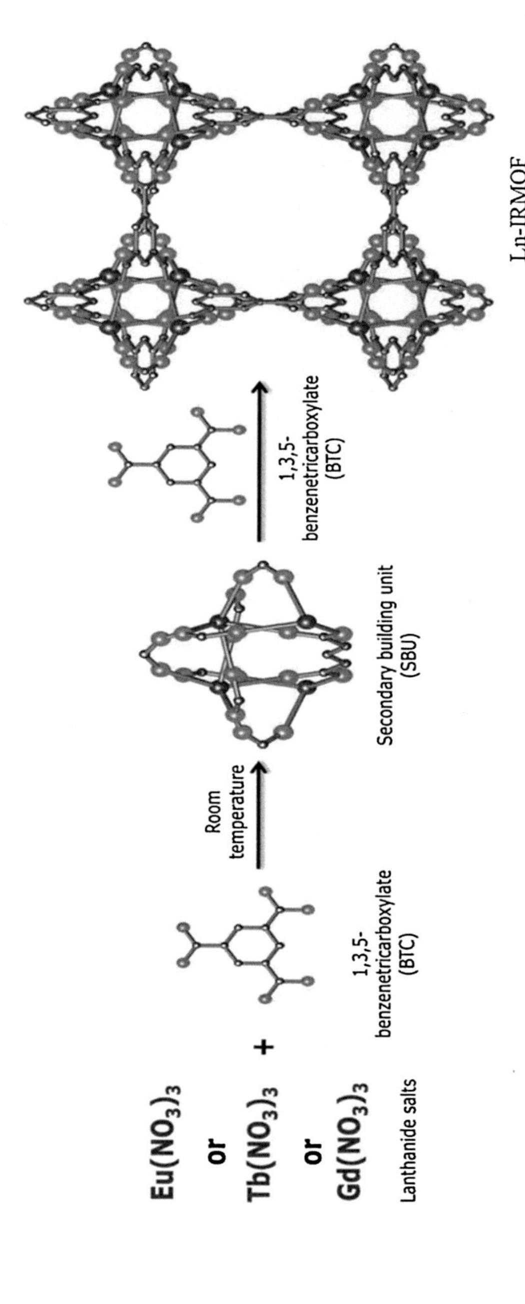

FIGURE 11.7 Schematic preparation method of Eu-, Tb-, and Gd-IRMOF by water-based precipitation methods [32].

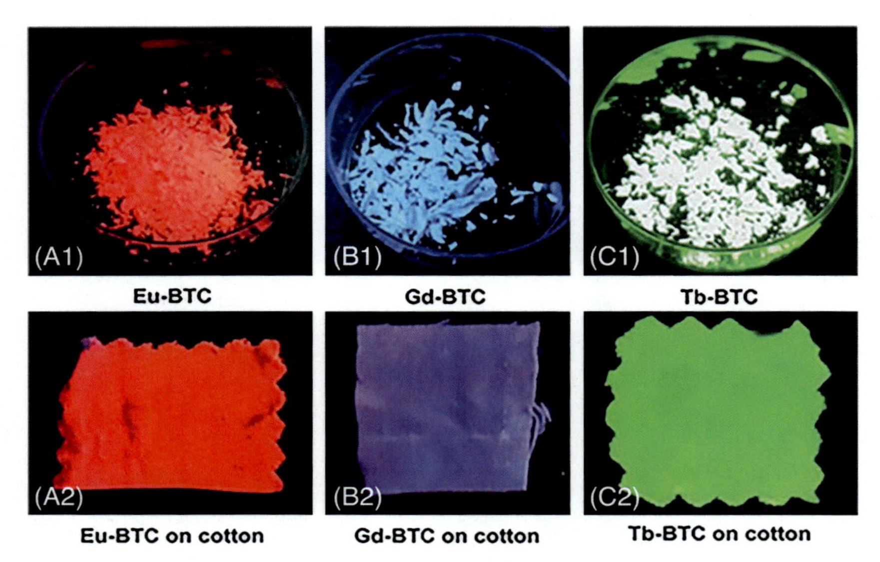

FIGURE 11.8 Red, blue, and green profiles of Eu-, Gd-, and Tb-IRMOFs under the influence of ultraviolet-C exposure [32].

- IRMOFs possess ultrahigh BET-specific surface area and high pore volume, which is very important in both adsorption and separation.
- Availability of structural tunability in IRMOFs allowed increasing adsorption capacity at moderate conditions.
- Alteration of organic linkers in IRMOFs expands the size of pores without deforming topology.

Different types of gases such as carbon dioxide, hydrogen, benzene, sulfur dioxide, ammonia, chlorine, dichloromethane, and ethylene dioxide and water are adsorbed/absorbed in various kinds of water-based synthesized IRMOFs. The elevated surface area, high pore volume, and cage-like framework make IRMOFs the best adsorbent/absorbent for gases [13,24]. IRMOFs bearing the same characteristics as MOFs provide a vital platform for different types of gases to adsorb/absorb efficiently. The IRMOF such as HKUST-1 was first utilized for the adsorption of water in 2002 owing to its open-metal location that acquires a bimodal pore allotment and smaller pore pockets [33]. Küsgens et al. also announced the adsorption of water in HKUST-1 with intense study along with other types of IRMOFs [34]. Furthermore, both Wang and Küsgens et al. found water adsorption isotherm following type I form [33,34]. Open metal areas present in the HKUST-1 adsorb water into different types of hydrophilic pores at subsided relative pressure (P/Po) which subsequently attains a saturation of ~25 mol/kg (45 wt.%). This adsorption isotherm attitude is the same as microporous zeolites 5A and 13X as shown in Fig. 11.9 [35]. Furthermore, Liang et al., Henninger, Schoenecker, and their teams observed identical water adsorption

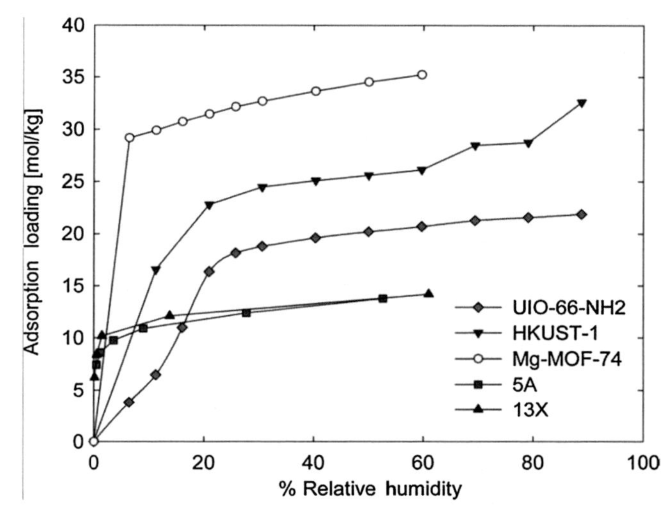

FIGURE 11.9 Comparison profile of water adsorption isotherm of UiO-66-NH_2, HKUST-1, and Mg-MOF-74 interconnected with zeolites 5A and 13X [35].

isotherm performances in HKUST-1 [36–39]. Li et al. observed that HKUST-1 is stable upon exposure to air for 7 days at ~40% RH. Moreover, at 40°C, no structural variation in HKUST-1 was found post-exposed to 40% RH for 2 weeks [40]. Decoste et al. reported the aging study of HKUST-1 embedding Mg-MOF-74. Moreover, in all circumstances, it was found that the structure decomposes [41].

Efficient CO_2 adsorption in IRMOFs is possible because of open metal sites, which leads to the existence of coordinated unsaturated metals as binding locations for different types of adsorbates. Moreover, the utilization of IRMOFs for CO_2 adsorption is interesting as compared to other materials due to lower regeneration costs. The presence of an exact amount of water present in the HKUST-1 structure is liable for efficient CO_2 uptake and selectivity. Yazaydın et al. reported experimental and simulation analysis of CO_2 adsorption of HKUST-1, which shows that the presence of little quantity of water, enhances CO_2 adsorption equilibrium [42]. Furthermore, also reported that selectivity over N_2 and CH_4 adsorption existed inside the material by enhanced electrostatic cooperation created by the quadrupole juncture of CO_2 collaborating with the electric field produced by bounded water at the open metal locations as shown in Fig. 11.10. The computational analysis performed by Yu et al. by analyzing water loading from zero to 8% into the similar sample found same results [43]. Another structure series of IRMOFs including M-MOF-74/M-DOBDC/M-CPO-27 exhibit coordinated unsaturated metal sites that show high CO_2 adsorption/absorption characteristics. Moreover, Mg-MOF-74 shows the highest CO_2 adsorption isotherm when measured at low pressure, which is more

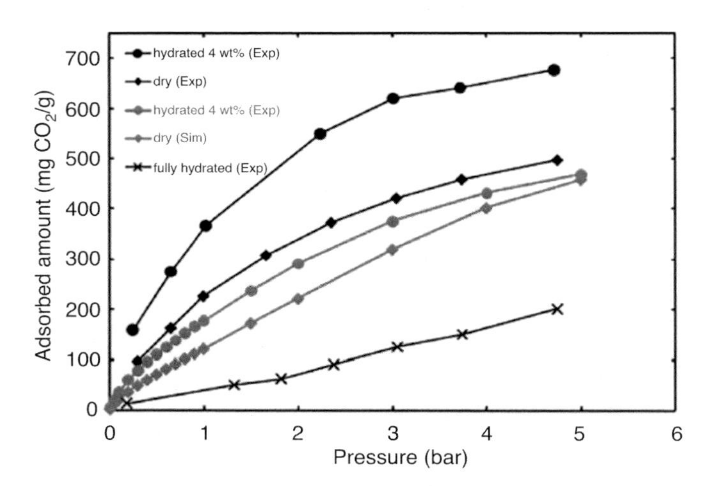

FIGURE 11.10 The simulated and experimental CO_2 adsorption isotherms performed at 298 K in HKUST-1 at various levels of water [42].

applicable to post-combustion flue gases. It was observed that experimental results found moisture present in the gas, which had negative effects on M-MOF-74 CO_2 capture efficiency [44–47]. The adsorption characteristics of ZIF-93 prepared in an aqueous medium confirmed the new dynamic development experiments performed on a CO_2/N_2 mixture. Various kinds of experiments performed in these adsorbed materials are exposed to a mixture of gases to determine the duration in which adsorbent can judiciously capture CO_2 while the remaining gases pass over. The holding of CO_2 longer time suggests an outstanding effective separation ability. The CO_2 and CH_4 uptake and separation of their mixtures in IRMOF-1 at RT utilizing atomistic simulation were performed. It was revealed that IRMOF-1 stores more CO_2 than CH_4 measured by using the gravimetric method. IRMOF-1 shows higher CO_2 and CH_4 adsorption capacity as compared to silicates and C_{168} Schwarzite performed under similar experimental conditions. However, all three adsorbents IRMOF-1, silicates, and C_{168} Schwarzite showed similar results when compared based on the selectivity of CO_2 over CH_4 [5].

The utilization of IRMOFs in adsorption and separation is considered the best solution for industrial instrument fabrication and environmental protection. Zehan et al. reported the usage of different types of IRMOFs-n (n = 1, 3, and 6) for adsorption and separation. This kind of IRMOF is considered a potential candidate for adsorption and separation owing to its elevated specific surface area and high pore volume. As shown in Fig. 11.11, where different types of IRMOFs adsorption of CO_2 are displayed shows that IRMOF-n possesses the highest sorption capacity at moderate conditions. Moreover, IRMOF-1 was found to be an excellent adsorbent of CO_2 adsorption capacity [17].

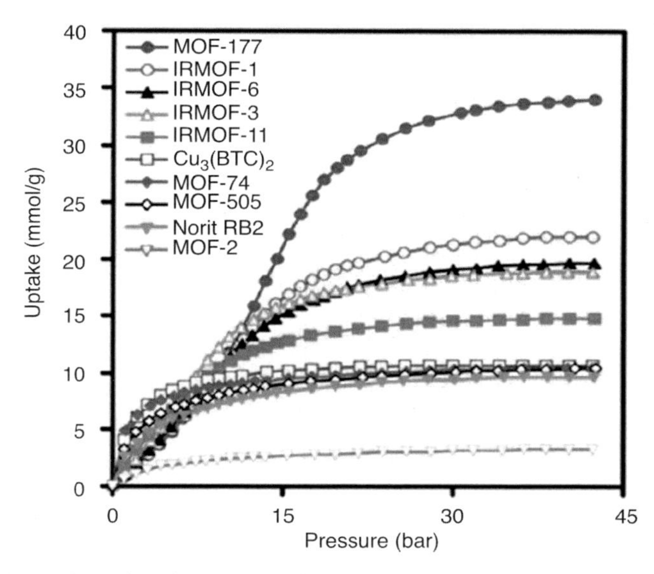

FIGURE 11.11 CO_2 adsorption isotherm of IRMOF-n (n = 1,3,6) and other MOFs at moderate conditions [17].

The hydrogen adsorption in water-based synthesized IRMOFs shows promising results owing to its elevated surface area and high pore volume. Hydrogen storage in highly porous IRMOFs material is critical in terms of low cost and environmental friendliness when utilized for onboard applications. Hydrogen uptake of HKUST-1 series including as-synthesized, activated at 160°C and 200°C performed at 303 K and 353 K, and 35 bar by gravimetric method showed very low hydrogen uptake [23]. However, measurement at 77 K and 1.0 bar enhances many folds up to 1.95 wt.% as shown in Fig. 11.12. Poirier et al. reported an experimental analysis of hydrogen uptake of IRMOF-1 performed at various low temperatures (50–100 K) and 40 bar. Moreover, hydrogen uptake of IRMOF-1 was found to be 9 wt.% at 50 K and 10 bar. Hydrogen uptake enthalpy was obtained as a function of temperature and the fractional filling was found to be 3–6 kJ/mol [48]. Many IRMOFs show low hydrogen storage capacity at moderate conditions, however after modification in terms of doping or decorating IRMOFs with transitional metal elements show many fold enhancement of hydrogen adsorption. Moreover, the introduction of carbon bridges into IRMOFs improves hydrogen uptake further. The enhancement of doped/embedded IRMOFs is due to the spillover mechanism phenomena and ease of spilling of hydrogen atoms dissociated by metals from primary to secondary receptions [40].

The utilization of methane gas as a successful fuel has been a challenging task for many years owing to its problems in its transport and storage. To be commercially viable, methane sorption into highly porous materials including

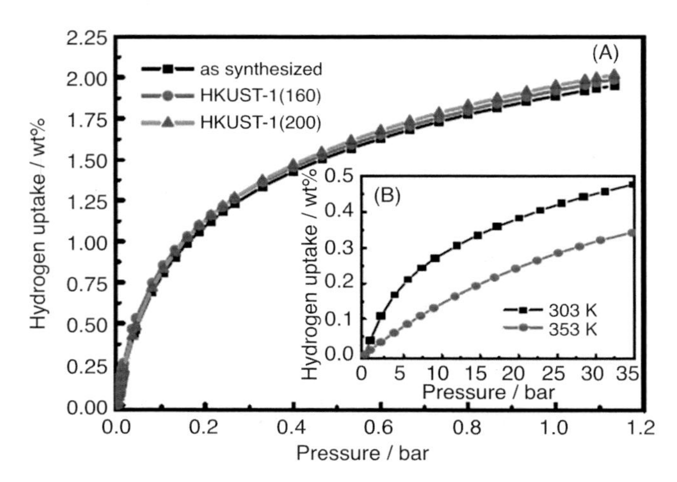

FIGURE 11.12 Hydrogen storage of HKUST-1. Profile (A) low pressure hydrogen isotherms of as-produced, activated at 160°C and 200°C measured at 77 K. Inset profile (B) high pressure hydrogen isotherms performed at 303 K and 353 K respectively of as-produced HKUST-1 [23].

IRMOFs must be feasible at moderate temperature and pressure. Among many IRMOFs, IRMOF-6 has a van der Waals size of 5.9 Å, the preferred parameter for efficient methane adsorption [17,18]. Different types of gas sorption isotherms performed at moderate conditions indicate that IRMOFs have an inflexible framework and maintain porosity without guests. The chloroform present as a guest molecule was eliminated from pores after chloroform-exchange IRMOF-6 was inserted into the microbalance instrument followed by evacuation at RT. Moreover, no more weight loss was found when evacuated overnight followed by elevation to 150°C. The XRD confirms that evacuated and as-synthesized IRMOF-6 are the same indicating the architectural stability of the evacuated IRMOF-6. The N_2 sorption displayed in Fig. 11.13 performed at 78 K shows reversible type I isotherm, which is an indication of a microporous type material. The plateau of isotherm was obtained at little pressure with no further adsorption at ambient pressure indicating the homogeneous nature of pores. According to Langmuir and Dubinin-Raduskhvich's (DR) model, the high surface area and pore volume of the IRMOF-6 was found to be 2630 m^2/g and 0.60 cm^3/g. The IRMOF-6 sample post evacuation was exposed to different types of gases including CH_2Cl_2, C_6H_6, CCl_4, and C_6H_{12} which shows type I reversible isotherm as presented in Fig. 11.13. The IRMOF-6 was also exposed to methane gas and measurement was performed at RT and pressure range of 0–42 atm. The methane adsorption was found to be 240 cm^3 as shown in Fig. 11.14 at RT and 36 atm, which exceeds the adsorption of zeolite 5A and other coordination frameworks.

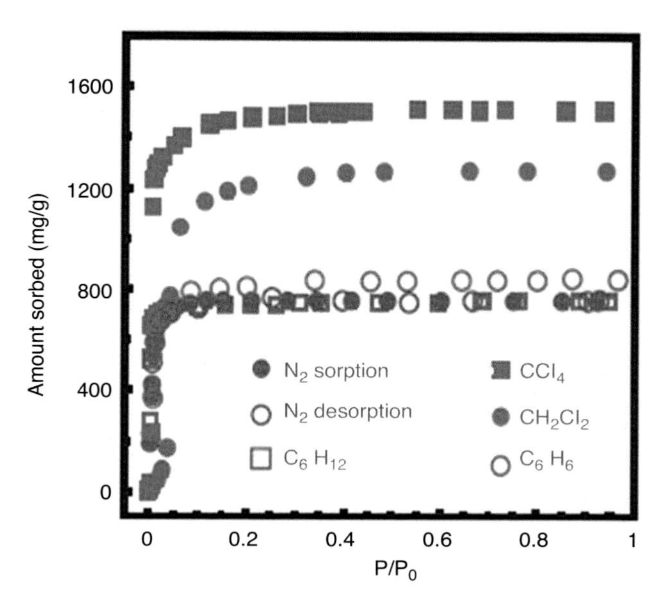

FIGURE 11.13 Gas (N_2) and organic gases (CCl_4, CH_2Cl_2, C_6H_6, and C_6H_{12}) adsorption/desorption of IRMOF-6 performed at 78 K [18].

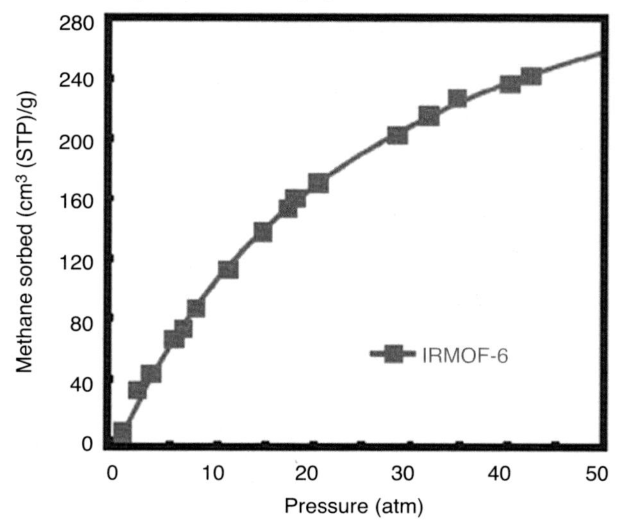

FIGURE 11.14 The methane gas adsorption of IRMOF-6 at room temperature fitted with Langmuir equation [18].

11.3.2 Catalysts

From the beginning of the synthesis of MOFs, the heterogeneous catalyst was considered as one of the important applications owing to their presence of active sites, high porosity, high surface area, and easily functionalized. Generally, zeolites are vastly used as a catalyst for many decades because of their large internal surface area, uniform pores, and cavity size. However, IRMOFs as one of the forms of MOFs differ from zeolites in certain ways as a catalyst. IRMOFs can be prepared in greater chemical variety as compared to zeolites. Moreover, IRMOFs show high thermal stability, as some of them go beyond 500°C. These characteristics of IRMOFs indicate that IRMOFs can be utilized as a catalyst like zeolites. Currently with so much known chemistry of IRMOFs, it is possible to explore a variety of catalyst ideas in the commercial area. The concept of IRMOFs as a catalyst is employed in different areas including heterogenization of homogeneous catalyst, framework-equilibrium of short-span catalysts, framework-enclosing of molecular catalysts, combination of catalyst to chemical separations and post-preparation of embedding of catalytic metal sites [10,49,50].

11.3.3 Sensors

The modification of IRMOFs as a composite is utilized as a detector of variety of items including environmental and industrial based applications and detecting special analytes including cations, anions, biomolecules, gas molecules, temperature, humidity and organic compounds. IRMOFs based composites are involved in the electrochemical and luminescent detection performances. Different methods involved in sensing character include photoluminescence (lanthanide and non-lanthanide-based frameworks), electrochemical, gas sensing, surface-enhanced Raman spectroscopy sensing, and refractive index sensor. Linag reported that UiO-66-NH_2 in fluorescent selective detection of mercury (II) in living cells. Moreover, UiO-66-NH_2 is used as an electrochemical sensor for cadmium ion detection. IRMOF-1 is employed for fluorescent detection of phosphate and gas sensing particularly CO_2. HKUST-1 or Cu-BTC IRMOF with the combination of silica-colloidal crystals for vapor sensing. Moreover, HKUST-1 composite is also utilized for refractive index sensor and determination of dihydroxybenzene isomers in water, hydroquinone and catechol. Formation of IRMOF-3 with magnetic nanoparticles of cobalt ferrite composite are used to detect the SERS sensing platform for revelation of N-terminal pro-brain natriuretic peptide [12].

11.3.4 Biomedical

Isoreticular metal–organic frameworks-based material emerged as a potential application in the biomedical field including stability, toxicology, drug delivery, and biocompatibility. Toxicology is very important especially when IRMOFs are used in health and biological applications. Iron-fumarate-based IRMOFs are approved for oral iron supplements, which is related to toxicology. Another very important challenge in drug delivery is the transmission of drugs inside the body utilizing non-toxic nanoparticles. There are some conditions mandatory for efficient therapy with nanoparticle carriers including (1) control of the discharge and dodging the "burst effect," (2) manage matrix deterioration and design its surface, (3) evident by different imaging methods, and (4) implicate drugs with immense embedding quantity. Moreover, biocompatibility and toxicology are other critical conditions associated with materials treated as possible novel drug carriers [11].

11.4 Conclusion

Different types of IRMOFs synthesized by a variety of methods utilizing conventional expensive organic solvents and precursors incorporate several disadvantages such as production cost, energy, and toxicity. However, using water as compared to toxic organic solvents for the preparation of IRMOFs provides a variety of advantages including stability, low production cost, minimum energy requirement, disposability, and zero environmental hazards. Moreover, organic solvents used for the preparation of IRMOFs are difficult to eliminate from pores and channels that existed in the framework while water is disposed of easily post-synthesis upon exposure to the activation procedure. The structure and thermal stability and other property enhancement were observed in the water-based synthesis of IRMOFs. For toxic or hazardous-free IRMOFs, several synthesis routes utilizing water as a solvent including microwave irradiation, ultrasound-assisted, microwave-assisted solvothermal, and in situ growth methods utilized to boost porosity tenability, stability, and production rate. Isoreticular MOFs extension of general MOFs because of framework structure emerges as an important pillar for many applications including biomedicine, catalysts, storage, separation, catalyst, sensors, toxicology, and drug delivery. Moreover, IRMOFs possess several properties including elevated surface area and high pore volume, tunable pore structure, topology, and channels provides a basic platform for different applications. The main application such as adsorption and separation of IRMOFs is possible because of thermal, chemical, and physical stability and structural tenability. The gases including CO_2, hydrogen, benzene, sulfur dioxide, ammonia, chlorine, dichloromethane and ethylene dioxide and water are adsorbed/absorbed in various kinds of water synthesized IRMOFs. High surface area, high pore volume and cage like framework makes IRMOFs as a best adsorbent/absorbent for gases. Hydrogen storage and CO_2 capture in highly porous IRMOFs material are very important in terms of low cost and environmental friendliness when utilized for onboard and other applications. The IRMOFs usually show low hydrogen adsorption at moderate conditions, however after careful modification in terms of doping, embedding or decorating with transitional metals furnish many folds enhancement of hydrogen storage capacity. The high CO_2 capture is feasible owing to its open metal sites, which leads to the continuation of coordinated unsaturated metals as binding locations for different types of adsorbates.

References

[1] E.B. Avery, A.B. David, L. Bingqian, V.S Thoi, Metal-organic framework functionalization and design strategies for advanced electrochemical energy storage devices, Commun. Chem. 86 (2019) 1–14.

[2] N. Srinivasan, M. Partha, Metal-organic framework structures: how closely are they related to classical inorganic structures, Chem. Soc. Rev. 38 (2009) 2304–2318.

[3] K.G. Kranthi, M. Suresh, B.M. Saratchandra, B.J. Sreekantha, A review on contemporary metal-organic framework materials, Inorg. Chim. Acta 446 (2016) 61–74.

[4] J.M. Leslie, D. Mircea, J.R. Long, Hydrogen storage in metal–organic frameworks, Chem. Soc. Rev. 38 (2009) 1294–1214.

[5] B. Ravichandar, H. Zhongqiao, J. Jianwen, Storage and separation of CO_2 and CH_4 in silicalite, C168 Schwarzite, and IRMOF-1: a comparative study from Monte Carlo simulation, Langmuir 23 (2007) 659–666.

[6] T. Li, D.-L. Chen, J.E. Sullivan, M.T. Kozlowski, J.K. Johnson, N.L. Rosi, Systematic modulation and enhancement of CO_2: N_2 selectivity and water stability in an isoreticular series of bio-MOF-11 analogues, Chem. Sci. 4 (2013) 1746–1755.

[7] R.M. Andrew, M.Y. Omar, Metal-organic frameworks with exceptionally high capacity for storage of carbon dioxide at room temperature, J. Am. Chem. Soc. 127 (2005) 17998–17999.

[8] E.V. Ramos-Fernandez, A. Grau-Atienza, D. Farrusseng, S. Aguado, A water-based at room temperature synthesized ZIF-93, for CO_2 adsorption, J. Mater. Chem. 6 (2018) 5598.

[9] L. Jeong, W. Kecheng, S. Yujia, T.L. Christina, L. Jialuo, Z. Hong-Cai, Recent advances in gas storage and separation using metal–organic frameworks, Mater. Today 21 (2018) 108–121.

[10] Y.L. Jeong, J.R. Hiroyas, A.S. Karl, T.N. SonBinh, T.H. Joseph, Metal–organic framework materials as catalysts, Chem. Soc. Rev. 38 (2009) 1450–1459.

[11] K. Seda, K. Seda, Biomedical applications of metal organic frameworks, Ind. Eng. Chem. Res. 50 (2011) 1799–181.

[12] A. Ali, K. Sima, S. Vahid, Metal-organic framework-based nanocomposites for sensing applications: a review, Polyhedron 177 (2020) 114260–114297.

[13] K.F. Omar, E. Ibrahim, C.J. Nak, G.H. Brad, E.W. Christopher, A.S. Amy, Q.S. Randall, T.N. SonBinh, A.Özgür Y, T.H. Joseph, Metal−organic framework materials with ultrahigh surface areas: is the sky the limit? J. Am. Chem. Soc. 134 (2012) 15016–15021.

[14] H. Furukawa, Y.B. Go, N. Ko, Y.K. Park, F.J. Uribe-Romo, J. Kim, M. O'Keeffe, O.M. Yaghi, Isoreticular expansion of metal-organic frameworks with triangular and square building units and the lowest calculated density for porous crystals, Inorg. Chem. 50 (2011) 9147–9152.

[15] D.J. Tranchemontagne, J.R. Hunt, O.M. Yaghi, Room temperature synthesis of metal-organic frameworks: MOF-5, MOF-74, MOF-177, MOF-199, and IRMOF-0, Tetrahedron 64 (2008) 8553–8557.

[16] M.Y. Omar, O. Michael, W.O. Nathan, K.C. Hee, E. Mohamed, K. Jaheon, Reticular synthesis and the design of new materials, Nature 423 (2003) 705–713.

[17] M. Zehan, L. Dingxin, Synthesis and applications of isoreticular metal−organic frameworks IRMOFs-n (n = 1, 3, 6, 8), Cryst. Growth Des. 19 (2019) 7439–7462.

[18] E. Mohamed, K. Jaheon, R. Nathaniel, V. David, W. Joseph, O. Michael, M.Y. Omar, Systematic design of pore size and functionality in isoreticular MOFs and their application in methane storage, Science 295 (2002) 469–472.

[19] D. Chongxiong, Y. Yi, X. Jing, Z. Xuelian, L. Libo, Y. Pengfei, W. Junliang, X. Hongxia, Water-based routes for synthesis of metal-organic frameworks: a review, Sci. China Mater. 63 (2020) 667–685.

[20] C. Zhijie, W. Xingjie, C. Ran, B.I. Karam, L. Xinyao, C.W. Megan, K.F. Omar, Water-based synthesis of a stable iron-based metal–organic framework for capturing toxic gases, ACS Mater. Lett. 2 (2020) 1129–1134.

[21] C.B. Nicholas, J. Himanshu, S.W. Krista, Water stability and adsorption in metal–organic frameworks, Chem. Rev. 114 (2014) 10575–10612.

[22] S. Norbert, B. Shyam, Synthesis of metal-organic frameworks (MOFs): routes to various MOF topologies, morphologies, and composites, Chem. Rev. 112 (2012) 933–969.

[23] L. Kuen-Song, K.A. Abhijit, K. Chi-Nan, C. Chao-Lung, K. Hua, Synthesis and characterization of porous HKUST-1 metal organic frameworks for hydrogen storage, Int. J. Hydrogen Energy 37 (2012) 13865–13871.

[24] S. Carsten, X. Mergime, E. Sebastian-Johannes, S. Alexa, T. Niels, J. Christoph, Solid-solution mixed-linker synthesis of isoreticular Al-based MOFs for an easy hydrophilicity tuning in water-sorption heat transformations, Chem. Mater. 31 (2019) 4051–4062.

[25] J. Huo, M. Brightwell, S.E. Hankari, A. Garaia, D. Bradshaw, A versatile, industrially relevant, aqueous room temperature synthesis of HKUST-1 with high space-time yield, J. Mater. Chem. 1 (2013) 15220–15223.

[26] W.Y. Siew, N.H.H. Abu Bakar, M. Abu Bakar, The influence of green synthesis on the formation of various copper benzene-1,3,5-tricarboxylate compounds, Inorg. Chim. Acta 482 (2018) 53–61.

[27] M. Gerardo, P.R. Javier, Scalable room-temperature conversion of copper (II) hydroxide into HKUST-1 ($Cu_3(btc)_2$), Adv. Mater. 25 (2013) 1052–1057.

[28] S. Cadot, L. Veyre, D. Luneau, D. Farrusseng, E.A. Quadrelli, A water-based and high space-time yield synthetic route to MOF Ni_2(dhtp) and its linker 2,5-dihydroxyterephthalic acid, J. Mater. Chem. 2 (2014) 17757–17763.

[29] L. Garzon-Tovar, A. Carne-Sanchez, C. Carbonell, I. Imaz, D. Maspoch, Optimised room temperature, water-based synthesis of CPO-27-M metal-organic frameworks with high space-time yields, J. Mater. Chem. 3 (2015) 20819–20826.

[30] T. Didriksen, A.I. Spjelkavik, R. Blom, Continuous synthesis of the metal–organic framework CPO-27-Ni from aqueous solutions, J. Flow Chem. 7 (2017) 13–17.

[31] A.J. Patrick, U. Krunoslav, D.K. Athanassios, A.J.K. Simon, W. Timothy, K.F. Omar, Z. Yuancheng, C. José, S.G. Luzia, E. Martin, E.D. Robert, L.J. Stuart, H. Ivan, F. Tomislav, In situ monitoring and mechanism of the mechanochemical formation of a microporous MOF-74 framework, J. Am. Chem. Soc. 138 (2016) 2929–2932.

[32] R.R. Ozer, J.P. Hinestroza, One-step growth of isoreticular luminescent metal–organic frameworks on cotton fibers, RSC Adv. 5 (2015) 15198–15204.

[33] Q.M. Wang, D. Shen, M. Bülow, M.L. Lau, S. Deng, F.R. Fitch, N.O Lemcoff, J. Semanscin, Metallo-organic molecular sieve for gas separation and purification, Microporous Mesoporous Mater. 55 (2002) 217–230.

[34] P. Küsgens, M. Rose, I. Senkovska, H. Fröde, A. Henschel, S. Siegle, S. Kaskel, Characterization of metal-organic frameworks by water adsorption, Microporous Mesoporous Mater. 120 (2009) 325–330.

[35] Y. Wang, M.D. LeVan, Adsorption equilibrium of carbon dioxide and water vapor on zeolites 5A and 13X and silica gel: pure components, J. Chem. Eng. Data 54 (2009) 2839–2844.

[36] Z. Liang, M. Marshall, A.L. Chaffee, CO_2 adsorption-based separation by metal organic framework (Cu-BTC) versus zeolite (13X), Energy Fuels 23 (2009) 2785–2789.

[37] P.M. Schoenecker, C.G. Carson, H. Jasuja, C.J.J. Flemming, K.S. Walton, Effect of water adsorption on retention of structure and surface area of metal–organic frameworks, Ind. Eng. Chem. Res. 51 (2012) 6513–6519.

[38] S.K. Henninger, F.P. Schmidt, H.M. Henning, Characterization and improvement of sorption materials with molecular modeling for the use in heat transformation applications, Adsorption 17 (2011) 833–843.

[39] S.K. Henninger, F. Jeremias, H. Kummer, C. Janiak, MOFs for use in adsorption heat pump processes, Eur. J. Inorg. Chem. 2012 (2012) 2625–2634.

[40] Y. Li, R.T. Yang, Hydrogen storage in metal-organic and covalent-organic frameworks by spillover, AIChE J. 54 (2008) 269–279.

[41] J.B. DeCoste, G.W. Peterson, B.J. Schindler, K.L. Killops, M.A. Browe, J.J. Mahle, The effect of water adsorption on the structure of the carboxylate containing metal–organic frameworks Cu-BTC, Mg-MOF-74, and UiO-66, J. Mater. Chem. 1 (2013) 11922–11932.

[42] A.O. Yazaydın, A.I. Benin, S.A. Faheem, P. Jakubczak, J.J. Low, R.R. Willis, R.Q. Snurr, Enhanced CO_2 adsorption in metal-organic frameworks via occupation of open-metal sites by coordinated water molecules, Chem. Mater. 21 (2009) 1425–1430.

[43] J. Yu, Y. Ma, P.B. Balbuena, Evaluation of the impact of H_2O, O_2, and SO_2 on postcombustion CO_2 capture in metal—organic frameworks, Langmuir 28 (2012) 8064–8071.

[44] J. Liu, Y. Wang, A.I. Benin, P. Jakubczak, R.R. Willis, M.D. LeVan, CO_2/H_2O adsorption equilibrium and rates on metal-organic frameworks: HKUST-1 and Ni/DOBDC, Langmuir 26 (2010) 14301–14307.

[45] S.R. Caskey, A.G. Wong-Foy, A.J. Matzger, Dramatic tuning of carbon dioxide uptake via metal substitution in a coordination polymer with cylindrical pores, J. Am. Chem. Soc. 130 (2008) 10870–10871.

[46] A.C. Kizzie, A.G. Wong-Foy, A.J. Matzger, Effect of humidity on the performance of microporous coordination polymers as adsorbents for CO_2 capture, Langmuir 27 (2011) 6368–6373.

[47] T. Remy, S.A. Peter, S. Van der Perre, P. Valvekens, D.E. De Vos, G.V. Baron, J.F.M. Denayer, Selective dynamic CO_2 separations on Mg-MOF-74 at low pressures: a detailed comparison with 13X, J. Phys. Chem. 117 (2013) 9301–9910.

[48] E. Poirier, A. Dailly, Investigation of the hydrogen state in IRMOF-1 from measurements and modeling of adsorption isotherms at high gas densities, J. Phys. Chem. 112 (2008) 13047–13052.

[49] W. Songhai, M. Xin, R. Jingyu, Z. Yanfei, Q. Fengxiang, L. Yong, Application of basic isoreticular nanoporous metal–organic framework: IRMOF-3 as a suitable and efficient catalyst for the synthesis of chalcone, RSC Adv. 5 (2015) 14221–14227.

[50] Z. Xi, Z. Yi, Y. Xiangui, Z. Liangzhong, W. Gongying, Functionalized IRMOF-3 as efficient heterogeneous catalyst for the synthesis of cyclic carbonates, J. Mol. Catal. A Chem. 361–362 (2012) 12–16.

Chapter 12

Preparation and applications of water-based coordination pillared-layer

Atif Husain[a], Benjamin Siddiqui[a], Malik Nasibullah[a], Naseem Ahmad[a], Mohd. Asif[a] and Mohd. Sufian Abbasi[b]

[a]*Department of Chemistry, Integral University, Lucknow, Uttar Pradesh, India,* [b]*Department of Civil Engineering, Integral University, Lucknow, Uttar Pradesh, India*

12.1 Introduction

The possibility of creating a wide range of visually pleasing formations that may also be of considerable relevance for applications across a variety of sectors connected to porous materials has drawn a lot of attention to the development of metal–organic frameworks (MOFs) over the last two decades [1]. This includes more common domains like storing, separation, and catalysis that are influenced by pores size and form as well as host-guest interactions. Additionally, research into biological uses and sensor materials is ongoing. The topic of the biosynthesis of MOFs has been covered in in-depth reviews [2–4]. Coordination pillared layers are unique, systematically structured frameworks with variable pore sizes, geometries, and surface qualities that allow for flexibility [5].

One kind of MOF is pillared-layer structures, and their construction process is regarded as one of the most viable and readily controllable methodologies for constructing 3D porous frameworks [6]. The synthetic process promises to generate a wide range of porosity MOF materials, the architectures of which can be predicted by carefully selecting linker components. Shape organic linkers are utilized to systematically adjust pore size and shape, and functional groups contained in the ligand in columns and/or layers may be introduced into the pore to control their pore surface to get desired features [7,8].

It is also important to consider the usage of green synthesis protocols for the effective and efficient synthesis of these CPL by adhering to green pathways on a sustainable scale, such as by reducing the use of harmful solvents. A new area of study that may point the way to the creation of novel materials with exceptional structures and characteristics is represented by the preparation

Synthesis of Metal–Organic Frameworks via Water-Based Routes: A Green and Sustainable Approach.
DOI: https://doi.org/10.1016/B978-0-323-95939-1.00010-1

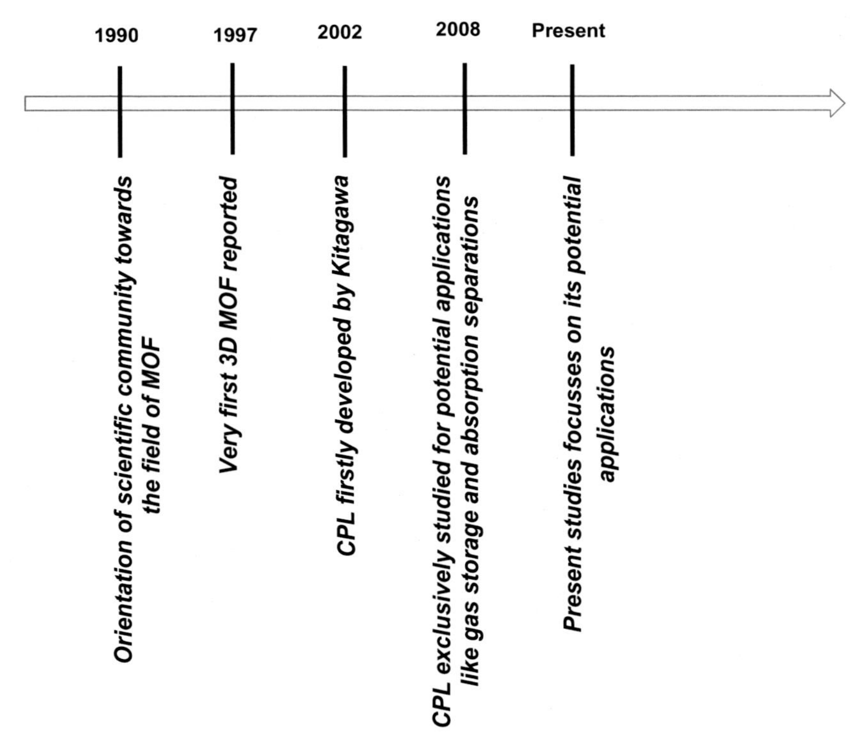

FIGURE 12.1 Historical developments in the field of multi–organic framework and selection and exclusion of coordination pillared layers for potential applications.

of water-based coordinated pillared layered (CPL) by adhering to water-based synthesis conditions, which is also noteworthy [9,10].

12.2 Brief history of the construction of CPL

The first CPL structure was designed as a function that supports the fabrication of a range of functionalized materials by Kitagawa et al. by constructing neutral 2D layers with Cu^{II} and 2,3-pyrazinedicarboxylate (pzdc) braced by multiple dipyridyl bridging ligands as pillars (Fig. 12.1) [11,12].

The CPL series used in their strategy has garnered a lot of interest due to its simple synthetic method, periodicity, great performance factor, and reduced assembly [13]. Due to its distinct characteristics, CPL has received a lot of study attention for pumped storage and adsorbent separations [14,15]. Chen and colleagues reported another historical investigation resulting in the adsorptive separation of propane by gate-opening action on adsorption over propylene. Similarly, Kishida et al. showed preferential ethylene adsorption over ethane in CPL-1 by a gas-induced structural change. Furthermore, acetylene can be substantially restricted in micropores by CPL-1, which effectively separates acetylene from the mixed gas that contains carbon dioxide [16,17].

12.3 Different synthesis procedures employed in the preparation of CPL

The many synthetic techniques used in MOF synthesis today by several organizations across the world are explained by these differing scientific backgrounds, particularly with regard to reactors and reaction temperatures. They also provide an explanation of the various synthetic processes and approaches used in MOF synthesis. Electrochemical, mechanochemical, and ideas like the precursor method or in situ, linker synthesis have all been made possible by coordination chemistry. However, zeolite chemistry has also been used to explain framework agents, hydrothermal method process conditions, microwave-assisted synthesis, mineralizers, and steam-assisted conversion [18].

Techniques like electrochemistry, mechanochemistry, microwave heating, ordinary electric heating, and ultrasonic heating have all been used in addition to room temperature synthesis. To effectively exploit the geometry of MOF products, many objectives have been met.

In addition to the dimensions or shape of the crystal, nanoplatelets, membrane, and several other structures formed of transition metal oxides have also been described, and they need the use of diverse synthesis techniques. One of them, the solvothermal approach, was the first method for building pillared layer MOFs that was publicly published (Fig. 12.2) [19,20].

The study of pillar-layered MOFs has advanced significantly over the past two decades, both in terms of the identification of new structural properties and the creation of fresh functional aspects. However, little emphasis has been made on informing new synthetic techniques with improved efficiency. Although constructing self-assembly is the primary method for producing pillar MOF, components under optimal circumstances, not all desirable structures are built in a single step. Post-synthesis modification (PSM) is an alternate technique for creating unachievable frameworks. The topology, application, and functionality of produced MOFs may be changed to allow for the chemical alternation of organic moieties [21]. The same group also looked at the olefin/paraffin adsorption and separation on bendable CPL structures.

12.4 Preparation of CPL on a greener and sustainable scale

The CPL preparation is comparable to that of other related structures such as zeolitic imidazolate frameworks (ZIFs), porous coordination networks (PCN), isoreticular MOFs (IRMOFs), MIL, and UiO. To address the needs of functionalization on a sustainable scale, they must be produced in water. However, only a few CPL have indeed been synthesized using liquid synthesis. However,

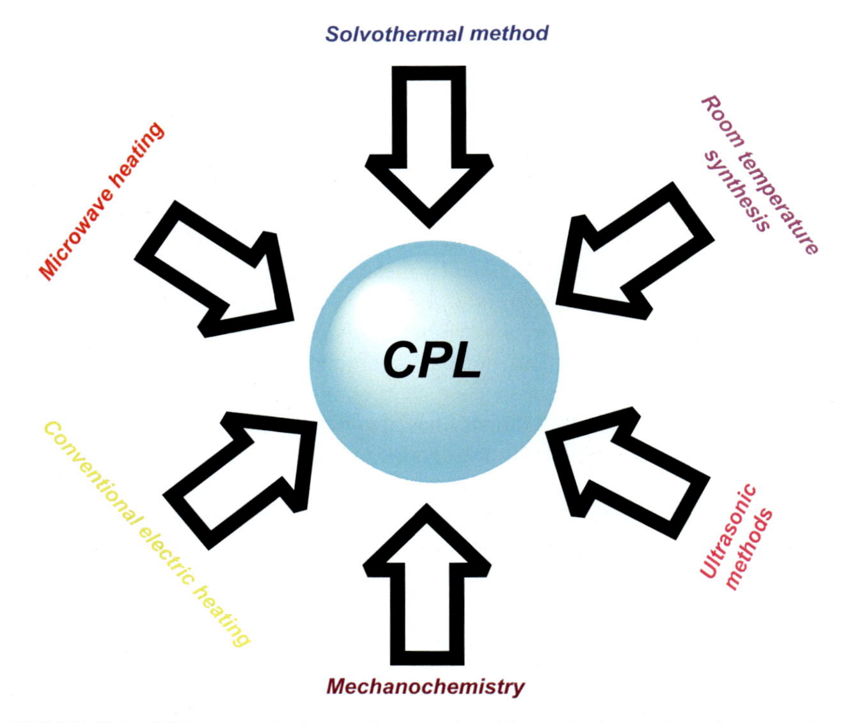

FIGURE 12.2 Different synthesis procedures employed for synthesis apart from the conventional procedures.

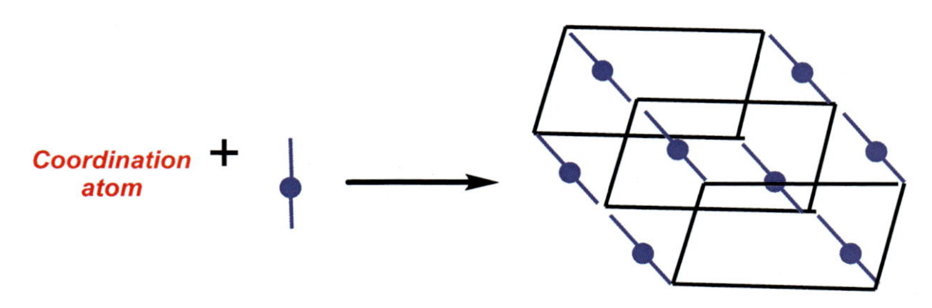

FIGURE 12.3 Basic schematic illustration of synthesis of coordination pillared layers from coordination atoms and pillared ligands.

this ratio is steadily growing. Chen and coworkers reported the effective production of CPL-1 ([[$Cu_2(pzdc)_2(L)$]$\cdot xH_2On$]) by reacting $Cu(ClO_4)_2 \cdot 6H_2O$ with Na_2pzdc and pyrazine in aqueous solutions [22]. Mechanochemical approaches can enhance other criteria suitable for the sustainable scale, such as cheaper cost, increased safety, ease of operation, efficient reaction rate, and excellent selectivity when compared to conventional solvothermal synthesis (Fig. 12.3) [23,24].

Sakamoto et al. described the mechanochemical production of CPL-1 in humid circumstances without the use of organic solvents [25]. Apart from the advances achieved in this sector, several challenges, such as solubility using water as the solvent (some pillared ligands are extremely insoluble in water), cannot be overlooked. Although The additional water is important throughout the mechanochemical process because of the fact that it serves as a materials transporter and promotes species diffusion inside an inter-organizational and intra-particle pattern.

12.5 Properties of CPL leading to a diversity of applications

The existence of major properties in coordination pillared layers opens the door to a plethora of applications in a variety of sectors. Flexible frameworks and simple parameters for synthesis like ambient pressure and temperature play an important role in its use in a wide range of applications [26,27].

Through the contraction and expansion of the framework, flexibility can result in high porosity and thermal stability [28]. Applying flexible linkers is a potential strategy for managing flexibility. This strategy is most effective in pillar-layered MOFs due to the availability of two linkers with customizable flexibility. The degree of conformational isomerism is made easier and has an impact on the control and selectivity of the porosity when long flexible pillars are used instead of stiff ones to create porous frameworks [29,30]. Innovative solutions can result from the interaction between the stiffness and flexibility of organic ligands.

12.6 Potential applications of pillared layered MOFs

By altering the pillar foundation or carrying out straightforward converts inside the interlayer or functional areas in pillar bridges, researchers can create structures with the ideal pore volume and effectiveness, trying to make them excellent contenders in a variety of applications such as adsorption at the surface, detachment, catalyst supports, sensing, and so on.

Pillar-layered MOFs are remarkably versatile and can be used in a wide range of applications, including the adsorption process [31], catalytic [32], sensing [33,34], drug delivery [35], and conduction [36]. The existence of multiple fusions inside one MOF could improve pore environment tunability, which is the most essential property of pillar-layered MOFs. The impact of supplementary linker backbone on the inherent properties of frameworks shows unpredictability. Substantial binding effects on structural traits, size, and topology are mostly to blame for this. The development of pillar-stacked MOFs of a future generation will require managing these consequences [37,38], given the importance of such features in defining MOF performance.

The design of the pillars is just efficient, and an increase in their length, the addition of functions, and the management of their flexibility can have an impact on the generated MOFs and improve the application of these frameworks. Possible uses are also being created by synthesizing these unique frameworks using solvent-free green synthesis techniques, which will ultimately reduce environmental pollution and make sustainable scales more practical [39].

12.7 Conclusion and outlook

The CPL is an exclusive structure belonging to the class of MOF having a wide scope for potential applications. Their selective properties give rise to new explorations made in their structure and synthetic procedures made over the past few decades. Due to the flexibility in their properties pillared layer MOFs are one of the most thriving fields of coordination compounds that may be comprehensively researched in different aspects. As the importance of research in this subject grows, there is a pressing need to investigate its structure and properties. This can help scientists gain a better grasp of the field and the relevant fields for the efficient construction of these structures.

Moreover, keeping in view, the environmental and economic perspectives related to the preparation and applications of this class of MOF the development of Wherever it is technically and economically feasible, a green synthetic method is essential. This will not only overcome the persistent challenges but will also cut back on the expensive, hazardous organic solvents used in MOF synthesis. Also, for future prospects, utilizing new synthetic methods for pillar-layered MOFs assemblies such as spray drying, sol-gel [40,41], or other cost-effective methods has not been studied up to now, which can be major research interest for upcoming researchers opening new gateways for potential applications.

References

[1] S. Kitagawa, R. Kitaura, S.-I. Noro, Functional porous coordination polymers, Angew. Chem. Int. Ed. 43 (2004) 2334.

[2] J.L.C. Rowsell, O.M. Yaghi, Metal–organic frameworks: a new class of porous materials, Microporous Mesoporous Mater. 73 (2004) 3.

[3] G. Férey, Hybrid porous solids: past, present, future, Chem. Soc. Rev. 37 (2008) 191.

[4] C. Janiak, J.K. Vieth, MOFs, MILs and more: concepts, properties and applications for porous coordination networks (PCNs), New J. Chem. 34 (2010) 2366.

[5] O.M. Yaghi, M. O'Keeffe, N.W. Ockwig, H.K. Chae, M. Eddaoudi, J. Kim, Reticular synthesis and the design of new materials, Nature 423 (2003) 705–714.

[6] M. Kondo, T. Okubo, A. Asami, S.-i. Noro, T. Yoshitomi, S. Kitagawa, T. Ishii, H. Matsuzaka, K. Seki, Rational synthesis of stable channel-like cavities with methane gas adsorption properties: $[\{Cu_2(pzdc)_2(L)\}n]$ (pzdc = pyrazine-2,3-dicarboxylate, L = a pillar ligand), Angew. Chem. Int. Ed. 38 (1999) 140−143.

[7] X.-L. Luo, Z. Yin, M.-H. Zeng, M. Kurmoo, The construction, structures, and functions of pillared layer metal−organic frameworks, Inorg. Chem. Front. 3 (2016) 1208−1226.

[8] H. Chun, D.N. Dybtsev, H. Kim, K. Kim, Synthesis, x-ray crystal structures, and gas sorption properties of pillared square grid nets based on paddle-wheel motifs: implications for hydrogen storage in porous materials, Chemistry 11 (2005) 3521−3529.

[9] B. Lim, Y. Xiong, Y. Xia, A water-based synthesis of octahedral, decahedral, and icosahedral Pd nanocrystals, Angew. Chem. Int. Ed. 46 (2007) 9279–9282.
[10] J. Ren, X. Dyosiba, N.M. Musyoka, et al., Review on the current practices and efforts towards pilot-scale production of metal-organic frameworks (MOFs), Coord. Chem. Rev. 352 (2017) 187–219.
[11] R. Ohtani, S. Kitagawa, M. Ohba, Coordination pillared layers using a dinuclear Mn(V) complex as a secondary building unit, Polyhedron 52 (2013) 591–597.
[12] H. Sakamoto, R. Kitaura, R. Matsuda, S. Kitagawa, Y. Kubota, M. Takata, Systematic construction of porous coordination pillared-layer structures and their sorption properties, Chem. Lett. 39 (2010) 218–219.
[13] S. Kitagawa, R. Matsuda, Chemistry of coordination space of porous coordination polymers, Coord. Chem. Rev. 251 (2007) 2490–2509.
[14] R. Kitaura, K. Fujimoto, S. Noro, M. Kondo, S. Kitagawa, A pillared-layer coordination polymer network displaying hysteretic sorption: $[Cu_2(pzdc)_2(dpyg)]n$ (pzdc=pyrazine-2,3-dicarboxylate; dpyg=1,2-di(4-pyridyl) glycol), Angew. Chem., Int. Ed. 41 (2002) 133–135.
[15] R. Kitaura, K. Seki, G. Akiyama, S. Kitagawa, Porous Coordination-polymer crystals with gated channels specific for supercritical gases, Angew. Chem., Int. Ed. 42 (2003) 428–431.
[16] Y. Chen, Z. Qiao, D. Lv, C. Duan, X. Sun, H. Wu, R. Shi, Q. Xia, Z. Li, Efficient adsorptive separation of C_3H_6 over C_3H_8 on flexible and thermoresponsive CPL-1, Chem. Eng. J. 328 (2017) 360–367.
[17] K. Kishida, Y. Watanabe, S. Horike, Y. Watanabe, Y. Okumura, Y. Hijikata, S. Sakaki, S. Kitagawa, DRIFT and theoretical studies of ethylene/ethane separation on flexible and microporous $[Cu_2$(2,3-pyrazinedicarboxylate$)_2$(pyrazine)]n, Eur. J. Inorg. Chem. 2014 (2014) 2747–2752.
[18] S.S.-Y. Chui, S.M.-F. Lo, J.P.H. Charmant, A.G. Orpen, I.D. Williams, A chemically functionalizable nanoporous material $[Cu_3(TMA)_2(H_2O)_3]_n$, Science 283 (1999) 1148.
[19] O.M. Yaghi, Z. Sun, D.A. Richardson, T.L. Groy, Directed transformation of molecules to solids: synthesis of a microporous sulfide from molecular germanium sulfide cages, J. Am. Chem. Soc. 116 (1994) 807–808.
[20] O.M. Yaghi, H. Li, Hydrothermal synthesis of a metal-organic framework containing large rectangular channels, J. Am. Chem. Soc. 117 (1995) 10401–10402.
[21] S. Jeong, D. Kim, X. Song, M. Choi, N. Park, M.S. Lah, Post-synthetic ligand exchange by mechanochemistry: toward green, efficient, and large-scale preparation of functional metal–organic frameworks, Chem. Mater. 25 (2013) 1047–1054.
[22] Y. Chen, Z. Qiao, D. Lv, et al., Efficient adsorptive separation of C_3H_6 over C_3H_8 on flexible and thermoresponsive CPL-1, Chem. Eng. J. 328 (2017) 360–367.
[23] Y. Li, J. Miao, X. Sun, et al., Mechanochemical synthesis of Cu-BTC@Go with enhanced water stability and toluene adsorption capacity, Chem. Eng. J. 298 (2016) 191–197.
[24] Y. Chen, H. Wu, Z. Liu, et al., Liquid-assisted mechanochemical synthesis of copper based MOF-505 for the separation of CO_2 over CH_4 or N_2, Ind. Eng. Chem. Res. 57 (2018) 703–709.
[25] H. Sakamoto, R. Matsuda, S. Kitagawa, Systematic mechanochemical preparation of a series of coordination pillared layer frameworks, Dalton Trans. 41 (2012) 3956–3961.
[26] O.J. García-Ricard, A.J. Hernández-Maldonado, Cu_2(pyrazine-2,3-dicarboxylate)2(4,4′-bipyridine) porous coordination sorbents: activation temperature, textural properties, and CO_2 adsorption at low pressure range, J. Phys. Chem. 114 (2010) 1827–1834.
[27] F. Zheng, L. Guo, B. Gao, et al., Engineering the pore size of pillared- layer coordination polymers enables highly efficient adsorption separation of acetylene from ethylene, ACS Appl. Mater. Interfaces 11 (2019) 28197–28204.

[28] X.-L. Li, G.-Z. Liu, L.-Y. Xin, L.-Y. Wang, Two topologically new trinodal cobalt(ii) metal–organic frameworks characterized as a 1D metallic oxide and a 2D → 3D penetrated porous solid, CrystEngComm 14 (2012) 5757–5760.
[29] S. Parshamoni, S. Sanda, H.S. Jena, K. Tomar, S. Konar, Synthesis and characterization of two lanthanide (Gd^{3+} and Dy^{3+})-based three-dimensional metal organic frameworks with squashed metallomacrocycle type building blocks and their magnetic, sorption, and fluorescence properties study, Cryst. Growth Des. 14 (2014) 2022–2033.
[30] L.-Z. Yang, R. Fang, W. Dou, A.M. Kirillov, C.-l. Xu, W.-S. Liu, Structural diversity in new coordination polymers modulated by semi rigid ether-linked pyridine phthalate building block and ancillary ligands: syntheses, structures, and luminescence properties, CrystEngComm 17 (2015) 3117–3128.
[31] D.-M. Chen, N. Xu, X.-H. Qiu, P. Cheng, Functionalization of metal–organic framework via mixed-ligand strategy for selective CO_2 sorption at ambient conditions, Cryst. Growth Des. 15 (2015) 961–965.
[32] L. Liu, C. Huang, X. Xue, M. Li, H. Hou, Y. Fan, Facile synthesis of one-dimensional organometallic–organic hybrid polymers based on a diphosphorus complex and flexible bipyridyl linkers, Cryst. Growth Des. 15 (2015) 4507–4517.
[33] C.-D. Si, D.-C. Hu, Y. Fan, Y. Wu, X.-Q. Yao, Y.-X. Yang, J.-C. Liu, Polytorsional-amide/carboxylates-directed Cd(ii) coordination polymers exhibiting multi-functional sensing behaviors, Cryst. Growth Des. 15 (2015) 2419–2432.
[34] B. Chen, Y. Ji, M. Xue, F.R. Fronczek, E.J. Hurtado, J.U. Mondal, C. Liang, S. Dai, Metal−organic frameworks with functional pores for recognition of small molecules, Inorg. Chem. 47 (2008) 5543–5545.
[35] M. Delavar, B. Afzalian, B. Notash, A novel mixed-ligand coordination polymer with pillared-layer & ladder like structure: synthesis, crystal structure, properties study, and application as sorbent for acetaminophen extraction, Int. J. Basic Appl. Sci. 4 (2015) 183.
[36] X. Li, X. Sun, X. Li, Z. Fu, Y. Su, G. Xu, Defect control to enhance proton conductivity in a metal–organic framework, Cryst. Growth Des. 15 (2015) 4543–4548.
[37] N.C. Burtch, K.S. Walton, Molecular-level insight into unusual low pressure CO_2 affinity in pillared metal–organic frameworks, Acc. Chem. Res. 48 (2015) 2850–2857.
[38] D. Singh, C.M. Nagaraja, Auxiliary ligand-assisted structural variation of Cd(II) metal–organic frameworks showing 2D → 3D polycatenation and interpenetration: synthesis, structure, luminescence properties, and selective sensing of trinitrophenol, Cryst. Growth Des. 15 (2015) 3356–3365.
[39] A.L. Garay, A. Pichon, S.L. James, Solvent-free synthesis of metal complexes, Chem. Soc. Rev. 36 (2007) 846.
[40] M. Rubio-Martinez, C. Avci-Camur, A.W. Thornton, I. Imaz, D. Maspoch, M.R. Hill, New synthetic routes towards MOF production at scale, Chem. Soc. Rev. 46 (2017) 3453–3480.
[41] M. Asif, T. Azaz, B. Tiwari, M. Nasibullah, Propagative isatin in organic synthesis of spirooxindoles through catalysis, Tetrahedron (2023) 133308. https://doi.org/10.1016/j.tet.2023.133308.

Chapter 13

Preparation and applications of water-based porous coordination network

Priyanka Singh and Kafeel Ahmad Siddiqui
Department of Chemistry, National Institute of Technology Raipur, Raipur, Chhattisgarh, India

13.1 Introduction

In the context of coordination polymers (CPs), porosity is one of the most investigated individual features. This is not surprising given the wide range of industrial applications for synthetic and natural zeolites, which include gas separation, ion exchange, purification, and petrochemical cracking. CP, like zeolites, is composed of an infinite network of inter-connected structural units capable of holding nanometer-sized gaps. As a result of these applications, commercial interest has emerged in areas for example, heterogeneous catalysis, solvent and gas storage, and molecular separations. Porous coordination gains new applications. These polymers (ligands + metal ions) [1a–e] are an exciting family of porous materials for gas sorption [2a–e], separation [3a–c], catalysis [4a–f], sensors [5a,b], electronics [6a,b], and biological applications [7a,b] due to their framework topologies and pore sizes may be controlled by an adequate selection of metal geometries and organic ligand assemblies. One of the most effective techniques in crystal engineering for both solid analysis and design is to reduce crystal structures to networks. Networks can help with the comprehension of complex structures, or they can serve as a blueprint for the targeting of specific packing arrangements and their related features [8].

Covalent bonds or weak non-covalent interplay-based metal–organic porous CPs cannot participate with zeolites in terms of thermal stability. As a result, oxide-based systems offer unique thermal and chemical stability and could be exploited for potential advantage.

Despite their exceptional porosity and structure tunability, coordination networks' overall stability, and performance can be inclined by a number of aspects, including the structure itself, working environment, moisture exposure, pH, hydrophobicity, metal ions, and a slew of other synthesis and operational

Synthesis of Metal–Organic Frameworks via Water-Based Routes: A Green and Sustainable Approach.
DOI: https://doi.org/10.1016/B978-0-323-95939-1.00008-3

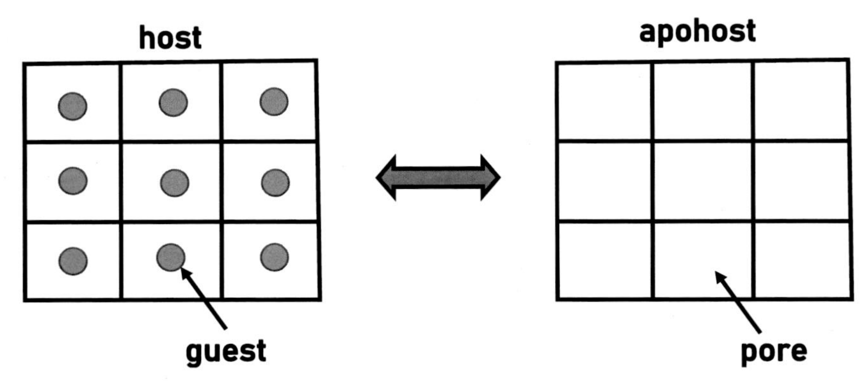

FIGURE 13.1 Schematic representation of permanent porosity.

parameters. Depending on the desired uses, a variety of synthesis techniques can be utilized to produce porous coordination networks [9]. To obtain the recommended functionalities from metal sites on the pore surface, it must be permanently accessible to guests. Though, because of a lack of electronic requirements, the unsaturated state of metal ions is extremely reactive and frequently inconstant [10], consequential in a loss of framework integrity or permanent porosity, and guest molecules are unable to access the metal sites [11]. Metal–organic porous CPs have a structural topology with a controlled cavity size (small to very large), availing the potency usability of different metal ions (node) geometry and versatility, size, and shape of the organic bridging ligands (spacer) [12].

13.1.1 Terminology

When referring to pores, it is well defined as "minute openings through which fluids or gases can pass" [13]. Isotherm studies demonstrate permanent porosity, which is known as the retention of architectural integrity during guest species adsorption and desorption from the host material. Permanent porosity is a type of porosity in which the guest molecules can be removed reversibly from the host material's pores to generate a stable apohost (Fig. 13.1). Permanent porosity can also be established directly through structural characterization of the empty host material (termed the apohost), such as single-crystal or powder X-ray diffraction experiments, but should be validated with parallel gas sorption studies [8].

13.1.2 Designing permanent porosity

While designing permanent porosity, a critical step is to first target either a widget framework, where there are essentially no differences observed between empty and full host structures, or a flexible framework, which may exhibit a significant movement of the host structure with desolvation but without porosity loss.

Seven very general sub-classes will be described to aid in the description of flexible through rigid frameworks: (1) 1D chains, (2) interdigitated 2D layers, (3) 2D stacked layers, (4) 3D inter-penetrated, (5) 3D pillared layer, (6) 3D networks, and (7) pseudo-3D materials are all possible. Porosity can be obtained in 1D, 2D, and 3D materials.

We have compiled a list of 3D porous polymer complexes that were synthesized based on water or that contain lattice water or coordinated water. Despite the abundance of paper, we attempted to compile the last 10 years' worth of reports.

13.2 Preparation

The synthesis of metal–organic CPs is a subfield of supramolecular chemistry that is being investigated in order to understand how crystalline and ordered materials can be created, as well as to elucidate the underlying process of self-assembly. CPs are typically synthesized in a single step via a self-assembly process like gradual evaporation of the reactant solution, layering of two solutions, or one component slowly diffusing to another. Other methods, however, are also used to obtain crystalline CPs. Ligand, metal cation, and metal-to-ligand ratio properties all influence coordination polymer formation, as do temperature, solvent, pH, concentration, time, inorganic counterion, and similar crystallization conditions. Normal synthetic processes can take weeks or months to produce a product, limiting the research of these materials to some extent. Hydrothermal or solvothermal approaches have been tremendously successful in lowering the time required to crystallize coordination polymer to a few days. The solubility of minerals in a solvent or a solvent mixture under high temperature and pressure is crucial in the synthesis of single crystals of CPs utilizing hydrothermal techniques. The history of hydrothermal synthesis may be traced back more than a century to the work of Schafhäutel, Wöhler, and Sénermont [14]. The word hydrothermal customarily represents any heterogeneous reaction that takes place under high pressure and temperature settings in aqueous solvents in order to dissolve and recrystallize materials comparably intractable under normal circumstances. Reactants that would otherwise be difficult to dissolve as complexes become easier to dissolve under hydrothermal conditions. Changes in critical criteria like concentrations, temperatures, and pH can result in a wide variety of network topologies [15] (Table 13.1).

In addition to these conventional methods, the literature describes numerous additional artificial methodologies, including a non-miscible solvent mixture [34], an electrochemical way [35], and a high-throughput route [36]. Microwave irradiation is one of the most promising alternatives because it provides access to a broad temperature range and may be utilized to reduce crystallization time while managing the distribution of particle size and face morphology [37]. However, a significant disadvantage of this method is the common lack of production of large enough crystals to provide excellent structural data.

TABLE 13.1 Synthesis of metal coordination polymer.

S. no.	Ref code	Synthesis	Synthesis method	Geometry	References
1.	BONNEY	$Cu(OH)_2$ (0.50 mmol) + H_2pydc (0.50 mmol) + bpp (0.50 mmol) + 20 mL H_2O + 10 mL MeOH	Room temperature	Distorted square pyramidal	[16]
2.	BODWIC10	(a) HBTC (0.095 mmol) + bpp (0.095 mmol) + DMF (4.8 mL) (b) $Cd(NO_3)_2.4H_2O$ (0.095 mmol) + 3.6 mL H_2O + H_2SO_4 (1 M, 0.2 mL). Each solution was mixed and heated.	Room temperature	Distorted octahedral	[17]
3.	REZFAG	H_4TPT (0.1 mmol) + bib (0.2 mmol) + $CdSO_4{\cdot}8/3H_2O$ (0.2 mmol) + NaOH (0.2 mmol) + H_2O (12 mL) + DMF (2 mL)	Hydrothermal method	Distorted octahedral	[18]
4.	FOSYAP	$Cu(OH)_2$ (0.50 mmol) + H_2pydc (0.50 mmol) + bpp (0.50 mmol) + 20 mL H_2O + 2 mL triethylamine (TEA)	Room temperature	Distorted square pyramidal	[16]
5.	NOVWAN	Orotic acid potassium salt (OAK) (15 mg) + 1, 3 dpp (15 mg) + 4 mL H_2O + 1 mL methanol. Added to 1 mL aqueous solution of $Zn(OAc)_2{\cdot}2H_2O$ (40 mg)	Room temperature	Distorted square pyramidal	[19]
6.	FENHIS	$CdCl_2.2.5H_2O$ (0.1 mmol) + H_4BTTB (0.05 mmol) + H_2O (4.0 mL) + DMA (1.0 mL) + 0.5 mol NaOH	Hydrothermal method	Dodecahedral	[20]
7.	COKBIQ	Hatz (0.20 mmol) + $Zn(NO_3)_2{\cdot}6H_2O$ (0.20 mmol) + H_4btc (0.20 mmol) + H_2O (5 mL) + MeCN (5 mL)	Hydrothermal method	Distorted tetrahedral	[21]
8.	WIWGAM	$Pb(NO_3)_3$ (0.20 mmol) + H_4TETA (0.10 mmol) + NaOH(0.2 mL, 1M) + d/w (10 mL)	Hydrothermal method	Distorted pentagonal pyramidal	[22]

(continued on next page)

9.	QICSED	$NiCl_2.6H_2O$ (0.1 mmol) + pytpy (0.1 mmol) + H_2sfdb (0.2 mmol) + CH_3OH (2.5 mL) + H_2O (10.0 mL) + NaOH (0.5 M)	Hydrothermal method	Distorted octahedral	[23]
10.	FAFSUE	$Zn(NO_3)_2{\cdot}6H_2O$ (0.20 mmol) + H_3TCMB (0.05 mmol) + 1H-Benzotriazole (0.1 mmol) + HBF_4 (30 μL) + H_2O (4 mL) + ethanol (4 mL)	Hydrothermal method	Square pyramidal	[24]
11.	$\{[Zn(Or)(Bimb)(H_2O)]_2{\cdot}6H_2O\}_n$	$Zn(NO_3)_2{\cdot}6H_2O$ (0.135 mmol) + H_2O (1 mL) + 4 mL $H_2O{:}CH_3OH$ (3:1) + OAK (0.077 mmol) + Bimb (0.063 mmol)	Room temperature	Distorted trigonal bipyramidal	[25]
12.	COKBOW	$Zn(NO_3)_2{\cdot}6H_2O$ (0.2 mmol) + Hatz (0.2 mmol) + 5-H_2nipa (0.20 mmol) + 10 mL H_2O	Hydrothermal method	Zn1 = distorted tetrahedral Zn2 = distorted octahedral Zn3, Zn4 = distorted tetrahedral	[21]
13.	TIMZIB	$Er(NO_3)_3.5H_2O$, (0.1 mmol) + sulfosalicylic acid (0.1 mmol) + 20 mL H_2O + 4–5 drops NaOH	Hydrothermal method	Distorted square-antiprismatic	[26]
14.	RIXBEI	$Ni(NO_3)_2{\cdot}6H_2O$ (0.04 mmol) + H_5L (0.02 mmol) + 4,4′-bipy (0.02 mmol) + NaOH (0.04 mmol) + H_2O (5 mL)	Hydrothermal method	Ni1 and Ni2 = octahedral	[27]
15.	FAFSOY	$Co(NO_3)_2{\cdot}6H_2O$ (0.20 mmol) + H_3TCMB (0.05 mmol) + 1H-Benzotriazole (0.1 mmol) + H_2O (4 mL) + EtOH (4 mL)	Hydrothermal method	Co1 = octahedral Co2= distorted octahedral	[24]
16.	WEPRUH	NMP (30 mL) + $Zn(OAc)_2{\cdot}2H_2O$ (0.625 mmol) + H_2icdc (0.625 mmol) + dabco (0.31 mmol).	Hydrothermal method	Tetragonal pyramidal	[28]
17.	WEFTUZ01	$Zn(NO_3)_2{\cdot}6H_2O$ (0.2 mmol) + H_2L (0.1 mmol) + 4,4-bipy (0.05 mmol) + DMF (10 mL)	Hydrothermal method	Square pyramidal	[29]

(continued on next page)

TABLE 13.1 Synthesis of metal coordination polymer—cont'd

S. no.	Ref code	Synthesis	Synthesis method	Geometry	References
18.	RIXMAP	$Co(NO_3)_2{\cdot}6H_2O$ (11.6 mg, 0.04 mmol) + H_5L (0.02 mmol) + 4,4′-bipy (0.02 mmol) + NaOH (0.04 mmol) + H_2O (5 mL)	Hydrothermal method	Co1= tetrahedral Co2= distorted octahedral	[27]
19.	PUZWIS	$Ni(NO_3)_2{\cdot}6H_2O$ (0.19 mmol) + TBIB (0.094 mmol) + H_3BTC (0.094 mmol) + EtOH:H_2O (6 mL, 2:1)	Hydrothermal method	Distorted octahedral	[30]
20.	PUZWOY	Crystals of **PUZWIS** were immersed in DCM	Room temperature	Distorted octahedral	[30]
21.	DIGDUV	$Zn(NO_3)_2{\cdot}6H_2O$ (0.2 mmol) + bibp (0.1 mmol) + H_4L (0.1 mmol) + H_2O (3 mL) + DMF (6 mL)	Hydrothermal method	Tetrahedral	[31]
22.	FIMYAD	(a) DEF/H_2O/TEA (2 mL:1 mL:0.125 mL) + H_4BuTC (0.1 mmol) (b) DEF/H_2O(2 mL: 0.9 mL) + (NiLallyl)$(ClO_4)_2$ (0.2 mmol). Both solutions were added slowly.	Room temperature	Square planar	[32]
23.	FIMYEH	(a) [NiLallyl]$(ClO_4)_2$ (0.2 mmol) + acetonitrile (1.0 mL) + H_2O (40 mL) (b) acetonitrile (1 mL) +H_4BuTC (0.1 mmol) + TEA (0.120 mL)	Room temperature	Square planar	[32]
24.	CAJVOC	(a) H_3BTC (0.095 mmol) + Bimb (0.095 mmol) + DMF (4.8 mL) (b) $Cd(NO_3)_2{\cdot}4H_2O$ (0.095 mmol) + H_2O (3.6 mL) + H_2SO_4 (1 M, 0.2 mL) each solution heated and poured in a vial again heated at 80°C for 24 h	Room temperature	Distorted octahedral	[33]
25.	REBJEO01	Urazole (15 mg) + 4,4′ -dipyridyl (15 mg) + $Cd(CH_3COO)_2{\cdot}2H_2O$ (30 mg) + H_2O + methanol	Hydrothermal method	Distorted octahedral	[33]

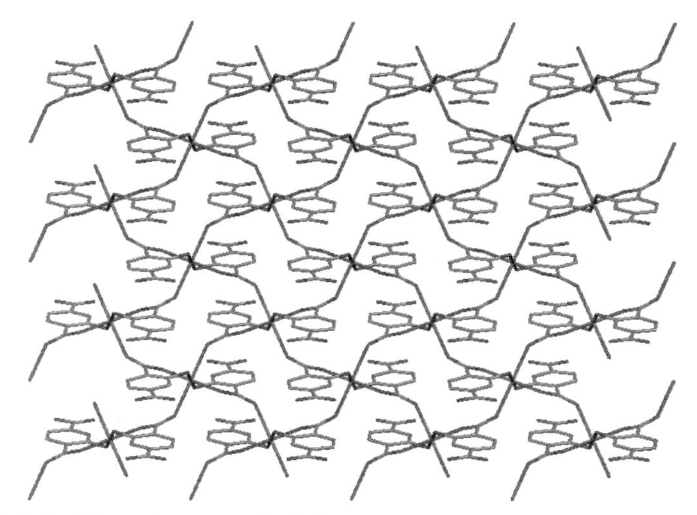

FIGURE 13.2 3D crystal structure of BONNEY.

13.3 Structural analysis

BONNEY's crystal structure contains a five-coordinating Cu(II) atom and it has a coordination environment of distorted square-pyramidal nature, coordinated by two oxygen atoms and one nitrogen atom through two ligands (2,3-pydc^{2-}), one nitrogen atom from one of the ligands (bpp), and one nitrogen atom from another ligand (bpp) [*H_2pydc = pyridine-2,3-dicarboxylic acid, bpp = 1,3-bis(4-pyridyl)propane)*]. The complex's 3D structure exhibits porosity, having a void of 15.3% per unit cell volume. Porosity is evidenced by the 3D structure of the complex; it retains a void of 15.3% per unit cell volume (Fig. 13.2).

BODWIC10 has a six-coordinated coordination environment with distorted octahedral geometry around Cd(II), which is ligated by four oxygens (O1, O1′, O3, O4) from carboxyl groups of three HBTC^{2-} ligands and two nitrogen (N1, N2) from two coordinated BPP ligands and has N,N-Dimethylformamide (DMF) and two aqua ligands in the lattice [*HBTC = 1,3,5-benzenetricarboxylic acid*]. The complex's 3D structure exhibits porosity (Fig. 13.3).

The crystal structure REZFAG features a distorted octahedral Cd(II) ion, which is surrounded by four oxygen atoms from two TPT^{4-} ligands and two nitrogen atoms from two distant bib linkers, generating a hexa-coordinated environment, as well as contains half of the lattice water molecules [*H_4TPT = p-terphenyl-2,2″,5″,5‴-tetracarboxylate acid*, and *bib = 1,3-bis(imidazol-1-yl) benzene*]. Porosity has been shown in the 3D structure of the complex; it retains a void of 20.4% per unit cell volume (Fig. 13.4).

FOSYAP's crystal structure contains Cu(II), which is five-coordinated and has a coordination environment of distorted square-pyramidal nature, coordinated by two oxygen atoms from two ligands (*2,3-pydc^{2-}*), one nitrogen atom

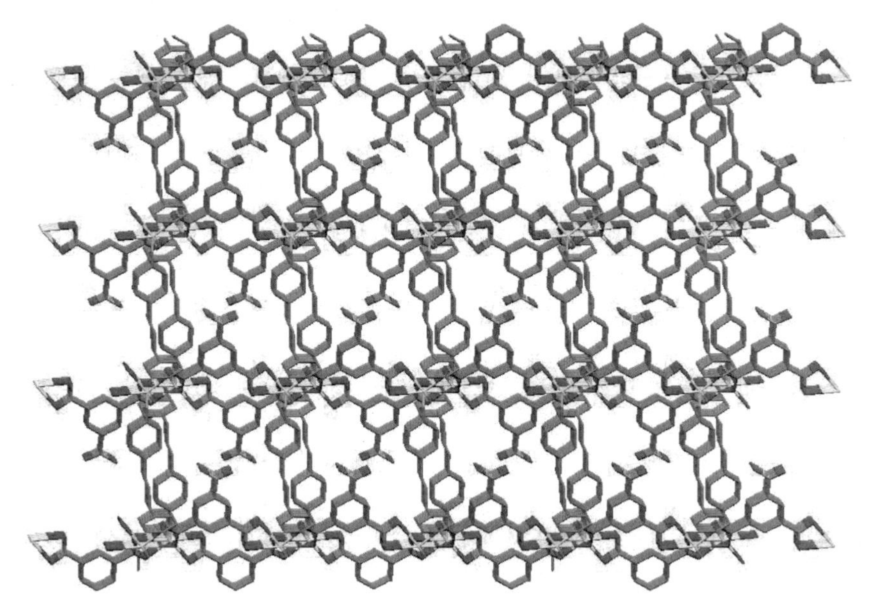

FIGURE 13.3 3D crystal structure of BODWIC10.

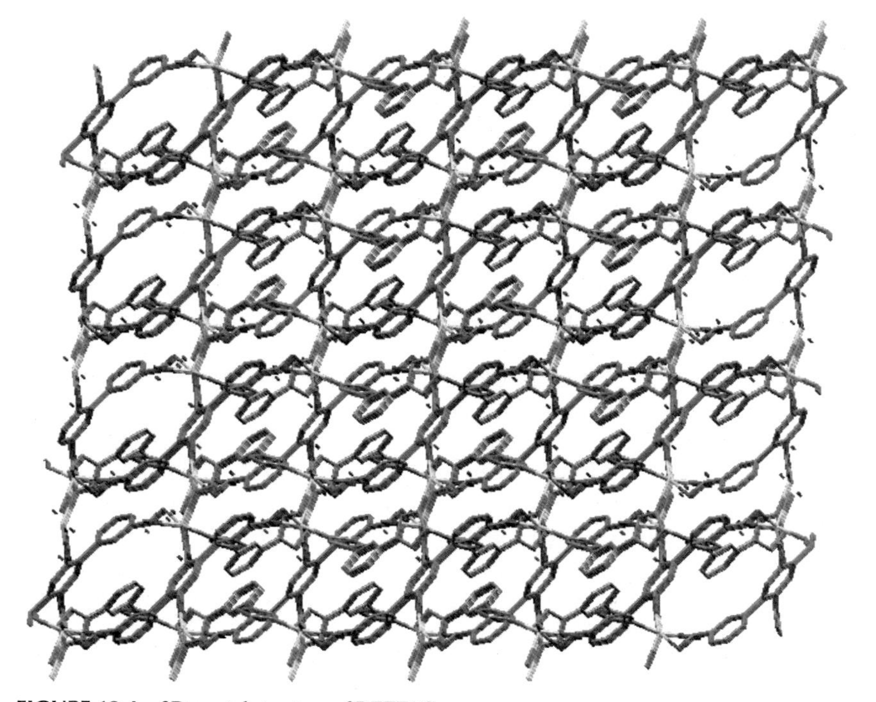

FIGURE 13.4 3D crystal structure of REZFAG.

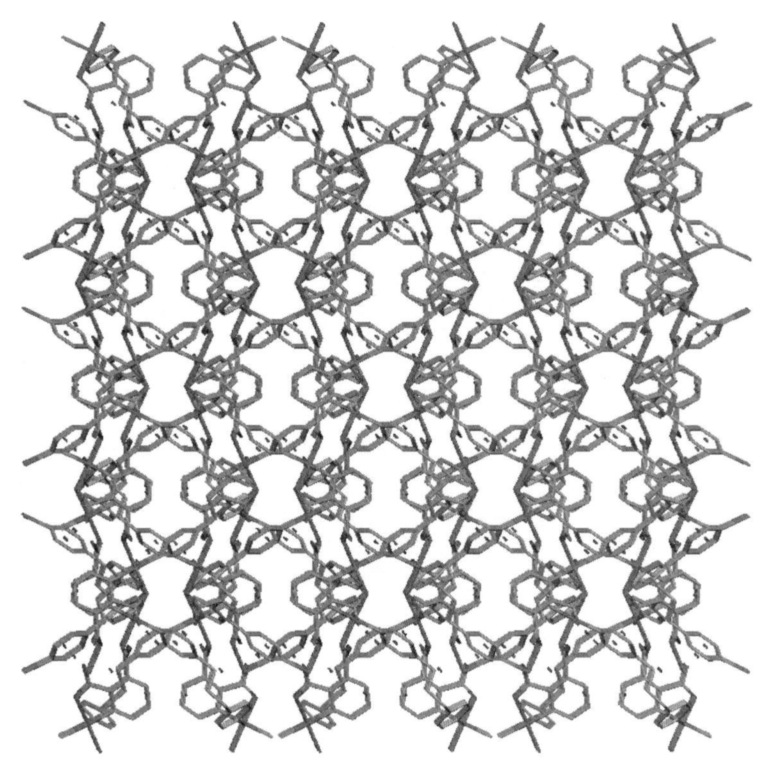

FIGURE 13.5 3D crystal structure of FOSYAP.

of ligand (bpp), and one nitrogen atom from another bpp ligand. Porosity has been shown in the 3D structure of the complex; it retains a void of 22.4% per unit cell volume (Fig. 13.5).

The structure of the crystal NOVWON is made up of two independent Zn(II) ions that are bonded by two orotate ligands (OrK), $2.\frac{1}{2}$ dpp, and two coordinated water molecules with distorted square pyramidal [*dpp = 1,3 di(4-pyridyl)propane, OrK = potassium orotate*]. In a 2D complex structure, porosity may be seen (Fig. 13.6).

The crystal structure FENHIS has a Cd1 core with dodecahedral geometry and is coordinated to eight oxygen (O1, O2, O1A, O2A, O3, O4, O3A, and O4A) atoms, thus, making it octa-coordinated through four distinct anions groups of $(H_2BTTB)^{2-}$ and it has half lattice water [*H_4BTTB = 4,4′,4″,4‴ benzene-1,2,4,5-tetrayltetrabenzoic acid*]. Porosity has been shown in the 3D structure of the complex; it retains a void of 50.6% per unit cell volume (Fig. 13.7).

COKBIQ's crystal structure contains Zn1, which is five-coordinated, depicting distorted tetrahedral geometry, by three oxygen atoms from two different btc^{4-} ligands and two nitrogen atoms from two different atz^{-} ligands, as well as

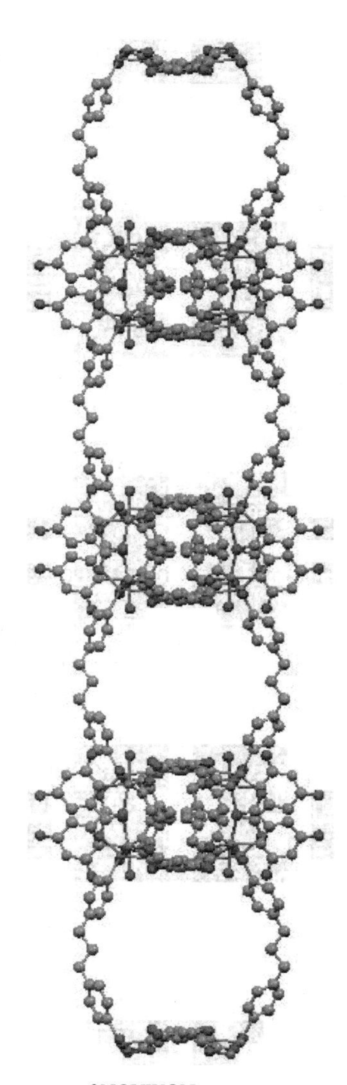

FIGURE 13.6 2D crystal structure of NOVWON.

$1.\frac{1}{2}$ lattice water molecule [*H_4btc =1,2,4,5-benzenetetracarboxylic acid, Hatz = 3-amino-1,2,4-triazole*]. The complex's 3D structure exhibits porosity, with a void accounting for 18.5% of the unit cell volume (Fig. 13.8).

WIWGAM is a distorted pentagonal pyramidal crystal structure that contains a Pb1 atom that is coordinated by two amino nitrogen and two carboxylates oxygen from a single $TETA^{4-}$ ligand, two carboxylate oxygens from the other two $TETA^{4-}$ ligands, showing hexa-coordination environment, and it also has three independent lattice water molecules [*H_4TETA = 1,4,8,*

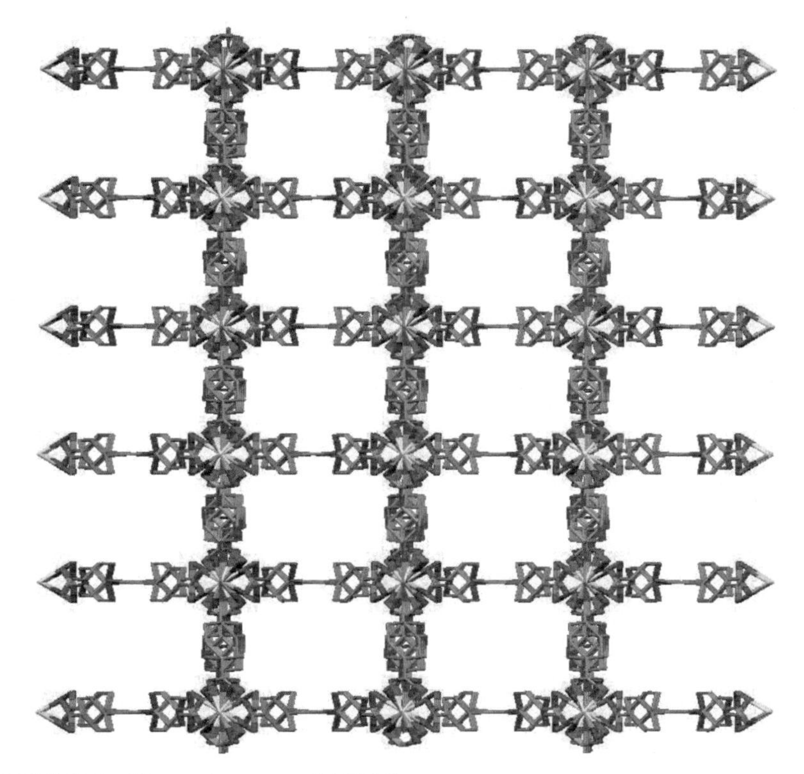

FIGURE 13.7 3D crystal structure of FENHIS.

11-tetraazacyclotetradecane-1,4,8,11-tetraacetic acid]. Porosity has been shown in the 3D structure of the complex; it retains a void of 30.8% of the unit cell volume (Fig. 13.9).

The crystal arrangement QICSED is a distorted octahedral comprises of Ni of six coordinated, and has a geometry of distorted octahedral with a coordination sphere made up of two nitrogen atoms (N1, N4A) and 4 carboxylate oxygen (O1, O2, O5A, and O6A) atoms from different pytpy ligands and a lattice water molecule [*pytpy = 4′-(4-pyridyl)-4,2′:60,400-terpyridine, H_2sfdb = 4,40–sulfonyldibenzoic acid*]. Porosity has been shown in the 3D structure of the complex (Fig. 13.10).

The structure of the crystal FAFSUE includes Zn has a geometry of square–pyramidal, is bonded by three hydroxyl oxygens (O2B, O2A, and O2) as well as two carboxylic acid oxygens (O3 and O1A), and contains half a molecule of lattice water [*H_3TCMB = 1,3,5-tris(carboxymethoxy)benzene*]. Porosity has been shown in the 3D structure of complex (Fig. 13.11).

$\{[Zn(Or)(Bimb)(H_2O)]_2 \cdot 6H_2O\}_n$ has a distorted trigonal bipyramidal geometry and comprises two Zn ions, with three nitrogen atoms (N1, N4, and N5) from one ligand orotate and two ligands bimb and also two oxygen atoms (O1, O5)

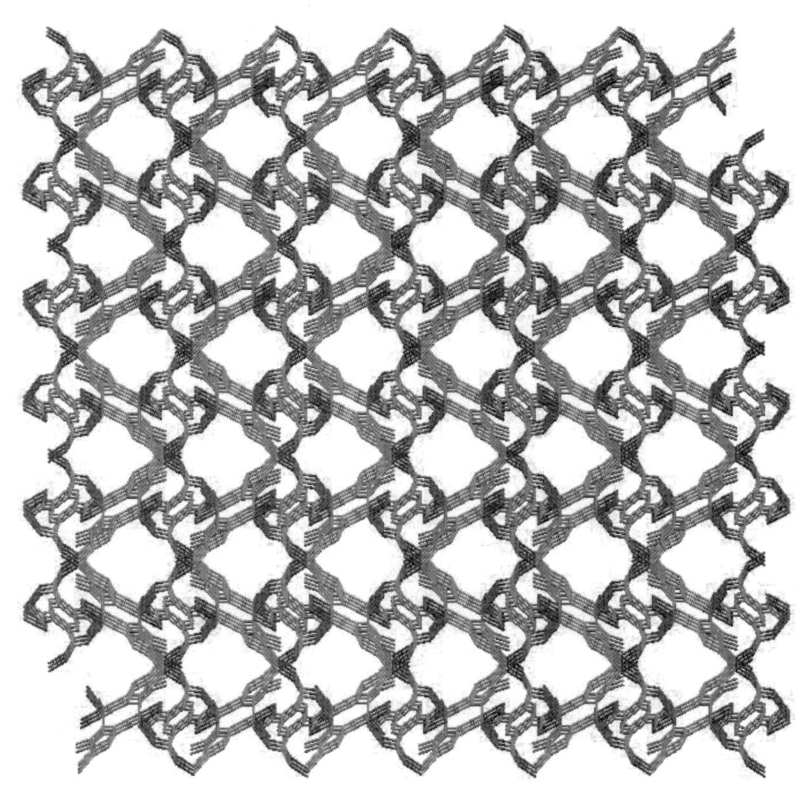

FIGURE 13.8 3D crystal structure of COKBIQ.

from orotate ligand and one aqua ligand surrounding both of the zinc centers. Six lattice water molecules capture the complex's vacancies. Both Zn ions are bonded to an aqua ligand and have six lattice water molecules [*Bimb = 1,4-bis[(1H- imidazole-1-yl)methyl]benzene)*]. Porosity has been shown in the 3D structure of the complex (Fig. 13.12).

The crystal arrangement COKBOW includes four separate Zn^{2+} ions, with Zn1 in a four-coordinated environment, enclosed by three nitrogen atoms out of three ligands (atz^-) and one oxygen atom from $nipa^{2-}$ ligand, resulting in a distorted tetrahedral shape. Zn2 depicts a distorted octahedral coordination structure that is bonded by four oxygen atoms from one $nipa^{2-}$, two nitrogen atoms from two atz^- ligands, 2 μ_3-OH^-, and one coordinated water molecule. One oxygen atom from one $nipa^{2-}$ ligand and three nitrogen atoms from three ligands (atz^-) bonded to Zn3. Zn4 is four-coordinated by three oxygen atoms from two $nipa^{2-}$ ligands and one nitrogen atom from one atz^- ligand and 1 μ_3-OH, and it depicts distorted tetrahedral geometry [*5-H_2nipa = 5–nitroisophthalic acid*]. Porosity has been shown in the 3D structure of the complex (Fig. 13.13).

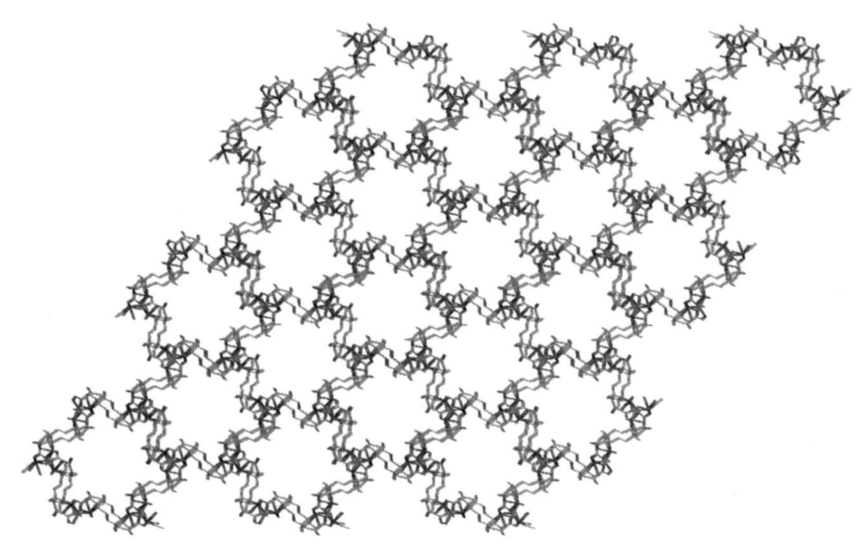

FIGURE 13.9 3D crystal structure of WIWGAM.

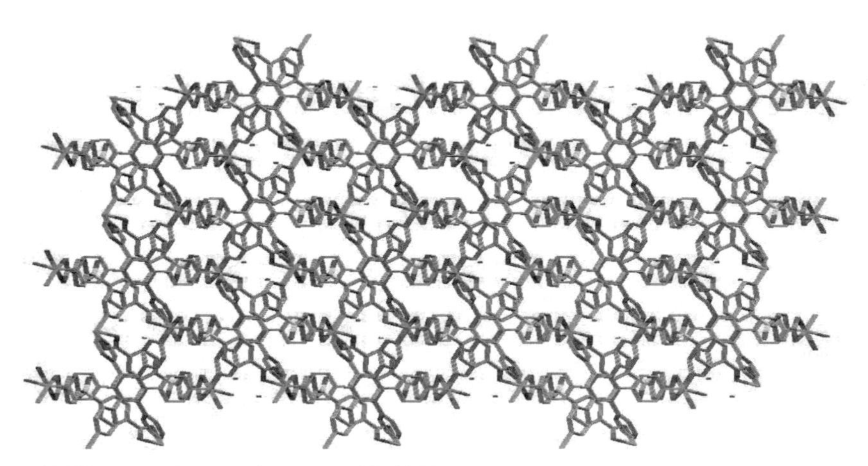

FIGURE 13.10 3D crystal structure of QICSED.

The structure of the crystal TIMZIB has distorted square anti-prismatic geometry and comprises an erbium atom that is bound to two coordinated water molecules and eight oxygen atoms by four SSA ligands, as well as a lattice water molecule [*SSA = 5-sulfosalicylic acid*]. Porosity has been shown in the 2D structure of the complex (Fig. 13.14).

RIXBEI has an octahedral geometry and comprises $1\frac{1}{2}$ nickel ions, one and a $\frac{1}{2}$, 4,4′-bipy, 1 H_2L^{3-} negative ion, three-coordinated water molecules coordinated, and two water molecules in lattice [H_5L = *3,5-(di(2′,5′-dicarboxylphenyl)*

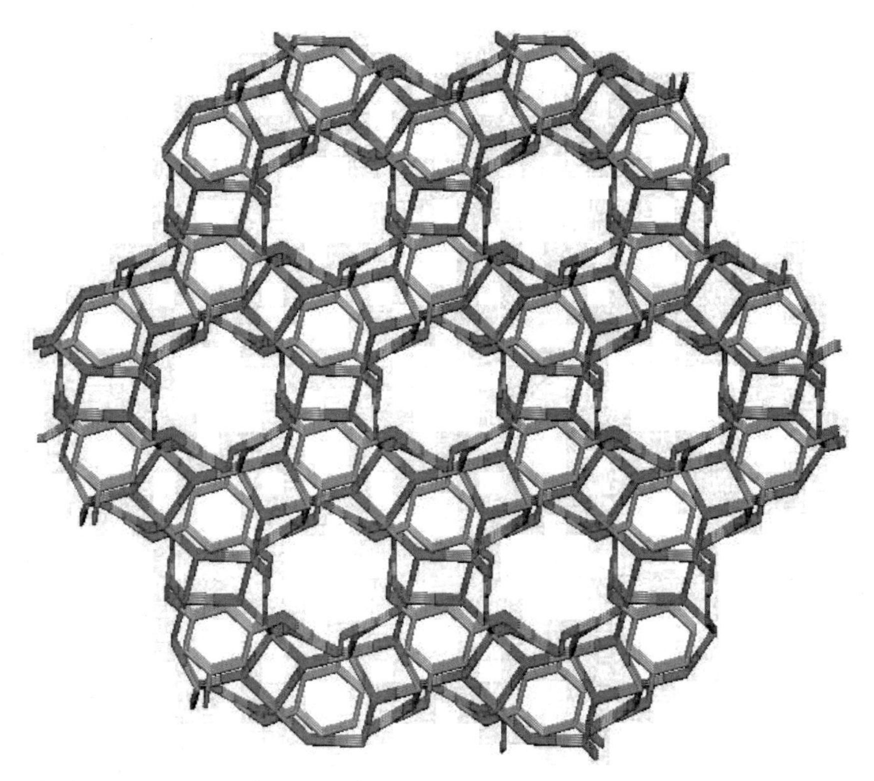

FIGURE 13.11 3D crystal structure of FAFSUE.

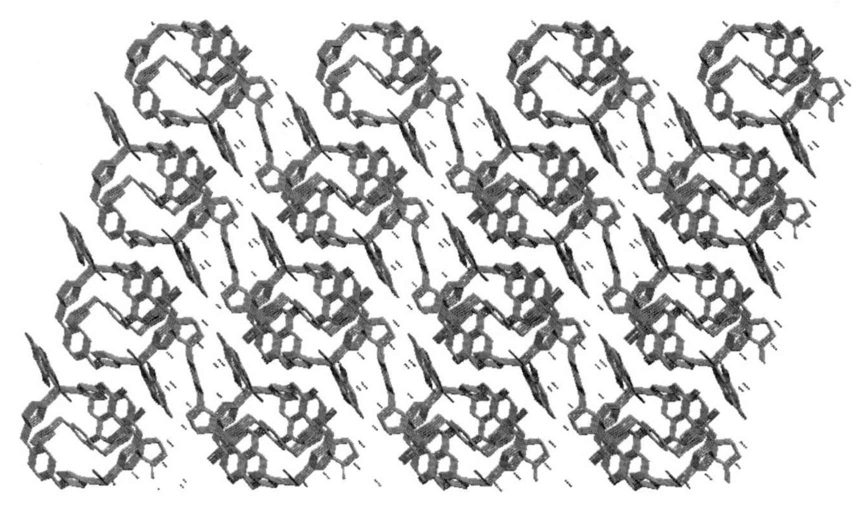

FIGURE 13.12 3D crystal structure of $\{[Zn(Or)(Bimb)(H2O)]_2 \cdot 6H_2O\}_n$.

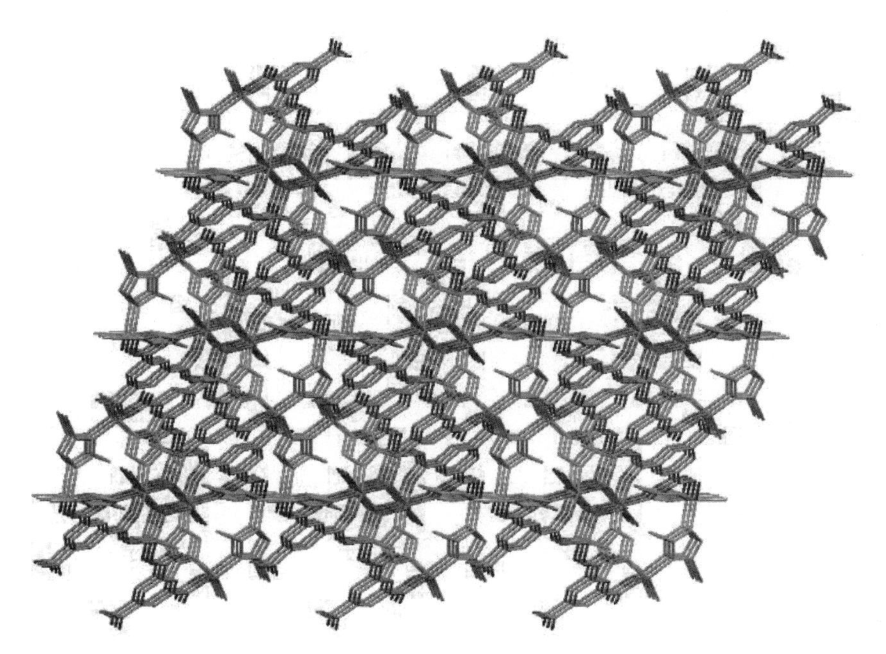

FIGURE 13.13 3D crystal structure of COKBOW.

benzoic acid, 4,4′bipy = 4,4′-dipyridyl]. Complex's 2D structure exhibits porosity (Fig. 13.15).

FAFSOY's crystal structure features Co1, which has an octahedral symmetrical geometry and is surrounded by four carboxylic acid oxygen molecules and two water molecules (O11A and O11). Co2 is linked to five carboxylic acid oxygens and atom O4 from water molecules via the distorted geometry of an octahedron [*H_3TCMB = 1,3,5-tris(carboxymethoxy)benzene*]. Four molecules of lattice water are also present in the compound. Porosity has been shown in the 3D structure of the complex (Fig. 13.16).

Structure of the crystal WEPRUH is a zinc cation with a tetragonal pyramidal coordination environment with a dabco ligand nitrogen atom and four oxygen atoms from four carboxylate groups, and it contains four NMP as well as an aqua ligand in the lattice [*dabco = 1,4-diazabicyclo[2.2.2]octane, NMP = N-methyl-2-pyrrolidone*]. Porosity has been shown in the 3D structure of the complex; it retains a void of 45% per unit cell volume (Fig. 13.17).

WEFTUZ01 has a square-pyramidal shape and comprises three Zn atoms, three L^{2-}, and one and a half BIPY ligands, as well as four DMF and one and a half water molecules as solvates [*BIPY = 4,4′-bipyridine*]. Porosity has been shown in the 3D structure of the complex; it has a void of 43.6% per unit cell volume (Fig. 13.18).

The RIXMAP crystal structure has two Co(II) ions, where Co1 is four-coordinated to three oxygen atoms (O1, O3, and O8) from 1 HL^{4-} carboxylate

FIGURE 13.14 3D crystal structure of TIMZIB.

group and one nitrogen atom from a 4,4′-bipy ligand, exhibiting coordination geometry tetrahedral. Co2 exhibited a distorted octahedral coordination shape with six-coordinated by a coordinated water molecule's oxygen atom, four oxygen (O4, O11, O5, and O6) atoms of one HL^{4-} carboxylate group, and one nitrogen of a ligand (4,4′-bpy) [*H_5L = 3,5-(di(2′,5′-dicarboxylphenyl)benzoic acid, 4,4′bipy = 4,4′-dipyridyl*]. This crystal structure also contains three lattice water molecules. Porosity has been shown in the 3D structure of RIXMAP (Fig. 13.19).

Morphology of the crystal PUZWIS includes two types of Ni(II) ions, one of which (Ni1 and Ni3) form the macrocyclic ring, while the other (Ni2) extends the chain via prolonged bridging with the BTC^{3-} ligand. This crystal structure contains distorted octahedral geometry, six molecules of coordinated water, and nine molecules of aqua ligand in the lattice, as well as five molecules of methanol [*TBIB = 1,3,5-tri(1H-benzo[d]imidazole-1-yl)benzene, H_3BTC = 1,3,5-benzene tricarboxylic acid)*]. Porosity has been shown in the 3D structure of the complex; it retains a void of 36% per unit cell volume (Fig. 13.20).

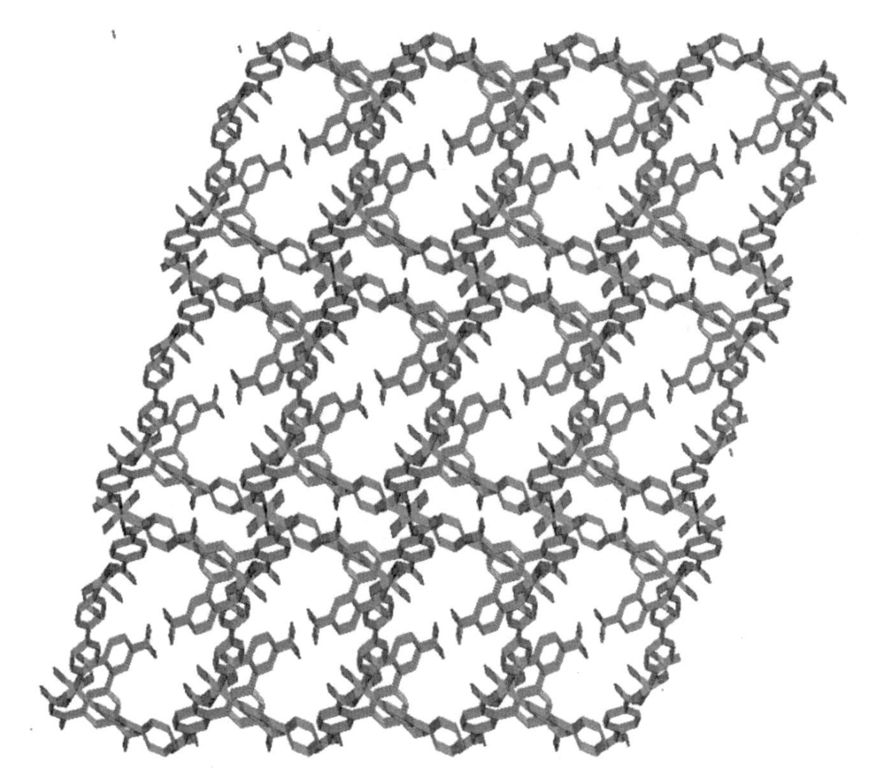

FIGURE 13.15 3D crystal structure of RIXBEI.

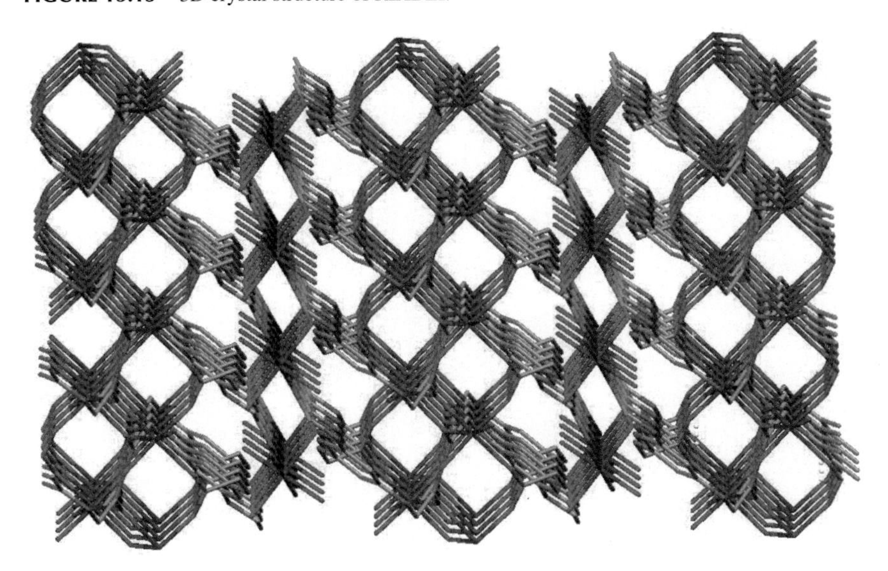

FIGURE 13.16 3D crystal structure of FAFSOY.

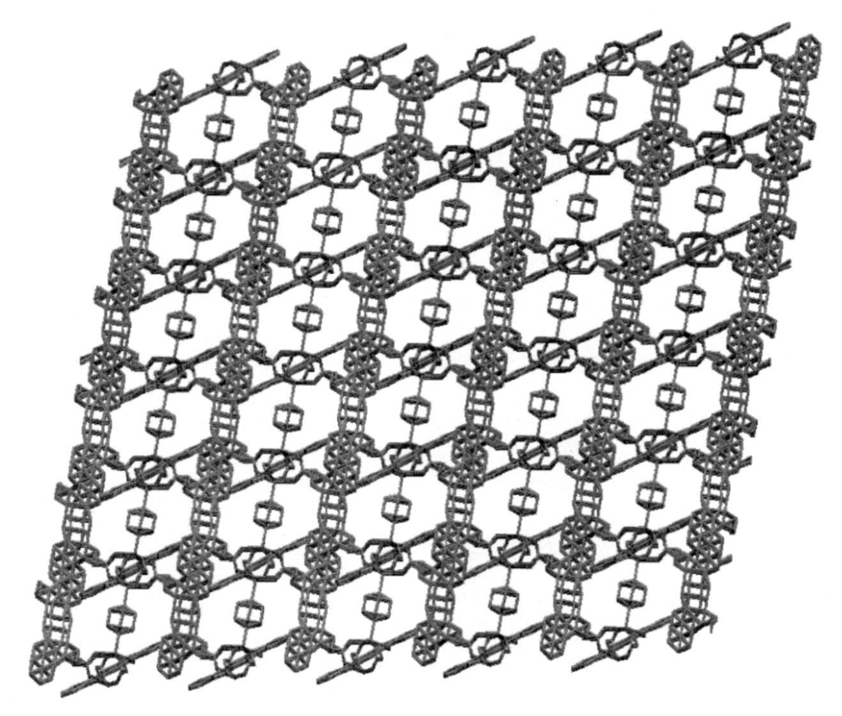

FIGURE 13.17 3D crystal structure of WERPUH.

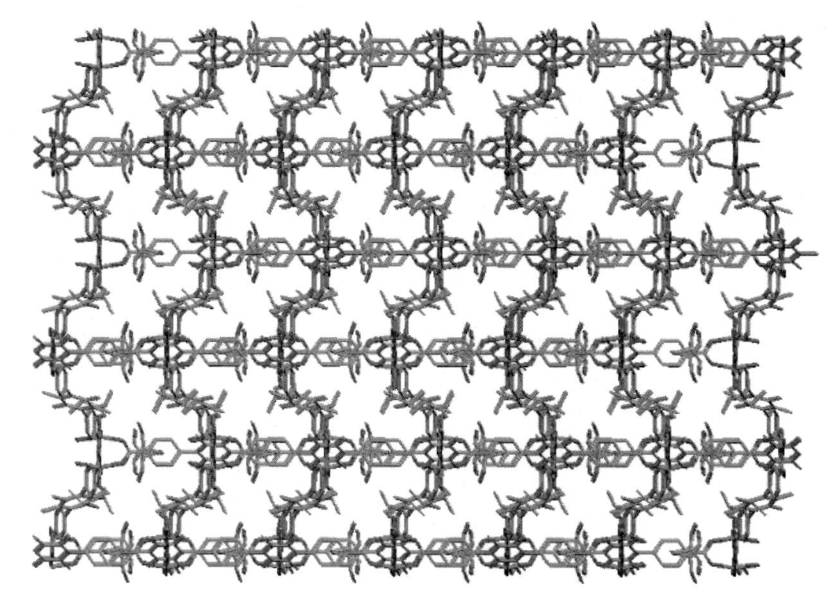

FIGURE 13.18 3D crystal structure of WEFTUZ01.

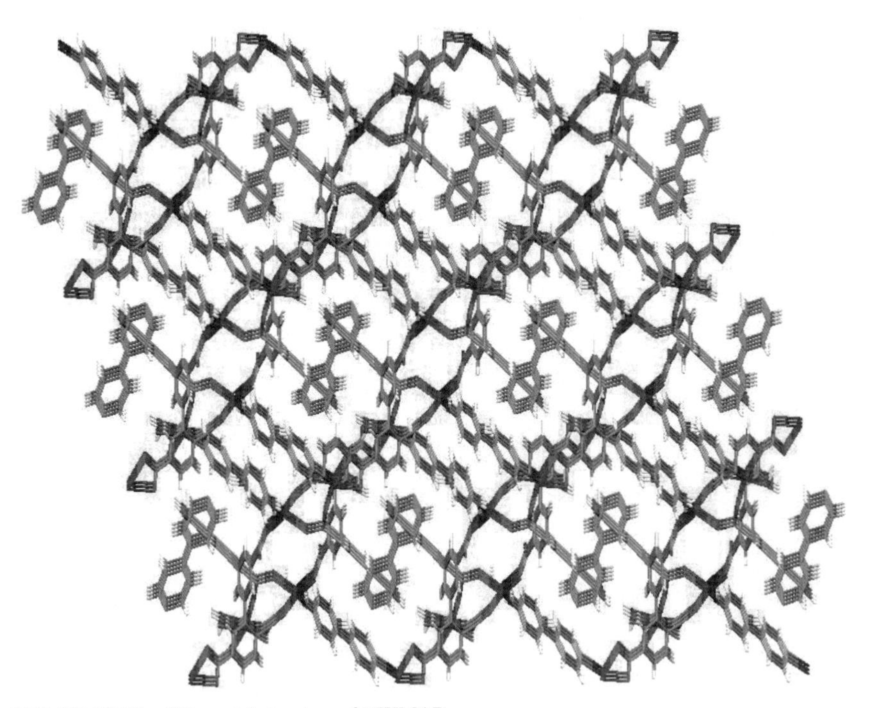

FIGURE 13.19 3D crystal structure of RIXMAP.

The PUZWOY crystal structure has distorted octahedral geometry and comprises two kinds of Ni(II) ions, one fully occupied and the other with a half occupancy factor, one BTC^{3-}, one TBIB, three DCM molecules, three coordinated water molecules, and has three lattice water molecules. Porosity has been shown in the 3D structure of complex; it retains a void of 35% per unit cell volume (Fig. 13.21).

The crystal arrangement DIGDUV includes two Zn ions, the Zn(II) center has a tetrahedral coordination sphere formed by two, O atoms from two distinct ligands (L^{4-}), and two N atoms from two different ligands bibp, and lattice comprises DMF and aqua ligands [*(H_4L = 5,5′-(biphenyl-4,4′-diylbis(methylene))bis(oxy)diisophthalicacid, bibp = 4,4′-bis(imidazolyl)biphenyl)*]. Porosity has been shown in the 3D structure of the complex (Fig. 13.22).

FIMYAD has square planar geometry, two crystallographically distinct Ni(II) macrocycles that are bonded with groups of carboxylate positioned in two distinct environments, the butane chain of the 1,2,3,4-$BuTC^{4-}$ ligand and two molecules of DEF, and two molecules of water present in the lattice [*H_4BuTC = 1,2,3,4-butanetetracarboxylic acid, DEF = N,N′-diethylformamide*]. Porosity has been shown in the 3D structure of the complex (Fig. 13.23).

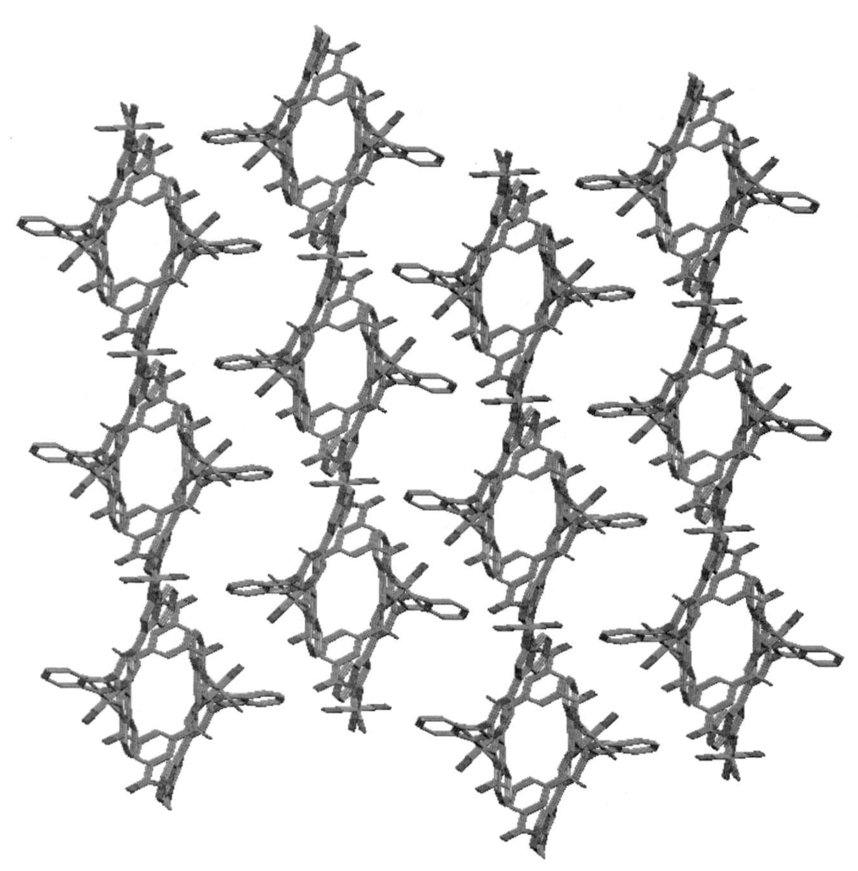

FIGURE 13.20 3D crystal structure of PUZWIS.

The crystal structure FIMYEH exhibits square planar geometry and comprises two crystallographically distinct Ni(II) macrocycles that are coupled with groups of carboxylate positioned in two separate environments, on the butane chain of 1,2,3,4-$BuTC^{4-}$ ligand and three molecules of aqua ligand in lattice [*H_4BuTC = 1,2,3,4-butanetetracarboxylic acid*]. Porosity has been shown in the 3D structure of the complex (Fig. 13.24).

CAJVOC's crystal structure consists of one Cd^{2+} ion, 0.5 $HBimb^+$ molecular ion, three water molecules, and one BTC^{3-} ligand [*H_3BTC = 1,3,5-benzenetricarboxylic acid*]. Cd1 is surrounded by a total of five number of oxygen (O1, O2, O4, O5, and O6) atoms from four different BTC^{3-} ligands with Cd(II), one oxygen atom (O7) from one of the molecules of water, completing six-coordinated distorted octahedral coordination geometry. Porosity has been shown in the 3D structure of the complex (Fig. 13.25).

The morphology of the crystal REBJEO01 includes four acetato groups, two 4,4′-dipyridyl ligands, and one molecule of water [*dpy = 4,4′-dipyridyl*].

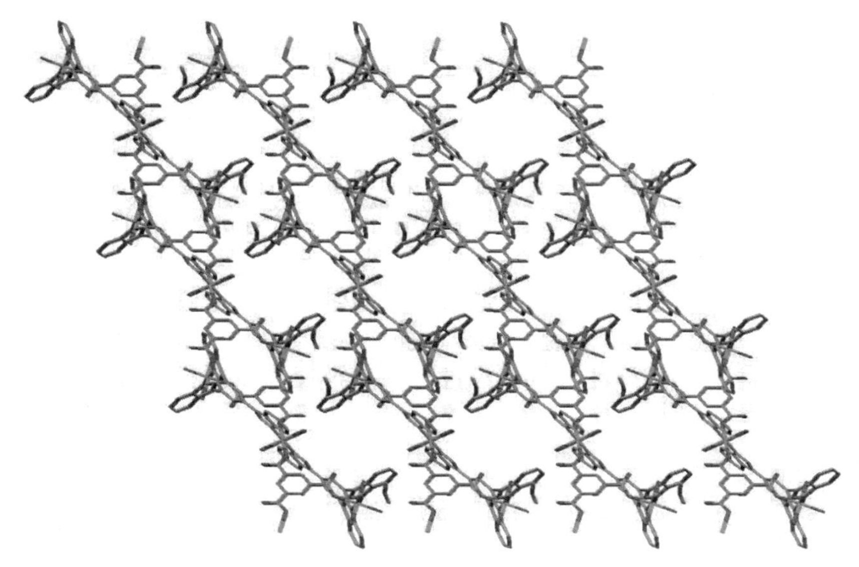

FIGURE 13.21 3D crystal structure of PUZWOY.

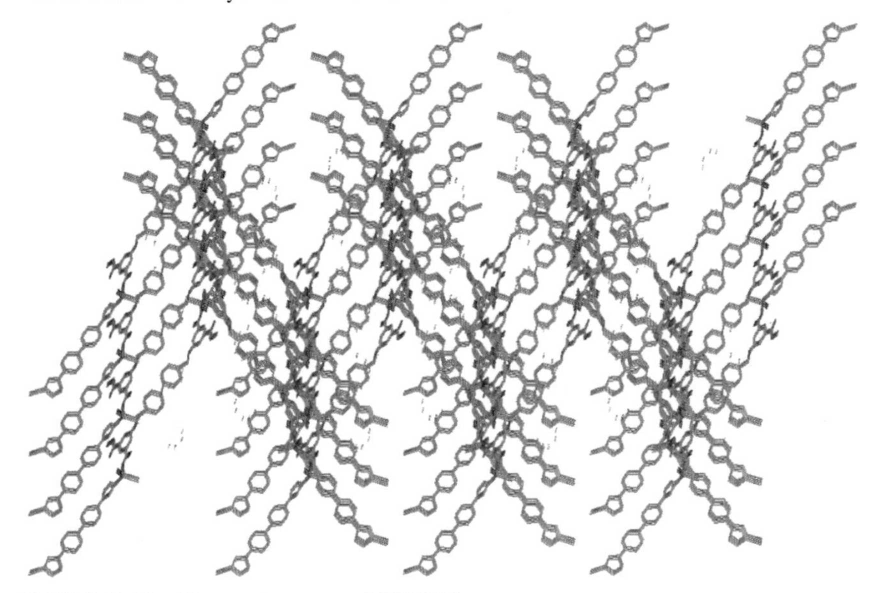

FIGURE 13.22 3D crystal structure of DIGDUV.

Six-coordinated Cd1 is showing distorted octahedral coordination form by the bonding of four of the oxygen (O1, O2) atoms from a single acetato group with chelating $\eta2$ mode and (O3, O7) from two different acetato groups, as well as two of the nitrogen (N3, N4) atoms from two distinctive 4,4′-dpy ligands complex's 3D structure reveals porosity (Fig. 13.26).

FIGURE 13.23 3D crystal structure of FIMYAD.

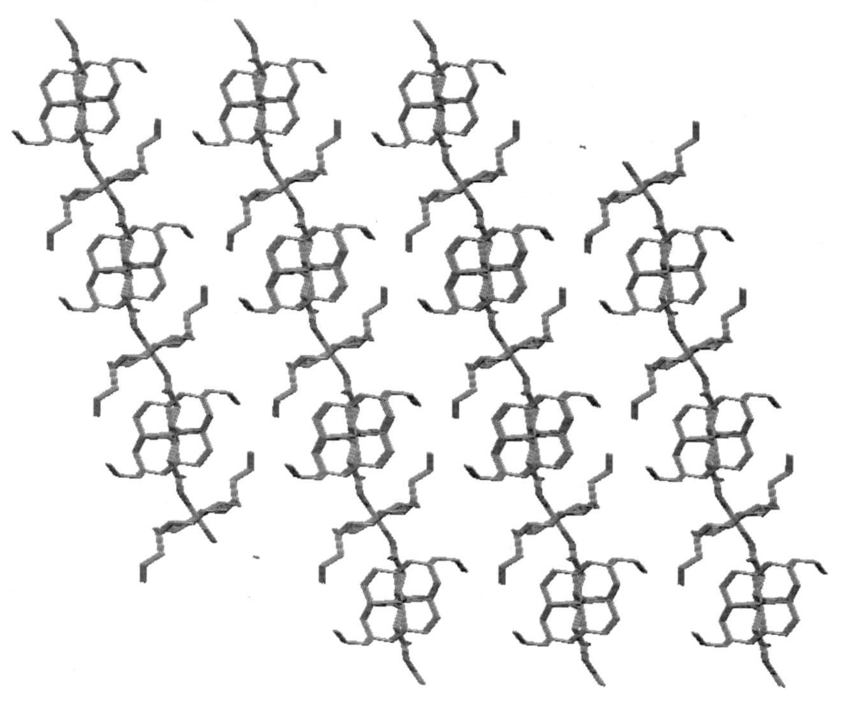

FIGURE 13.24 3D crystal structure of FIMYEH.

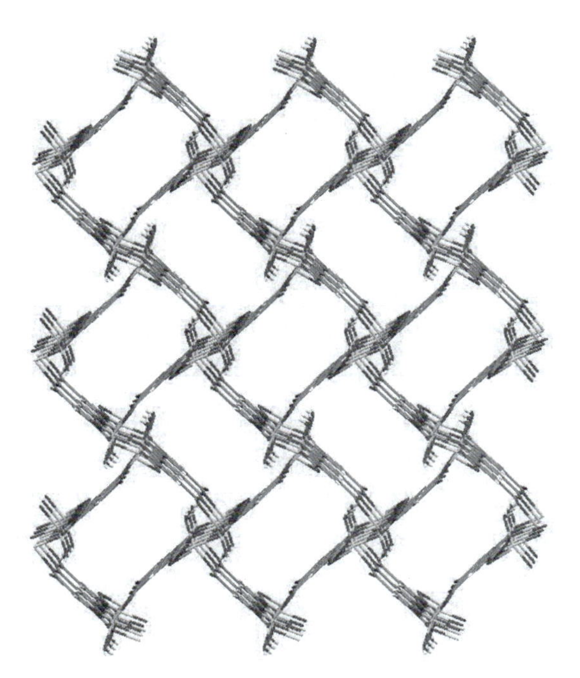

FIGURE 13.25 3D crystal structure of CAJVOC.

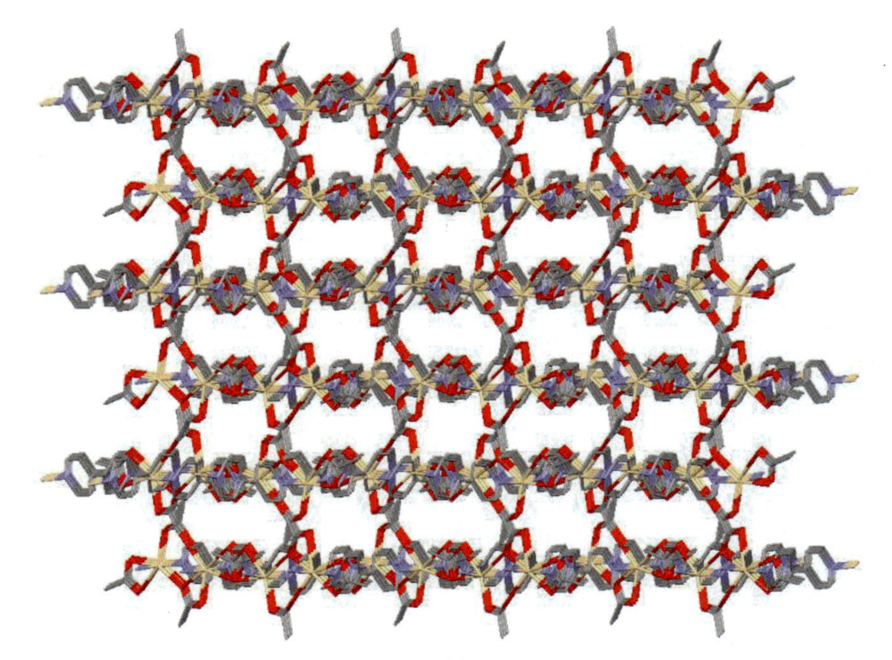

FIGURE 13.26 3D crystal structure of REBJEO01.

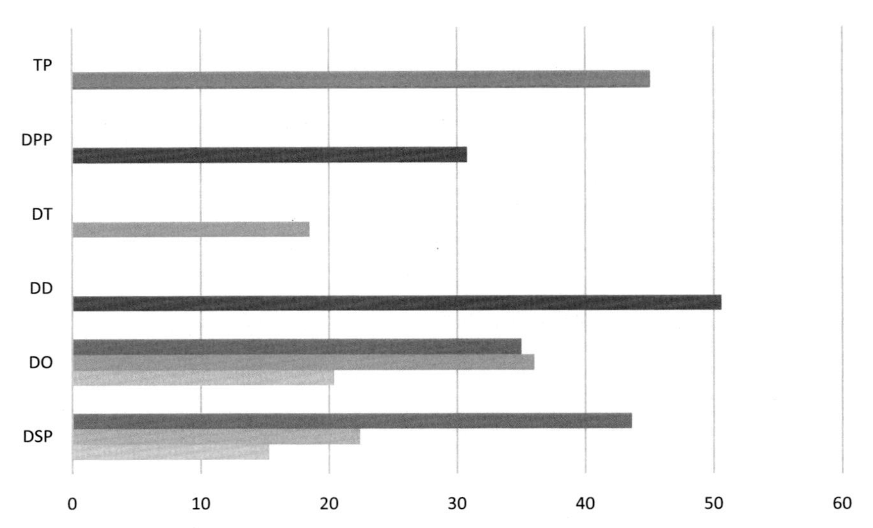

CHART 13.1 Complexes' void percentages based on their geometry.
TP, tetragonal pyramidal; *DPP*, distorted pentagonal pyramidal; *DT*, distorted tetrahedral; *DD*, dodecahedral; *DO*, distorted octahedral; *DSP*, distorted square pyramidal.

As a result of the void percentage detected in 50% of the reported complexes (Chart 13.1), concluded that distorted square pyramidal geometry complexes have porosity ranging from 15.3% to 43.6%, and distorted octahedral geometry complexes have porosity ranging from 20.4% to 36%. A distorted tetrahedral, tetragonal pyramidal and distorted pentagonal pyramidal complex have porosities of 18.5%, 45%, and 30.8%, respectively, whereas a dodecahedral complex has a porosity of 50.6%. As a result, dodecahedral and tetrahedral pyramidal geometries have the most porosity, whereas distorted tetrahedral has the least. Distorted square pyramidal can rise as high as well as have reduced porosity.

13.4 Application

13.4.1 Photoluminescence/magnetism

An excited state-to-ground state electronic transition caused by photo-excitation results in the emission of light [38]. Since Cd(II) ions are not easy to reduce or oxidize because of their d10 configuration, the emission is neither ligand-to-metal charge transfer (LMCT) nor metal-to-ligand charge transfer. After excitation at 270 nm, REZFAG observed a strong emission band at 417 nm. Luminescent carboxylate-based CPs based on d-block metal ions have also been developed. Both 2D and 3D FENHIS have been reported upon excitation at 348 nm it exhibits a sharp emission band at 400 nm. Thereby, the emission peak for both complexes can be attributed to intra-ligand ($\pi^*\rightarrow n$ or $\pi^*\rightarrow\pi$) emission. The carboxylic groups of H_4BTTB ligands remain protonated, suggesting that

TABLE 13.2 Characterization of properties of complexes.

Ref code	Excitation wavelength	Emission maxima	Emission state	χ_{mT}	θ	Magnetic property
BODWIC	390 nm	340 nm	Intra-ligand n→π* π→π*	–	–	–
REZFAG	270 nm	417 nm	Intra-ligand π*→n π*→ π	2.95	−9.95	AF
NOVWAN	290 nm	380 nm	Intra-ligand n→π* π→π*	–	–	–
FENHIS	348 nm	400 nm	Intra-ligand π*→n π*→ π	–	–	–
FAFSUE	280 nm	386 nm	Intra-ligand n→π* π→π*	–	–	–
{[Zn(Or)(Bimb) $(H_2O)]_2 \cdot 6H_2O\}_n$	340 nm	390 nm	Intra-ligand n→π* π→π*	–	–	–
DIGDUV	330 nm	382 nm	LLCT LMCT	–	–	–
TIMZIB	349 nm	450 nm	Intra-ligand n→π* π→π*	–	–	–
WEPRUH	360 nm	400–561 nm	Intra-ligand n→π* π→π*	–	–	–
RVXMAP	–	–	–	5.48	−2.81	AF
RIXBEI	–	–	–	4.05	−2.46	AF

DIGDUV's emission peak was found at 382 nm by causing excitation at 330 nm, resulting from LLCT and admixing with LMCT transition. Zn and Cd complexes mostly show the fluorescent property. From the above table, the χ_{mT} value of the complexes is between 3.00 cm^3/mol K and 5.49 cm^3/mol K and the θ value is between −2.5 K and −10.00 K. The majority of the complexes are anti-ferromagnetic.

they could be host–guest interaction sites that are active, and it leads to the luminescent sensor by opening the door of FENHIS being a candidate.

The luminescence activity in $\{[Zn(Or)(Bimb)(H_2O)]_2 \cdot 6H_2O\}_n$, NOVWON, FAFSUE, BODWIC, TIMZIB, WEPRUH leads toward the intra-ligand charge transfer or n → π^* & π → π^* emissions. Complexes cause excitation from 280 nm to 390 nm causing fluorescence with emission maxima about 340–561 nm. These kinds of fluorescent characteristics are determined by an intra-ligand (n→π^* and π→π^* transitions) emission state (Table 13.2).

TABLE 13.3 Sorption studies of porous complexes.

Ref code	Gas storage	Heat of adsorption (K)	Adsorption amount (cm^3/g)		
			CO_2	N_2	H_2
BODWIC	CO_2 and N_2	363	17.9	6.5	–
		453	17.4	5.8	–
		573	17.2	8.7	–
FENHIS	CO_2 and N_2	77	–	Cannot diffuse	–
		195	28.08		–
WEFTUZ	N_2	453	–	6.6×10^5	–
PUZWIS	CO_2, N_2, and H_2	77 (N_2 and H_2)	–	Cannot diffuse	Cannot diffuse
		195	29.76×10^5	–	–
		273	55	–	–
		298	40	–	–
FIMYAD	H_2 and N_2	77	–	3.2×10^{-2}	Cannot diffuse
FIMYEH	H_2 and N_2	77	–	8.1×10^{-2}	Cannot diffuse

13.4.2 Sorption

Compression for the gases of interest (e.g., H_2, N_2) at high pressures and at room temperature is required for the secure transport and storage of gases at high concentrations. Porous CPs with a large proportion of homogeneous microporosity and a large degree of tuneability have appeared as attractive alternatives for solid gas adsorbents.

The accompanying Table 13.3 shows that maximal sorption has occurred in the CO_2 and N_2 gases of interest. The pore volume absorbed declines as the temperature rises. As the temperature rises, the layers compress and the available space between them shrinks, leaving very little room for gas molecules to be caught between them. For CO_2 storage, experiments were conducted across a temperature range spanning from 195 K to 573 K. The corresponding measurements of pore volume absorption exhibited values between 17.2 and 28.08, with the exception of an anomalous result at 29.76×10^5. Similarly, N_2 storage was examined over a temperature range of 77 K to 573 K, revealing pore volume adsorption values ranging from 3.2×10^{-2}, with the exception of 6.6×10^5. These findings shed light on the adsorption behavior of CO_2 and N_2 and contribute to our understanding of their storage properties. Then H_2 does not show any diffusion to any of the above complexes (Table 13.4).

13.4.3 Variable temperature luminescence analysis

This particular investigation was carried out by our group, and the application in variable temperature photo-luminescence has recently been examined in these articles [17,25,33]. We evaluated the luminescence properties of complexes by

TABLE 13.4 Emission maxima at different temperatures of complexes.

Ref code	λ_{ex} (nm)	As synthesized λ_{max} (em)	90°C λ_{max} (em)	180°C λ_{max} (em)	300°C λ_{max} (em)	500°C λ_{max} (em)
$\{[Zn(Or)(Bimb)(H_2O)]_2 \cdot 6H_2O\}_n$	340	390	427	364	363	439
BODWIC10	340	390 and 364	390 and364	390 and 364	390 and 363	–
CAJVOC	350	391	391	390	390	–
REBJEO01	330	396	399	396	399	–

From the above table, it can be concluded that the emission intensity of as synthesized complexes as compared to the emission intensity of complexes heated at different temperatures shows changes in their property.

heating them at various temperatures such as 100°C, 120°C, and 200°C and studying the influence on luminescence properties.

13.5 Conclusion

We have discussed many of the 3D porous coordination networks, either there was the presence of water in synthesis or there was coordinated or lattice water in the complexes. The building block method by which porous frameworks are constructed is a significant factor in giving them strength relative to other porous materials. This approach has resulted in a number of significant porous material series. CPs have advanced beyond the level at which other porous materials held the records and, in certain cases, exceeded government objectives set for future years in such porous applications as gas storage and molecular separations. Indeed, among the vast range of porosity applications that CPs are currently finding, there are instances that demonstrate genuine potential future uses in the medical world, for example, targeted drug administration, and the physical world, for example, molecular sensing, gas separations, and electrical devices.

Acknowledgment

Dr. Kafeel Ahmad Siddiqui is thankful to the Council of Scientific and Industrial Research, New Delhi, India for awarding a research grant (01(2962)/18/EMR-II dated on May 1, 2018) to support his "crystal engineering" research.

References

[1] [1a] O. Yaghi, M. O'Keeffe, N. Ockwig, H.K. Chae, M. Eddaoudi, J. Kim, Reticular synthesis and the design of new materials, Nature 423 (2003) 705–714; [1b] S. Kitagawa, R. Kitaura, S. Noro, Angew, Functional porous coordination polymers, Chem. Int. Ed. 43 (2004) 2334–2375; [1c] G. Ferey, C. Mellot-Draznieks, C. Serre, F. Millange, Crystallized frameworks with giant pores: are there limits to the possible? Acc. Chem. Res. 38 (2005) 217–225; [1d]

D. Bradshaw, J.B. Claridge, E.J. Cussen, T.J. Prior, MJ. Rosseinsky, Design, chirality, and flexibility in nanoporous molecule–based materials, Acc. Chem. Res. 38 (2005) 273–282; [1e] C.Z.M. Wang, Photosynthetic modification of metal-organic frameworks, Chem. Soc. Rev. 38 (2009) 1315–1329.

[2] [2a] M. Kondo, T. Yoshitomi, K. Seki, H. Matsuzaka, S. Kitagawa, Three-dimensional framework with channeling cavities for small molecules: $\{[M_2(4,4'\text{-bpy})_3(NO_3)_4]\cdot xH_2O\}_n$ (M = Co, Ni, Zn), Angew. Chem. Int. Ed. 36 (1997) 1725–1727; [2b] J.L.C. Rowsell, A.R. Millward, K.S. Park, O. M Yaghi, Hydrogen sorption in functionalized metal-organic frameworks, J. Am. Chem. Soc. 126 (2004) 5666–5667; [2c] A.R. Millward, OM. Yaghi, Metal−organic frameworks with exceptionally high capacity for storage of carbon dioxide at room temperature, J. Am. Chem. Soc. 127 (2005) 17998–17999; [2d] R.E. Morris, PS. Wheatley, Gas storage in nanoporous materials, Angew. Chem., Int. Ed. 47 (2008) 4966–4981; [2e] A.M. Dinc, J.R. Long, Angew, Hydrogen storage in microporous metal-organic frameworks with exposed metal sites, Angew. Chem. Int. Ed. 47 (2008) 6766–6779.

[3] [3a] U. Mueller, M. Schubert, F. Teich, H. Puetter, K. Schierle-Arndt, J. Pastree, Metal–organic frameworks: prospective industrial applications, J. Mater. Chem. 16 (2006) 626–636; [3b] J.-R. Li, R.J. Kuppler, H.-C. Zhou, Selective gas adsorption and separation in metal–organic frameworks, Chem. Soc. Rev. 38 (2009) 1477–1504; [3c] S. Couck, J.F.M. Denayer, G.V. Baron, T. Rèmy, J. Gascon, F. Kapteijn, An amine-functionalized MIL-53 metal−organic framework with large separation power for CO_2 and CH_4, J. Am. Chem. Soc. 131 (2009) 6326–6327.

[4] [4a] M. Fujita, Y.J. Know, S. Washizu, K. Ogura, Preparation, clathration ability, and catalysis of a two-dimensional square network material composed of cadmium(II) and 4,4′-bipyridine, J. Am. Chem. Soc. 116 (1994) 1151–1152; [4b] S. Hasegawa, S. Horike, R. Matsuda, S. Furukawa, K. Mochizuki, K. Yoshinori, et al., Three-dimensional porous coordination polymer functionalized with amide groups based on tridentate ligand: selective sorption and catalysis, J. Am. Chem. Soc. 129 (2007) 2607–2614; [4c] S. Horike, M. Dincă, K. Tamaki, J.R. Long, Size-selective Lewis acid catalysis in a microporous metal-organic framework with exposed Mn^{2+} coordination sites, J. Am. Chem. Soc. 130 (2008) 5854–5855; [4d] J. Lee, O.K. Farha, J. Roberts, K.A. Scheidt, S.T. Nguyen, J.T. Hupp, Metal–organic framework materials as catalysts, Chem. Soc. Rev. 38 (2009) 1450–1459; [4e] L. Ma, C. Abney, W. Lin, Enantioselective catalysis with homochiral metal–organic frameworks, Chem. Soc. Rev. 38 (2009) 1248–1256; [4f] J. Juan-Alcañiz, E.V. Ramos-Fernandez, U. Lafont, J. Gascon, Building MOF bottles around phosphotungstic acid ships: one-pot synthesis of bi-functional polyoxometalate-MIL-101 catalysts, J. Catal. 269 (2010) 229–241.

[5] [5a] E. Biemmi, A. Darga, N. Stock, T. Bein, Direct growth of $Cu_3(BTC)_2(H_2O)_3{\cdot}xH_2O$ thin films on modified QCM-gold electrodes: water sorption isotherms, Micropor. Mesopor. Mater. 114 (2008) 380–386; [5b] Y. Takashima, V.M. Martínz, S. Furukawa, M. Kondo, S. Shimomura, H. Uehara, et al., Molecular decoding using luminescence from an entangled porous framework, Nat. Commun. 2 (2011) 168.

[6] [6a] S.T. Meek, J.A. Greathouse, M.D. Allendorf, Metal-organic frameworks: a rapidly growing class of versatile nanoporous materials, Adv. Mater. 23 (2010) 249–267; [6b] O. Shekhah, J. Liu, R.A. Fischer, C. Wöll, MOF thin films: existing and future applications, Chem. Soc. Rev. 40 (2011) 1081–1106.

[7] [7a] P. Horcajada, T. Chalati, C. Serre, B. Gillet, C. Sebrie, T. Baati, et al., Porous metal–organic-framework nanoscale carriers as a potential platform for drug delivery and imaging, Nat. Mater. 9 (2009) 172–178; [7b] K.M.L. Taylor-Pashow, J.D. Rocca, Z. Xie, S. Tran,

W. Lin, Postsynthetic modifications of iron-carboxylate nanoscale metal-organic frameworks for imaging and drug delivery, J. Am. Chem. Soc. 131 (2009) 14261–14263; [7c] A.C. Mckinlay, R.E. Morris, P. Horcajada, G. Fèrey, R. Gref, P. Couvreur, et al., BioMOFs: metal–organic frameworks for biological and medical applications, Angew. Chem. Int. Ed. 49 (2010) 6260–6266.

[8] S.R. Batten, S.M. Neville, D.R. Turner, Coordination Polymers Design, Analysis and Application, RSC Publishing, 2000.

[9] M.A. Abdelkareem, Q. Abbas, M. Mouselly, H. Alawadhi, A.G. Olabi, High-performance effective metal–organic frameworks for electrochemical applications, J. Sci. Adv. Mater. Devices 7 (2022) 100465.

[10] H. Nagashima, H. Kondo, T. Hayashida, Y. Yamaguchi, M. Gondo, S. Masudaet, et al., Chemistry of coordinatively unsaturated organoruthenium amidinates as entry to homogeneous catalysis, Coord. Chem. Rev. 245 (2003) 177–190.

[11] J.L.C. Rowsell, O.M. Yaghi, Strategies for hydrogen storage in metal–organic frameworks, Angew. Chem. Int. Ed. 44 (2005) 4670–4679.

[12] T.K. Maji, S. Kitagawa, Chemistry of porous coordination polymers, Pure Appl. Chem. 79 (2007) 2155–2177.

[13] L.J. Barbour, Crystal porosity and the burden of proof, Chem. 11 (2006) 1163–1168.

[14] G.W. Morey, E. Ingerson, The pneumatolytic and hydrothermal alteration and synthesis of silicates, Econ. Geol. 32 (1937) 607–761.

[15] X.-M. Chen, M.-L. Tonga, Solvothermal in situ metal/ligand reactions: a new bridge between coordination chemistry and organic synthetic chemistry, Acc. Chem. Res. 40 (2007) 162–170.

[16] H.-Q. Hao, C.-Y. Wang, Z.-C. Chen, Two different porous coordination polymers regulated by the conformations of flexible ligands, Inorg. Chem. Commun. 49 (2014) 151–154.

[17] L. Tyagi, S. Sahu, R. Singh, P. Lama, Gas sorption and luminescence properties of activated forms of a Cd(II)-coordination polymer, J. Coord. Chem. 74 (2021) 2227–2238.

[18] L. Fan, Y. Zhang, J. Wang, L. Zhao, X. Wang, T. Hu, et al., Modular construction, magnetic property, and luminescent sensing of 3D Mn(II) and Cd(II) coordination polymers based on p-terphenyl-2,2″,5″,5‴-tetracarboxylate acid, J. Solid State Chem. 260 (2018) 46–51.

[19] A.K. Bharati, L.P. Somnath, K.A. Siddiqui, A novel mixed ligand Zn-coordination polymer: synthesis, crystal structure, thermogravimetric analysis and photoluminescent properties, Inorg. Chim. Acta. 500 (2020) 119219.

[20] F. Yuan, C.M. Yuan, H.M. Hu, T.T. Wang, C.-S. Zhou, Luminescence, adsorption and topological analysis of a new two-fold interpenetrating 3D Cd(II) coordination polymer based on a tetracarboxylate ligand, Polyhedron 139 (2018) 257–261.

[21] S.Y. Zhang, B. Liu, S.H. Zhang, K.F. Yue, Z.Q. Huang, Crystal structures and thermal decomposition kinetics of three new Zn(II) coordination polymers based on 3-amino-1,2,4-triazole, J. Solid State Chem. 278 (2019) 120901.

[22] X.D. Zhu, T.X. Tao, W.X. Zhou, F.H. Wang, R.M. Liu, L. Liu, et al., A novel lead(II) porous metal–organic framework constructed from a flexible bifunctional macrocyclic polyamine ligand, Inorg. Chem. Commun. 40 (2014) 116–119.

[23] Y.-F. Zhao, B.-C. Wang, H.-M. Hu, X.-L. Yang, L.-N. Zheng, X. Ganglin, Three interpenetrating coordination polymers with 3D honeycomb networks derived from versatile ligand: 4′-(4-pyridyl)-4,2′:6′,4″-terpyridine, J. Mol. Struct. 1171 (2018) 38–44.

[24] Z.-Y. Li, H. Lin, L. Zhang, Y. Peng, Y. Jin, Z.-B. Pi, et al., Two transition metal coordination polymers: fluorescent sensing property and capability of treating neuropathic pain via regulating the ephrinBs-EphBs signaling pathway, J. Mol. Struct. 1228 (2021) 129697.

[25] S. Sahu, M. Ahmad, K.A. Siddiqui, 0D + 1D = 1D Zn-orotate-bimb polyrotaxane coordination polymer: synthesis, structure, thermogravimetric and variable temperature luminescence analysis, Polyhedron, 215 (2022) 115693.
[26] U. Erkarslan, A. Donmez, H. Kara, M. Aygun, M.B. Coban, Synthesis, structure and photoluminescence performance of a new Er^{3+}-cluster-based 2D coordination polymer, J. Clust. Sci. 29 (2018) 1177–1183.
[27] X.-K. Wang, J.-W. Tian, D.-D. Huang, Y.-P. Wu, L.-Q. Pan, D.-S. Li, Two novel Co(II)/Ni(II) coordination polymers based on 3,5-(di(2′,5′-dicarboxylphenyl)benozoic acid ligand: crystal structures, magnetic properties and oxygen evolution reaction, J. Solid State Chem. 269 (2018) 348–353.
[28] I.S. Khan, D.G. Samsonenko, R.A. Irgashev, N.A. Kazin, G.L. Rusinov, V.N. Charushin, et al., Synthesis, crystal structure and fluorescent properties of indolo[3,2-*b*]carbazole-based metal-organic coordination polymers, Polyhedron 141 (2017) 337–342.
[29] V. Lozan, G. Makhloufi, V. Druta, P. Bourosh, V.C. Kravtsov, N. Marangoci, et al., Synthesis and structure of zinc(II) and cobalt(II) coordination polymers involving the elongated 2′,3′,5′,6′ tetramethylterphenyl-4, 4″-dicarboxylate ligand, Inorg. Chim. Acta 506 (2020) 119500.
[30] R.A. Agarwal, Structural diversity and luminescence properties of the novel $\{[Cd_3(TBIB)_4(BTC)_2]\cdot 2C_2H_5OH\cdot 11H_2O\}_n$ and $\{[Zn(TBIB)(HBTC)]\cdot 3H_2O\}_n$ coordination polymers, Polyhedron 85 (2015) 745–747.
[31] X.M. Meng, Y.C. Li, K.R. Huang, J.Q. Long, L.S. Cui, A multifunctional four-fold interpenetrated coordination polymer: quantifiable evaluation of luminescent sensing for Cr(VI)/Cu(II) and photocatalytic properties, Inorg. Chim. Acta. 482 (2018) 284–291.
[32] S.M. Hyun, T.K. Kim, Y.K. Kim, D. Moon, H.R. Moon, Guest-driven structural flexibility of 2D coordination polymers: synthesis, structural characterizations, and gas sorption properties, Inorg. Chem. Commun. 33 (2013) 52–56.
[33] A. Yadav, M. Ahmad, S. Sahu, K.A. Siddiqui, Luminescent properties of activated forms of rigid ligand Cd(II)-coordination polymers sustained by hydrogen bond, Inorg. Chim Acta 529 (2022) 120660.
[34] X.M. Chen, M.L. Tonga, Solvothermal in situ metal/ligand reactions: a new bridge between coordination chemistry and organic synthetic chemistry, Acc. Chem. Res. 40 (2007) 162–170.
[35] P.M. Forster, M. Thomas, A.K. Cheetham, Biphasic solvothermal synthesis: a new approach for hybrid inorganic—organic materials, Chem. Mater. 14 (2002) 17–20.
[36] U. Mueller, M. Schubert, F. Teich, H. Puetter, K. Schierle-Arndt, J. Pastre, Metal–organic frameworks—prospective industrial applications, J. Mater. Chem. 16 (2006) 626–636.
[37] E. Biemmi, S. Christian, N. Stock, T. Bein, High-throughput screening of synthesis parameters in the formation of the metal-organic frameworks MOF-5 and HKUST-1, Microporous Mesoporous Mater. 117 (2009) 111–117.
[38] K.A. Siddiqui, A.K. Bharati, P. Lama, Zinc-orotate coordination polymer: synthesis, thermogravimetric analysis and luminescence properties, SN Appl. Sci. 2 (2020) 392.
[39] D. Kyriakou, A. Mavrogiorgou, J.C. Plakatouras, A novel 3D Pb(II) coordination polymer from a flexible dicarboxylate ligand which reversibly absorbs water, Inorg. Chem. Commun. 134 (2021) 109076.
[40] S. Jana, S. Karim, S. Paul, E. Zangrando, M.S. El Fallah, D. Das, et al., Carboxylato bridging Cu(II) coordination polymer: structure, magnetism and catalytic reduction of nitrophenols, J. Mol. Struct. 1245 (2021) 131058.

Chapter 14

Metal–organic frameworks for wastewater treatment

Akhtaruzzaman, Samim Khan, Basudeb Dutta and Mohammad Hedayetullah Mir
Department of Chemistry, Aliah University, New Town, Kolkata, West Bengal, India

14.1 Introduction

Water is, possibly, the biggest contributor to the existence of living bodies. It combines atmospheric carbon dioxide through photosynthesis to trap the sun's energy that flows to all living beings in the form of a food chain. In the human body, every part including bones is made of a high amount of water. The major part of the surface of the earth, i.e., nearly 71% is covered with water. At any given time, half of its surface is covered by clouds making the earth a "Water Planet" as seen from space. This enormous system of water cycle, powered by solar energy, continuously exchanges water among the oceans, the atmosphere, and the land (Fig. 14.1) [1].

So far, the physical properties are concerned, water with the chemical formula of H_2O behaves as a universal solvent. In comparison to some alike molecules in terms of chemical formula or molecular weight, it possesses distinguished physical characteristics like extraordinary boiling point, melting point, surface tension, and many more. Such specific properties are attributed to its asymmetric structure coupled with polar character. Although it is electrically neutral on the whole, it exhibits a polar nature as the oxygen atom carries a net negative charge, as well as two hydrogen atoms possess net positive charges resulting in hydrogen bonds among water molecules. The polar nature further contributes to innumerable physical and chemical phenomena that led to water as the most sought-after chemical to mankind.

As evident from the Fig. 14.2, saline water accumulated in seas and oceans constitutes the lion's share of the earth's total water content. Again, the bulk of the freshwater is captured as glaciers and ice caps. Another considerable part is blocked as groundwater pools created in the past geological history. As a result, only the water bodies from oceans, rivers, and lakes are being recycled through evaporation and subsequent precipitation keeping a major part of the water pool inaccessible from the water cycle. Human civilization has been rising by

Synthesis of Metal–Organic Frameworks via Water-Based Routes: A Green and Sustainable Approach.
DOI: https://doi.org/10.1016/B978-0-323-95939-1.00001-0

FIGURE 14.1 Continuous movement of the earth's water in the water cycle [1].

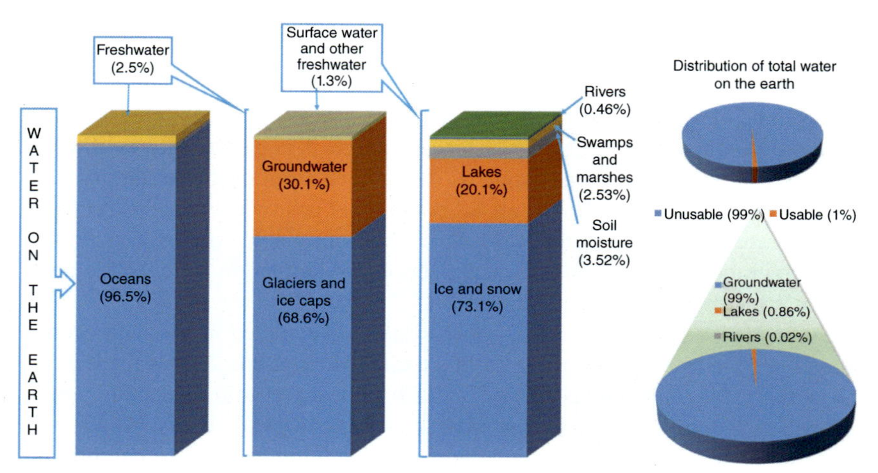

FIGURE 14.2 Earth's water reservoirs bar chart distribution and pie chart of usable water. (*From: United States Geographical Survey) [4]*).

employing this water in the area of food production, energy generation, industrial growth, and many others. Importantly, the growth of modern civilization has put a growing demand for more water. The demand for clean water is increasing because of the increasing population and simultaneous industrial growth all over the world. For example, an average domestic requirement in the United States is around 100 gallons per day per person, whereas, a person needs only around 1 gallon in a day to survive. The level of consumption of water does vary by the people in different countries as per their status of development. It is evident from the point of view of the production level that changing consumption patterns as well as shifting diets towards highly water-demanding foods is a warning of the state of water scarcity. For example, the increasing tendency of meat intake is a worsening signal as 1 kg of beef needs 15,000 L of water [2,3]. Consumption of water is increased from 0.24 billion to 3.8 billion in the last few years, which is nearly an amplification by four times [4,5].

The world is on the verge of enormous water stress owing to the crisis of sufficient clean water to its increasing population in many parts of the globe as illustrated in Fig. 14.3 [6]. It has been estimated that two-thirds population of the world, at present, inhabit in regions where scarcity of water occurs for more than a month in one year and nearly 500 million people survive in regions where consumption of water surpasses renewable water resources in the vicinity by a factor of two [2]. It has been predicted by the World Health Organization (WHO) that, by 2025, around half of the total population of the world is feared to come under water-stressed regions. When a high quantity of water is withdrawn in comparison to the total available water then that region is supposed to be under water stress, which necessitates immediate concern for a sustainable scheme as well as extensive execution of effective water remediation strategies [5]. The global population is increasingly facing water scarcity issues. The situation is getting worse in the low precipitation areas as well as in the densely populated areas leading to population migration and social inequality where poor people are liable to shrink access to clean water [6]. People in the coastal areas of the Gangetic Delta region of India are already experiencing a shortage of clean water due to the intrusion of saltwater during cyclonic storms and this problem is going to get worsened with rising sea levels due to global warming.

The quantity of clean water is dangerously depleting due to the contamination of substances that are potentially harmful to living bodies as well as the ecosystem. This phenomenon of contamination is largely termed water pollution and the contaminated water is otherwise called wastewater. Domestic effluent, industrial effluent, urban runoff, agricultural runoff, etc. are regarded as wastewater. Besides some natural processes like volcanic eruptions that lead to the generation of wastewater, the consistent surge in world population along with fast growth in manufacturing industries and urban sectors, are the common factors for increasing pollution on the earth. Human activities like the effluent generation in industries, resource mining, application of agrochemicals in agriculture, use of fossil fuels, and radioactive leftovers generated in the nuclear power sector are

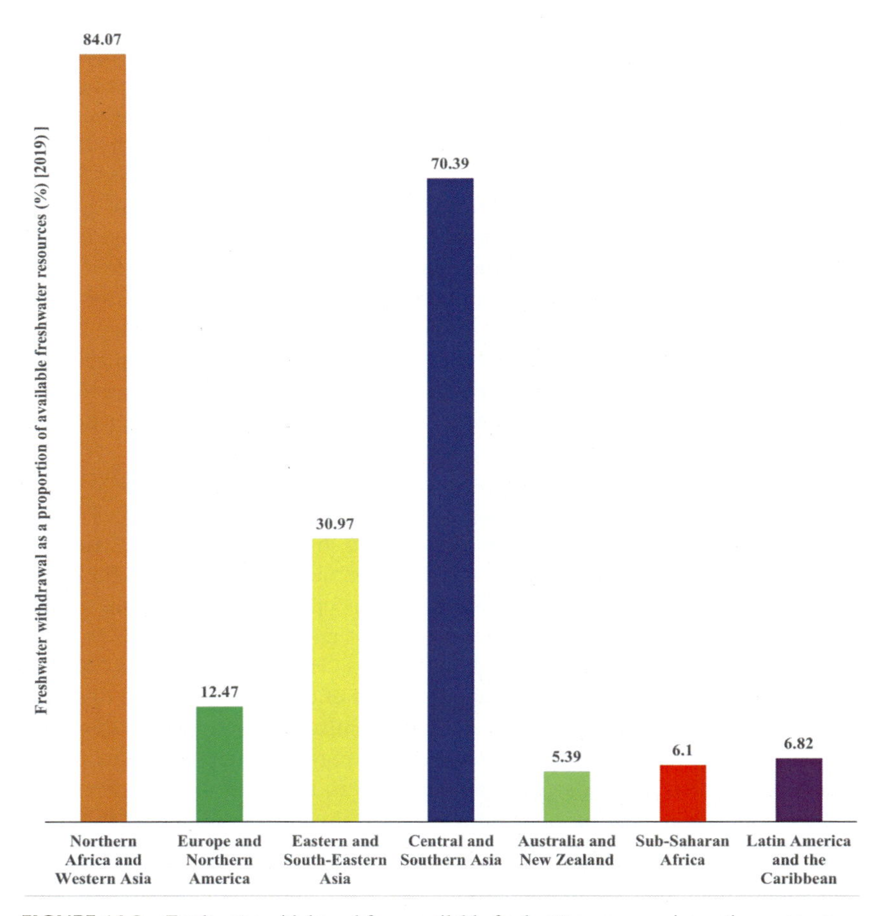

FIGURE 14.3 Freshwater withdrawal from available freshwater resources impacting water stress [6].

some of the examples of major causes of pollution. Since the beginning of the industrial revolution in Europe in the 18th century, lots of factories have been set up worldwide. Consecutive two world wars demanded urgent growth in the industrial sector to tackle the growing population and, in order to give production a fillip; the environmental aspects have been compromised a lot. Since then, an unequaled impact has been put upon the earth having restricted geological reserves [7]. Among the resources, freshwater sources are highly utilized as well as exploited to aid the growth of the industrial sector.

From the viewpoint of the survival of human civilization, we must focus globally to protect clean water and, most importantly, on the urgent need for the remediation of wastewater through efficient and sustainable treatment processes to remove emerging pollutants. Fast improvements in the management of wastewater encouraging reuse through the recycling of water now have become

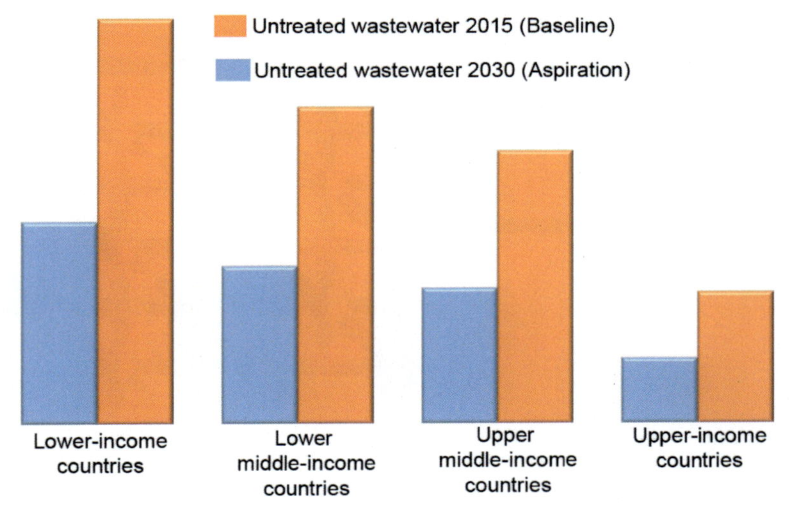

FIGURE 14.4 50% reduction in untreated wastewater for 2030 (aspirations) in comparison to that of 2015 (baseline) for countries with different income levels [2].

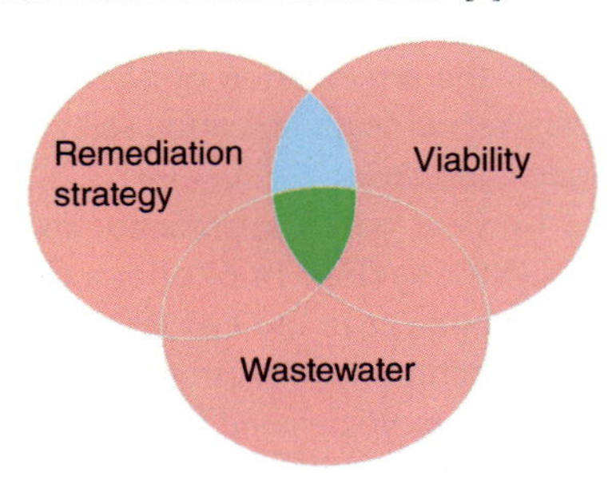

FIGURE 14.5 Viability of remediation strategy for wastewater [10].

a global issue to meet the ever-increasing demand for freshwater [8,9]. Scientists and engineers are extensively putting their efforts in quest of improved and sustainable technologies every now and then for recycling of wastewater to turn it fit again for consumption. In Fig. 14.4, the quantity (%) of untreated wastewater in countries having diverse income levels has been compared for 2015 and as targeted in 2030 and is expected to exhibit a 50% reduction [2].

In monitoring wastewater remediation strategies, mainly in developing nations, sustainable development goals remain a major challenge. Significant investments with upgraded infrastructure and appropriate technologies are the key to the success in the treatment and use of wastewater (Fig. 14.5).

In doing so, efficient and low-cost technologies are vital to remediation of wastewater [10,11]. Obviously, it is unwise to assume that any solitary approach would be the strategy to remove pollutants due to diversity in contaminants. Based upon the physical, chemical, and biological methods, several mechanisms have been so far engaged and effectively employed to treat wastewater. However, many such methods are complex in character in terms of construction

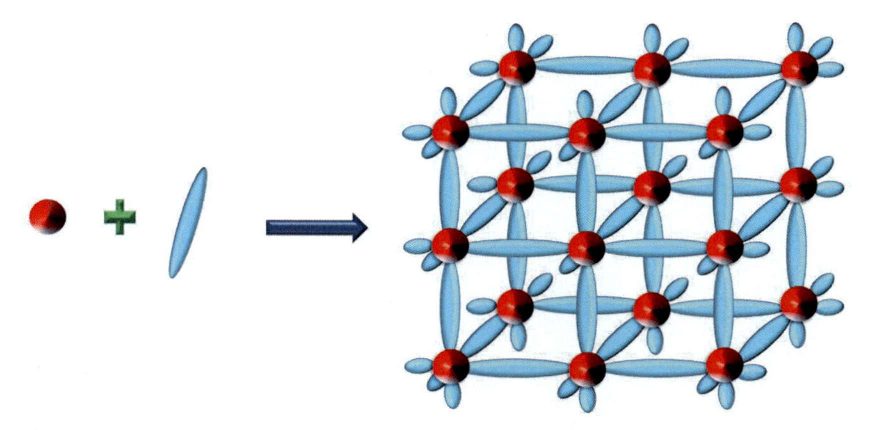

FIGURE 14.6 Schematic representation of synthetic route to MOF.

facilities and consume enough space with the involvement of huge start-up and maintenance expenses [12]. Porous adsorbent materials like iron oxides [13–16], zeolites [17], activated carbons [18,19], and aluminum oxide [20,21], etc. could be an alternative method for the remediation of wastewater. Moderate surface areas and tunability to design specific anion/cation selectivity are the key factors that are associated with such adsorbents [22].

In the area of such adsorbent based approaches in wastewater treatment, metal–organic frameworks (MOFs) (Fig. 14.6) offer a fascinating option. There has been an impressive development in the last few decades in the research of MOFs due to their sundry structural aspects in terms of virtually infinite crystalline networks with different topologies and porous nature. The constructions are made by connecting metal centers with organic linkers through self-assembly via coordination bonds to form several thousands of different MOFs [23–25].

In order to tune the properties and subsequent applicability of the MOFs, the organic linkers are tuned suitably incorporating functional groups. MOFs are constructed as three-dimensional lattices through the interconnection of secondary building units (SBUs) that consist of metal ions or the metal cluster embedded with organic ligands. The resultant topology of the Framework is controlled by the SBUs, whereas, the geometry of SBUs is dependent on several factors like the ligand's structure, the metal's category, the stoichiometric ratio of metal and ligand, the nature of solvent plus the source of anion that neutralize the positive charge of the metal ion. In this viewpoint, MOFs remain a strong contender for potential applications in extensive areas like catalysis [26–29], gas adsorption, accumulation as well as release [30–32], chemical sequestration [33], drug delivery [34], sensing [35], deactivation of chemical warfare agents [36,37], light harvesting [38,39] in addition to its relevance in the removal of toxic substances present in air and water [40]. Judicious selection of linkers and metal ions can lead to the tunability of the size of pores and their shape [41] leading to desired applications. Researchers have gracefully made use of

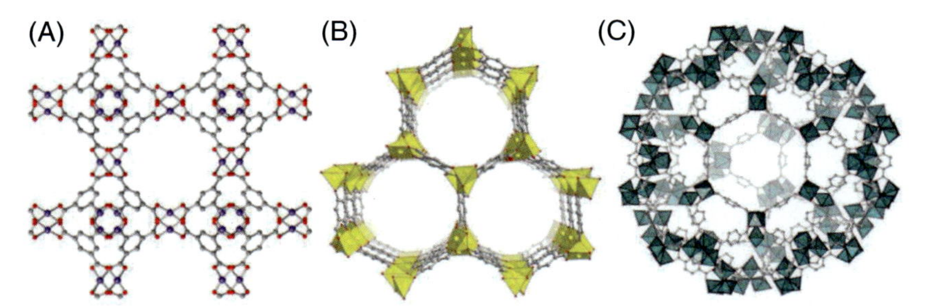

FIGURE 14.7 Structures with porosity of (A) HKUST-1, (B) Mg-MOF-74, and (C) MIL-101 (Cr). [C (*gray*), O (*red*), Cu (*purple*), Mg (*yellow polyhedra*), and Cr (*green polyhedral*)]. (*Reproduced with permission from Elsevier [53]*).

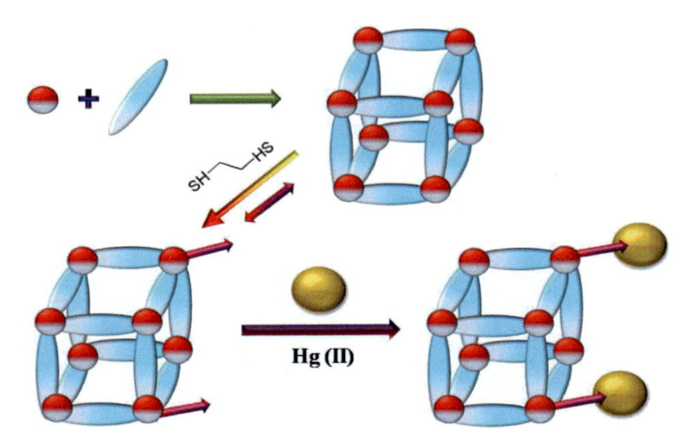

FIGURE 14.8 Illustration of the functionalization process of groups grafted on unsaturated metal centers in MOFs [55].

rational design, to tune the physical and chemical properties of MOFs to possess significant porosity and extraordinarily large surface areas with enough stability [42–46]. Among many advantages of MOFs over other materials, high surface area with porosity is one of the most striking features of MOFs, which has triggered their application as adsorbent materials exposing accessibility to the adsorption site and facilitating the diffusion of impurities of wastewater through the framework (Fig. 14.7). On account of the crystalline structure, the pores also possess highly ordered orientation. Farha et al. in 2012 reported a MOF with a very large BET area, ~7000 m^2/g, which is extraordinary for any porous material [47–52].

Another important aspect of the MOFs is their further tuning via post-synthetic modification (PSM) so that they can be customized for their interaction with the guest molecules [25]. In this regard, a design of a functionalization process of Thiol-HKUST-1 is mentioned in Fig. 14.8 where thiol groups have been embedded onto the unsaturated metal centers as reported by Ke et al. [54]. In regard to accommodating guest molecules, structural transformations are observed leading to the classification of MOFs into three classes by Kitagawa et al. (Fig. 14.8) [55]:

1. First-generation MOFs that suffer irreversible collapse upon elimination of guest molecules,
2. Second-generation MOFs, a type of non-conformationally changeable frameworks, that sustain rigidity even though the guest molecules are removed, and
3. Third-generation MOFs which show flexibility as well as dynamic functionalities resulting to guest responsive adsorption characteristics.

Owing to these advantages, the application of MOFs for removal of contaminants from wastewater is an emergent field of research. It is to keep in mind that, obviously, the applicability of MOFs in treating wastewater involves water-stable MOFs, which would maintain structural stability upon exposure to water. In general, durable coordination bonds or else substantial steric hindrance are exhibited by those water-stable MOFs to avert hydrolysis reaction leading to breakdown of metal-ligand bonds [56]. A good number of water-stable MOFs are known so far in the broad areas of frameworks based on metal carboxylates having high valence metal ions, metal azolate frameworks constructed with N-donor ligands and MOFs imparted with hydrophobic functionalities [56–59]. Metal centers with high valency [60] as well as greater coordination numbers, for example, Zr^{4+}, Fe^{3+}, and Cr^{3+} fabricate a rigid structure with bridging ligands of carboxylate-type to fabricate MOFs which are less vulnerable on exposure to water molecules [60,61]. However, metal ions with lower valency like Co^{2+}, Zn^{2+}, Cu^{2+}, and Ni^{2+} construct water-stable MOFs with azolate ligand types [61]. The robustness of MOFs in water medium could be introduced by imparting hydrophobic functionalities, which inhibit water molecules to approach the lattice plus to rupture the structure of the framework [61]. MOFs, thus, offer an amazing modular approach which can be designed and tuned as per the requirement for elimination of specific contaminants [62].

In this chapter of the book, recent developments focusing on the removal of different type of organic contaminants, heavy metals, pharmaceutical products, dyes, etc. have been discussed with a discussion on future scope of possible role by MOFs in remediation of wastewater through elimination of contaminants.

14.2 Wastewater treatments

The contaminants of wastewater bodies are easily separated out utilizing different chemical technology. Classical methods of wastewater treatment are the collective techniques of physical, chemical, and biological procedures. The main objective of this treatment is common to allow living systems and industrial outcomes to be disposed of without any danger to public health or unnecessary damage to the environment. Actually, the required quality of water is prescribed for its intense use, such as for drinking purposes, aquatic life, irrigation, etc. Therefore, to get an acceptable purity level of this precious commodity, noxious substances are to be removed from the water, which is done through cleaning and removal of toxic metal ions (iron, manganese, cadmium, mercury, etc.), sterilization, and desalination or softening. Besides, some substances are especially supplemented to upgrade the quality and, thereby, control the parameters such as pH value, boiling point, conductivity, etc. In this area of wastewater treatment, metal-coordinated frameworks are potential candidates. They can be strategically designed for employment in the detection, sorption, and destruction of various noxious agents.

14.2.1 Conventional methods

In order to remove solids, organic matters and some unwanted nutrients from wastewater, following procedures are obeyed:

- Physical treatment,
- Biological treatment,
- Chemical treatment,
- Sludge treatment, and
- Membrane treatment.

These classical ways utilized by people in removing pollutants present in the water system are discussed in the following.

- **Physical treatment:** Physical wastewater treatment methods are incorporated by involving racks, screens, comminutors, clarifies (sedimentation and flotation) along with filtration. Processes for mechanical preparation such as aeration, accumulation, or thermal influence are also involved.
- **Biological treatment:** This process of treatment is the subsequent step to the removal of unwanted solid materials. In this process, treatment of water requires involvement of various microorganisms, fungi, algae, or bacteria under aerobic or anaerobic conditions. In this method, large organic matters get oxidized or incorporated and thus can be eliminated. Basically, organic materials or unwanted nutrients are assimilated biological reaction through the microorganisms.
- **Chemical treatment:** This case of wastewater treatment is governed by utilization of common chemical technology. Chemical precipitation, i.e., coagulation and flocculation, is the fundamental technique of this process. In addition, ion exchange, adsorption, absorption, oxidation, and disinfection through chlorination, ozone, UV light, sun light, dichlorination, etc. are directly corresponding to this process. In spite of some limitation and advancement of recent technology, chemical wastewater treatment is also up to the minute.

- **Sludge treatment:** The main purpose of this process is to reduce the organic chemicals and disease-causing biological organisms. This method includes aerobic digestion, anaerobic digestion, and composting. Mainly sewage or industrial wastewater is treated using this process. This treatment decreases potential health risks that are caused through such polluted water.
- **Membrane treatment:** This is the very oldest technique of wastewater treatment. To separate numerous contaminates from water, this process is used. Size, charge, and nature of materials directly govern the method. Generally, this process comprises of filtration, microfiltration, as well as ultra-filtration plus nanofiltration along with reverse osmosis and electro dialysis. Many advanced materials are used in this process.

14.2.2 Recognition of pollutants in wastewater through MOFs

Recently huge attention has been employed for the recognition of noxious pollutants in water body. Thousands of classical methodologies have been designed to treat wastewater as well as to detect impurities in wastewater. Those methods have advantages as well as hassles, in addition, in few of the procedures, decontamination methods are executed sequentially to minimize its shortcomings. Actually, conventional methods of disinfection as well as decontamination processes are suffered from complicated operation technique, high cost for infrastructural maintenance and possibility of generation of secondary pollutants. In addition to the classical methods, there are some other procedures such as: gas chromatography, high performance liquid chromatography, capillary electrophoresis, flame atomic absorption spectroscopy, mass spectrometry, titrimetry, etc. Such methods are also time consuming, laboratory-based with moderate sensitivity and involve high cost of equipment as well as operational complexity. To overcome these limitations, luminescence and electrochemical detection methods are enormously adequate. Therefore, the problem of recognizing technology looks to be solved by these methods. Now, in search of detecting materials, various material components are coming to the surface: diverse fluorescent and redox active organic molecules, different form of nanoparticles, carbon nanotube and graphene, metal oxides, quantum dots, polymers, etc. However, most of these components are endured from some limitations: complex and multistep synthetic procedure, diminishing stabilities upon the introduction of analytes, high cost of materials, low selectivity and sensitivity, short-term constancy, poor recyclability, and reversibility. In many cases, it is hard to connect the structure and property relationship, which is imperative for the expansion in future technological aspect. Hence, it is a great incentive to improve the novel sensing materials with unique properties and excellent performance functionalities in diverse environmental situations. Those are extremely significant for real sample analysis and promotion of sensor research as well as wastewater treatment. In this context, MOFs or metal organic coordination networks (MOCNs) are impending candidates to serve the field. In these materials, effect of metal centers and some vital strategies of ligand selection such as donating atom, spacer length, conjugation in the linker, steric effect, and availability of π-electron on identifying activity are of major concerns. Recognition and treatment of various water pollutants and their corresponding mechanistic features are also significant. The challenges and the point of view for the future direction are well guided through the designing, characterization and resultant efficacy of MOF and MOF-based systems as sensors to sense groundwater contaminants.

14.2.2.1 Luminescence sensor

In MOFs, the functional linkers and different metal nodes play a major role as fluorescent sensors and their luminescence features can be controlled by sensible selection of metal centers and ligand variety like π-electron conjugation, steric hindrance, electronic nature of donor atom of the ligand, spacer length etc. Utilization of functionally decorated ligand moiety as linkers have an inclination to encompass supramolecular interactions with analyte and encapsulation of guest molecules in cavities can modulate the luminescent emission properties as well. Mainly, "turn-on" and "turn-off" type fluorescence sensing are observed. "Turn-on" type may have any of three different categories *viz.* normal fluorescence enhancement, ratio-metric enhancement and wavelength shift. However, in "turn-off" type, only quenching of fluorescence is observed.

Therefore, to remove contamination for a water body and to treat wastewater completely, porous hydride materials are highly efficient in a true manner. Heavy metal ions are extremely toxic and non-biodegradable in nature. Plants and animals are dangerously affected if these pollutants are incorporated in the biological systems. Overloading of some metal entities such as Hg, Cu, As, Pb, Cr, Cd, etc. is highly hazardous to human health and the aquatic ecosystem. A few real examples of wastewater were investigated in recent past for sensing of contaminants as noted here. Wang and his co-workers synthesized a cadmium-based MOF $\{[Cd_{1.5}(C_{18}H_{10}O_{10})]\cdot(H_3O)(H_2O)_3\}_n$, (Cd–EDDA) which displayed dual-emission signals and acted as MOF-implicated ratio-metric sensor for mercury(II) in pure water. In this report, the Cd–EDDA was found to be highly selective and sensitive to Hg^{2+} ions exhibiting fast response (~15 seconds), and the sensing of Hg^{2+} using Cd–EDDA was credited to the breakdown of the Cd–EDDA assembly [63]. Our research group recently synthesized a one-dimensional coordination polymer (1D CP), $\{[Cd_2(adc)_2(4\text{-}nvp)_6]\cdot(MeOH)\cdot(H_2O)\}_n$, (where, H_2adc = 9,10-anthracenedicarboxylic acid, 4-nvp = 4-(1-naphthylvinyl)pyridine). The CP, specifically detected Cr^{3+} ion in the presence of other biologically relevant metal ions by turn-on approach, which is very rare. The limit of detection (LOD) of Cr^{3+} was very impressive and determined by the $3\sigma/m$ method and was found a surprising low value as 0.31 μM in a semi-aqueous medium [64]. Cyanide ion (CN^-) being a chemical warfare agent is vital to regulate and control as well as it demands immediate detoxification. The main origin of cyanide ions is from effluents from industries.

Therefore, major attention needs to be drawn to quantifying and controlling its concentration in drinking water and wastewater. Karmakar et al. demonstrated guest-induced turn-on sensing of cyanide ions in an aqueous system. In this report, 3,6-diaminoacridinium cation (DAAC) was used as a marker of cyanide ion and bio-MOF-1 was employed as the host of DAAC molecules. Due to the anionic behavior of bio-MOF-1, DAAC was introduced inside the framework of bio-MOF-1 by cation exchange process to establish bio-MOF-1 DAAC. This dye plus MOF composite was used for the sensing studies of toxic CN^- ion in an aqueous medium and bio-MOF-1 DAAC showed a nearly 17-times increase in the fluorescence intensity even in the presence of coexisting anions with the detection limit of 5.2 parts per billion (ppb) [65].

14.2.2.2 Electrochemical sensor

The redox activity of MOFs is one of the momentous and fundamental attentions for the modern scientific community. The expansion of monomeric coordination motifs results in the corresponding polymeric compounds. Some CPs are highly attractive in recent times because of their emerging redox property. Appropriate selection of metal nodes or organic ligands can generate the electrochemical or redox activity in CP system. After the judicious synthesis of these materials, researchers utilized them for various potential applications. Electrochemical sensing of analytes or toxic materials for the environment is one of the challenging tasks. It is also important to note that most of these electrochemical sensors are to operate in a water medium. Baghayeri et al. reported a composite material of graphene oxide/zinc-based MOF (GO/MOF) prepared with Zinc chloride and histamine dihydrochloride through a simple solvothermal method. This material was utilized for electrochemical sensors and was found sensitive for the determination of As(III) because of the synergistic effect. The detection limit of the sensor for the detection of As(III) was 0.06 ppb. This advanced platform was engaged to monitor As(III) in environmental water samples [66]. CP-based material can also be applied in the electrochemical sensing of nitro compounds. Keeping in mind this concept, Gunasekaran and Prathap utilized ZIF-8 ($[Zn(MeIm)_2]$, (MeIM = 2-methylimidazolate) for electrochemical detection of TNT in water medium. Electrochemical study and Zeta potential exhibited that TNT molecules got adsorbed onto the nanosheet of ZIF-8. ZIF-8 containing electrode system was found to be capable to detect TNT and relevant nitro compounds like 2,4-DNT, 2,6-DNT, and 1,3-DNB. The differential pulse voltammetry (DPV) data determined the linear range of $1–460 \times 10^{-9}$ M maintaining a sensitivity of 6.94 $\mu A/nm/cm^2$ with LOD as 346×10^{-12} M of the sensor. In this work, the electron-transfer mechanism of the donor and acceptor facilitated the mechanism of electrochemistry [67]. In another work of our research group, Khan et al. synthesized two isotypical 1D CPs, $\{[Zn(adc)(avp)_2(H_2O)]\cdot(H_2O)_3\}_n$ and $\{[Cd(adc)(avp)_2(H_2O)]\cdot(H_2O)\cdot(CH_3OH)\}_n$ (H_2adc = acetylene dicarboxylic acid, avp = 4-[2-(9-anthryl)vinyl]pyridine) containing anthracene chromophore, which further constructed 3D structure and exhibited considerable detection

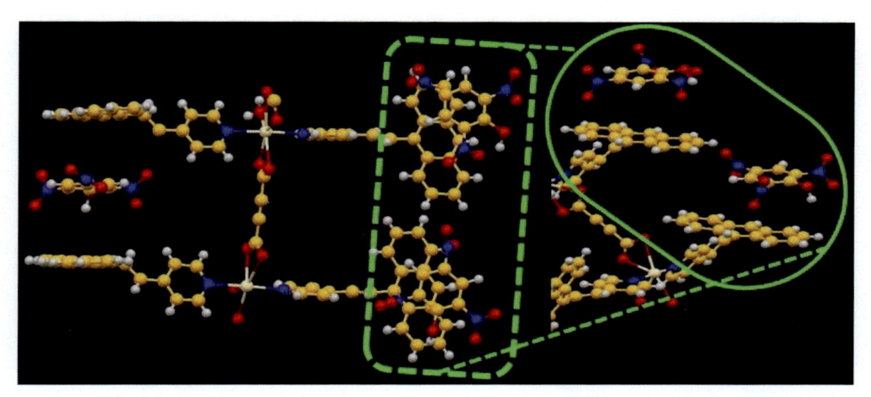

FIGURE 14.9 Probable $\pi\cdots\pi$ stacking in CP involving TNP and anthracene moiety [68].

of explosive nitroaromatic compounds (epNACs) with high selectivity toward TNP. Anthracene moiety being free electron-rich with rational spatial orientation assisted juxtaposition of sensor and analyte in the close vicinity for rational interaction in the molecular level through C–H$\cdots\pi$ and $\pi\cdots\pi$ stacking interactions (Fig. 14.9). It exemplified an innovative approach for preparation of functional materials for possible application in selective sensing of TNP [68].

14.3 MOFs in removal of wastewater pollutants

A sustainable strategy is the key to the success of the removal of contaminants from wastewater due to the diversity in the characteristics of contaminants. Wide ranges of physicochemical techniques are already in place as treatment methods. The application of water-stable MOFs is a good option for effective remediation of wastewater. Regarding employing MOFs, research is fast developing to understand and identify the possible mechanisms of working principles. Advancements of such research work are discussed in the following for different categories of pollutants.

14.3.1 Dyes removal

The detrimental effects of dyes are known even at very low concentrations. Before discharging to the water bodies, it is essential to treat the effluent thoroughly [69]. In the removal strategy of dyes from wastewater, presently, there are several technologies but each of them is having few deficiencies. One of the most sustainable techniques is the adsorption method owing to its economic viability, eco-friendly character, and easy regeneration. MOFs, in this regard, show huge potential in adsorption and catalysis to remove dyes from wastewater. In the following several developments involving MOFs for the removal of dyes are discussed.

Haque et al. in 2010 employed MOF-235, a Fe-terephthalate-based MOF, for adsorptive removal of large amounts of anionic methyl orange and cationic methylene blue from wastewater at various temperatures. It was proposed by

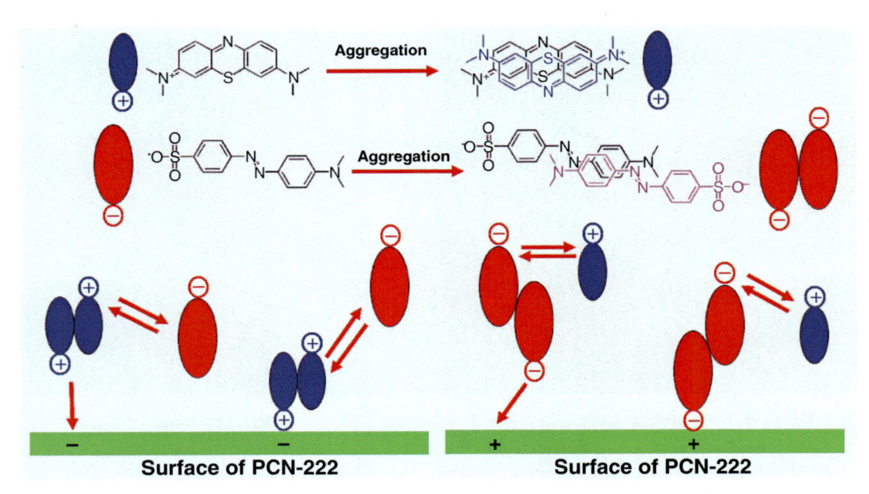

FIGURE 14.10 Push–pull mechanism for the adsorption process of PCN-222 [72].

them that the principle of electrostatic interaction was the driving force for such adsorption [69]. As MOF-235 could not adsorb nitrogen, the research team further pointed out that non-adsorbing MOFs could be beneficial in the removal of dyes in the liquid phase. For the removal of methylene blue dye present in wastewater, Soni et al. established an improvement in the adsorption output of Fe-Benzene dicarboxylic acid MOF through cobalt doping [70].

Abbasi et al. [71] investigated the role of methods of preparation in exhibiting probable effects on the adsorption properties. They prepared CuBTC MOFs through simple routes like mechanochemical and ultrasound techniques to compare their adsorption and desorption rates on crystal violet and methylene blue. As MOF prepared by ultrasound method adsorbed superior dye quantity, it was proposed that there could be the potential influence of method of fabrication on adsorption performance.

In 2017, Li et al. studied zirconium-based MOF PCN-222, a mesoporous porphyrinic MOF possessing large 1-D channels as well as suitable zeta potentials [72]. It demonstrated rapid diffusion to assist in improved adsorption capacity for cationic and anionic dyes in a water medium. It is more significant that the mutual enhancement in the adsorption capacity in a mixed dye solution was reported and it was explained by Li et al. as a push–pull mechanism, a unique phenomenon for MOFs (Fig. 14.10).

There are many others examples of zirconium based water stable MOFs that possess amazing chemical and thermal stability by retaining their crystalline and porous nature owing to strong Zr(IV)–O bonds in Zr-based nodes as well as between node and carboxylate group of the linkers as shown in Fig. 14.11. Their ability to remain stable over a range of pH allows it to act in wastewater remediation. In this respect, Cavka et al. in 2008, developed (UiO-66) which have been extensively studied by scientists in the area of removal of dyes for the purpose of wastewater treatment [73].

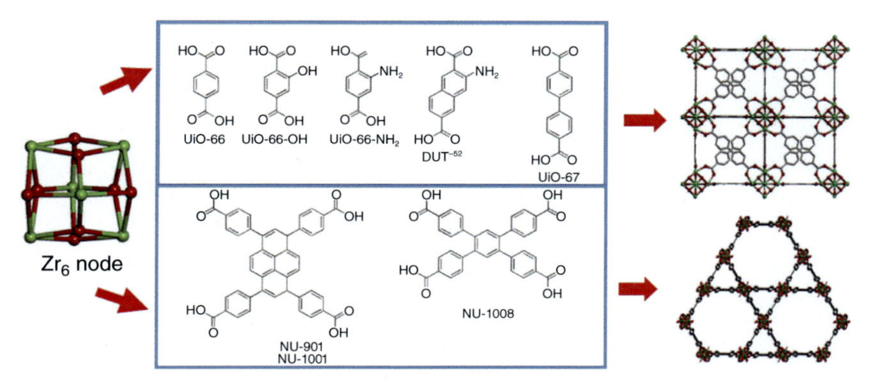

FIGURE 14.11 Structures of the Zr_6-node and different linkers with topologies for UiO series (top) and NU-1000 /NU-1008 (below) [74].

Chen et al. reported that both UiO-66 and UiO-66-NH_2 was capable of removing cationic dyes from the solution more efficiently than dyes of anionic nature due to selective adsorption facilitated by electrostatic attraction [74]. Furthermore, in comparison to UiO-66, UiO-66-NH_2 exhibited better adsorption behavior for cationic dyes and lesser affinity for anionic dyes. The reason behind this fact was explained by their microporous structure and high negative zeta potential possessed by UiO-66-NH_2. In an inclusive study for eight different dyes, Embaby et al. explored four parameters related to adsorption like concentration of the dye, amount of adsorbent, pH and the time of exposure at five levels each by involving a central composite design [75]. The study revealed excellent adsorptive ability of nano-porous and microwave-synthesized MOF UiO-66 towards anionic dyes in comparison to cationic types of dyes. In 2018, Molavi et al. examined UiO-66 for its long-term stability, even aging for one year in water, and periodic affinity for methyl red and methylene blue [76]. As per different characterization techniques including FESEM analysis, negligible changes were observed in the structure of UiO-66 on water-aging for 3 months (UiO-66-3), 9 months (UiO-66-9) and 12 months (UiO-66-12) as in Fig. 14.12 and the affinities got reduced to only 30%.

It was further revealed in the study that methyl orange adsorption was favored at low pH and methylene blue adsorption was preferred at high pH. The controlling by pH values suggested that both π–π stacking interactions and electrostatic interactions played the role in the adsorption mechanism. It has also been noticed that the adsorption of methyl orange was driven by enthalpy process, whereas adsorption of methylene blue was driven by entropy. In another approach akin to above, Mohammadi et al. found that in basic conditions around pH 9, maximum adsorption occurred, which could be due to favorable electrostatic interactions [77]. Another Zr-based MOF composite (UiO-66-P) was studied by J.M. Yang for adsorption as well as separation of organic dyes [78]. Utilizing, UiO-66, the adsorption of rhodamine B dye was examined by He et al. in 2014 [79]. Furthermore, the performances of zirconium-based MOFs were systematically

FIGURE 14.12 FESEM micrographs of as-synthesized UiO-66, UiO-66-3, UiO-66-9 and UiO-66-12. (*Reproduced with permission from Elsevier [76]*).

analyzed for their abilities to adsorb various dyes by the research group of O.K. Farha [80]. In Table 14.1, adsorption uptake capacities exhibited by some Zr-MOFs have been mentioned without specifying much importance to the wide ranges of conditions in respect to different studies.

In 2019, Tian et al. [81] performed large scale synthesis of water-stable cationic Fe-based MOF, CPM-97 Fe, having high specific surface area as well as large pore width to adsorb nine different anionic and cationic hazardous dyes efficiently with higher capacity than that of activated carbon. They advocated the involvement of factors like ion exchange, electrostatic interaction, π–π stacking interaction etc. in the adsorption mechanism. Zhang et al. studied Fe-loaded MOF-545(Fe) for methylene blue and methyl orange for catalytic mechanism in presence of H_2O_2 in 2020 [82]. The results showed that the ability of MOF-545(Fe) to catalyze the dyes in a similar way as done to peroxidase in a short span of time without any absorption of dye. Thereafter, in 2020, Abdi and

TABLE 14.1 Adsorption capacities of dyes on various Zr-MOFs [80].

Dye	MOF	MOF BET surface area (m^2/g)	MOF pore size (Å)	Adsorption capacity (mg/g)	References
Methylene blue (MB)	UiO-66	1280	17	70	[117]
	UiO-66	770	23 (BJH)	65	[118]
	UiO-66	840	~10 (BJH)	143	[115]
	UiO-66	980	6.1	23	[117]
	UiO-66-NH_2	620	~10 (BJH)	152	[115]
	UiO-66-P	710	5.6	91	[117]
	PCN-222	2340	32	910	[112]
Methyl orange	UiO-66	1280	17	84	[117]
	UiO-66	840	~10 (BJH)	48	[115]
	UiO-66	980	6.1	173	[117]
	UiO-66-NH_2	620	~10 (BJH)	37	[115]
	UiO-66-P	710	5.6	4	[117]
	PCN-222	2340	32	592	[114]
Rhodamine B	UiO-66	490	48.5 (BJH)	135 (at 25°C)	[118]
Alizarin Red S	UiO-66	Not reported	Not reported	Not reported	[116]

BJH, Barrett–Joyner–Halenda.

Abedini reported a new polymeric nanocomposite based on zeolitic imidazolate frameworks (ZIFs). It was prepared by means of a facile one step phase inversion (PI) technique and it was applied to remove malachite green present in colored wastewater. The framework could remove 99.2% dye molecules with a significant adsorption capacity of 613.2 mg/g. The removal principle was attributed to π–π stacking interactions involving the dye with the rings of imidazole ligand [83].

The assembled MOFs developed from metal/carbon aerogel are expected to act as an encouraging contender for adsorption of various oils and organic compounds. Such derived metal/carbon aerogels possessing same composition of elements as the original MOFs seem to be also effective for conversion through catalytic hydrogenation process. Those actions of the assembled MOFs could be beneficial for efficient application in the remediation of oily substances from wastewater. Su et al. investigated Ni-MOF hydrogel to study the adsorption efficiency for various oil plus organic compounds as well as the reusability was observed for up to 15 cycles. It was also proposed that the adsorption technique could be regarded as a capable way for effective separation of oil from water separation owing to many advantages in terms of its viability [84].

Zhu et al. reported fabrication of 3D hybrid melamine sponge created on ZIF-8, a mechanically stable MOF-based 3D structure, having excellent compressive behavior which could be employed as a suitable absorbent for separation of oil with sufficient reusability. It exhibited good hydrophobicity to absorb different organic solvents demonstrating an encouraging absorption up to 145 times of its own weight [85].

There have been numerous other studies that have added further evidences of the projected potential of MOFs as removal agent for organic pollutants from wastewater. Role of various physical parameters substantially influence the uptake capacity by MOF's. The relationship between MOFs' structure property relationships could be accurately demarcated on precisely controlled adsorption conditions which are guided by the research teams. In large-scale applications, however, it is not so easy to tune adsorption conditions. Thus, efforts are to be made in designing sorbents that would have the ability to perform well in wide sets of conditions.

14.3.2 Toxic agrochemicals removal

Ever-increasing need for food as demanded by world's population requires efficient use of agrochemicals, fertilizers, and pesticides. Such chemicals are notably increasing agricultural productivity and, in addition, they are entering into the water system. Poor biodegradability lead to the accumulation of those chemicals and it is a threat to human health as well as aquatic lives.

Atrazine is a triazine-based well-known herbicide and is suspected to show adverse effect on endocrine even at extremely low quantities like 0.1–25 ppb [86]. Although there are some available treatment methods, they often lead to more toxic by-products [87]. In terms of cost-effective methods, adsorption seems to be the efficient strategy. Use of MOFs started with UiO-67 through investigation of the adsorption of atrazine by Akpinar and Yazaydin [88]. They studied the affinity towards atrazine of some selected Zr-based MOFs possessing different types of linker π-system sizes along with chemical functionality, topologies of the frameworks, and pore dimensions (Fig. 14.13). All types of adsorbents could remove up to 98% atrazine from water and UiO-67 exhibited the fastest rate and was much superior to ZIF-8 and F400, an activated carbon. Moreover, they advocated the encouraging results for regenerated UiO-67.

Akpinar et al. continued their work to illuminate the mechanism of atrazine adsorption [89] by systematically tuning the abundance of prospective π–π interaction locations in UiO-67 and functionalized derivatives of UiO-66, DUT-52, NU-1000, NU-901, and NU-1008. It was reported that DUT-52 exhibited the maximum affinity for atrazine, which signified the $\pi - \pi$ interactions provided the primary role to adsorb atrazine (Table 14.2). The study indicated that the dimension of the π–system was bigger contributor than surface area or size of pore in the removal process.

In Fig. 14.14, atrazine adsorption uptake of above mentioned Zr_6-based MOFs has been mentioned. Adsorption uptake was reported as percentage of the

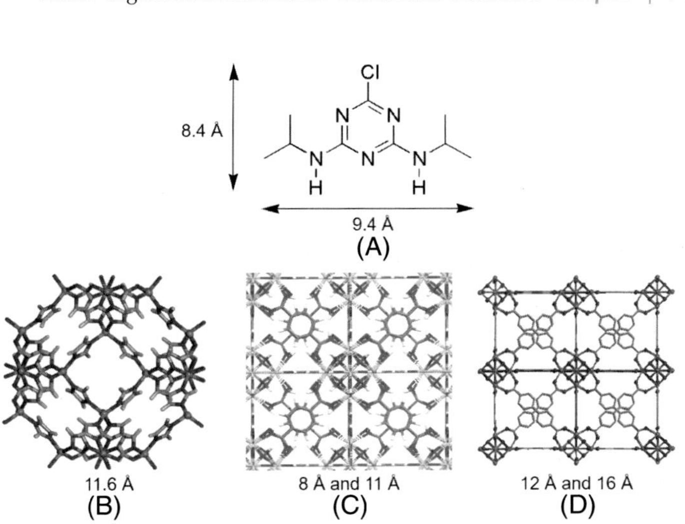

FIGURE 14.13 Structure of atrazine (A) and structural models of ZIF-8 (B), UiO-66 (C) and UiO-67 (D) showing pore sizes [Color of the atoms: carbon (*black*), oxygen (*red*), nitrogen (*blue*), zinc (*orange*), and zirconium (*cyan*)] [89].

TABLE 14.2 Parameters of surface area, pore size and pore aperture (approx.) of Zr_6-based MOFs screened for adsorption of atrazine [89].

MOF	Surface area (m^2/g)	MOF pore size (Å)	Pore aperature (Å)
UiO-66	1690	12 and 16	7.5
UiO-66-NH_2	1410	13	7.5
UiO-66-OH	1210	11	7.5
UiO-67	2510	13 and 23	12
DUT52	1960	12 and 20	9
NU-1008	1400	14 and 30	14 and 30
NU-901	2110	12	12
NU-1000	2210	12 and 30	12 and 30

total quantity of atrazine exposed to MOF samples where 3.5 mg was exposed to 10 mL of 10 ppm atrazine solution for one day (24 hours) under ambient conditions [89].

Seo et al. studied removal of methylchlorophenoxypropionic acid (MCPP), a phenoxyacid herbicide, from water by employing UiO-66 under various conditions [90]. In comparison to activated carbon, even at low MCPP concentrations, UiO-66 showed ~7.5 times higher adsorption with ~30 times faster uptake kinetics. It was further revealed in the study that, at low pH, removal mechanism was controlled by electrostatic interactions, whereas, π-π interactions was prevailing at higher pH. It was also confirmed that UiO-66 could be reused

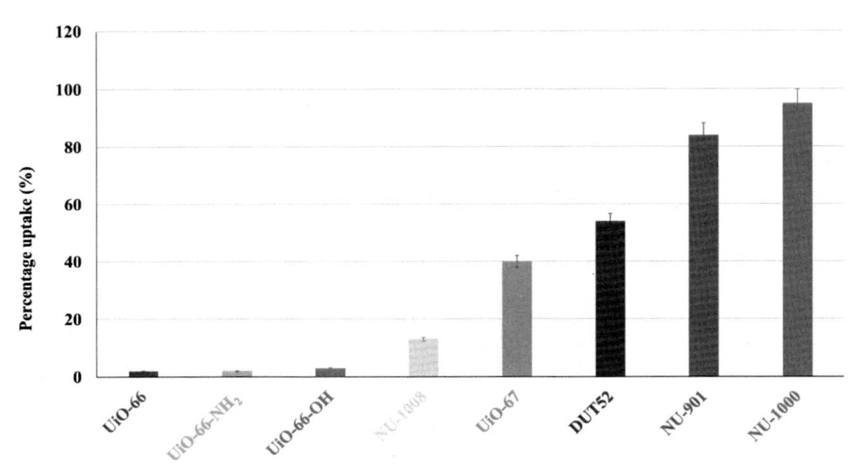

FIGURE 14.14 Atrazine adsorption uptake as a percentage of the total amount of atrazine exposed to MOF samples [89].

after simple washing with water or ethanol. Zhu et al. [91] worked on UiO-67 for removal of glyphosate up to 537 mg/g. They thoroughly optimized the adsorption conditions by analyzing several factors like glyphosate and sorbent concentrations, glyphosate adsorption and exposure time, pH, and ionic strength. Improved adsorption capability was also owing to high specific surface area as well as the adequate pore size. The observations were extremely encouraging in comparison to previously examined adsorbents making UiO-67 as a potent candidate for removal of such pesticides from wastewater. In continuation of exploration of Zr-based MOFs, Pankajakshan and co-workers synthesized NU-1000 and UiO-67 in the nanometer to micrometer ranges to investigate specific adsorption of the glyphosate [92]. The study revealed negative correlation between the MOF particle size and uptake kinetics owing to the rise in the exterior surface area and minimization of diffusion barriers and result of the study were better for NU-1000 than UiO-67, which could be due to shorter Zr···O—P distance as well as stronger interactions (Table 14.3). Another pesticide, 2,4-dichlorophenoxyacetic acid (2,4-D) was attempted for removal by Tan and Foo [93] by employing an iron-based MOF, MIL-100(Fe) that showed high adsorption capacity from aqueous solution owing to high specific surface area. The crystalline framework was credited for favorable host-guest interaction with the 2,4-D molecules for the thermodynamically favored rapid adsorption process which was also found recyclable through ethanol washing.

14.3.3 Pharmaceutical products

The persistence of pharmaceutical contaminants is an important matter as health hazards of those contaminants in low-level exposure is not fully known. Many MOFs allow simultaneous detection as well as removal of pharmaceutical

TABLE 14.3 Adsorption capacities of glyphosate by NU-1000 and UiO-67 at different sizes and comparison of the data with various other adsorbents [92].

Adsorbents	Pore size	q_{max} (mmol/g)
NU-1000	100–200 nm	8.97
NU-1000	250–300 nm	6.49
NU-1000	500–700 nm	6.37
NU-1000	1000–2000 nm	6.05
UiO-67	100–200 nm	7.90
UiO-67	250–300 nm	6.12
UiO-67	500–700 nm	5.44
UiO-67	1000–2000 nm	3,21
Fe_2O_3@SiO_2@UiO-67		1.52
$MnFe_2O_4$-graphite		0.23
MnO_2/Al_2O_3		0.69
dendro biochar		0.26
UiO-67		3.18
Chitosan/alginate membrane		4.73×10^{-5}
Polyaniline/ZSM-5		0.58
Montmorillonite		0.295
Alum sludge		0.67
Ni_2AlNO_3		1.02
α-FeOOH		0.23
MgAl-LDH		1.09

residues. Some reported works of removal of highly consumable pharmaceutical drugs from wastewater are described in brief.

Adsorptive removal from aqueous solution of a toxic sulfonamide antibiotic, sulfachloropyradazine (SCP), at first was examined on UiO-66 and ZIF-67 by Azhar et al. [94] in 2017. Superior adsorption capacity was shown by UiO-66 in comparison to ZIF-67, demonstrating it as a promising adsorbent and, even, superior to previously studied MOFs and adsorbents in wastewater treatment processes on account of fast kinetics and easy regeneration. Electrostatic interactions coupled with π–π interactions and hydrophobicity were the main contributor for such high adsorption. Chen et al. [95] studied the role of UiO-66 in water on adsorption behaviors of carbamazepine, an anticonvulsant and tetracycline hydrochloride antibiotic. Reported maximum adsorption levels as exhibited by UiO-66 for carbamazepine and tetracycline were 37.2 mg/g and 23.1 mg/g at 25°C, respectively. Adsorption capacities were reported to get diminished as the pH of the solution moved up. Such competitive adsorption advocated

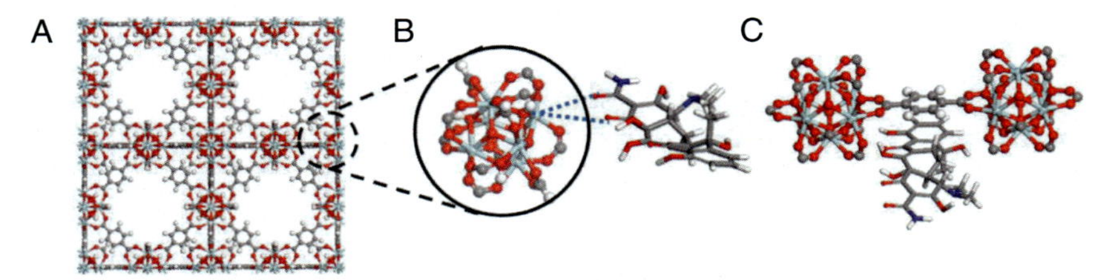

FIGURE 14.15 (A) Schematic illustration of H-UiO-66 structures, (B) Interaction between Zr_6 cluster of H-UiO-66 and O atoms in tetracycline, (C) π–π stacking between ligand of H-UiO-66 and tetracycline [Color of atoms: C (*gray*), H (*white*), N (*ultramarine blue*), O (*red*), and Zr (*cyan*)]. (*Reproduced with permission from Elsevier [96]*).

that the adsorption sites for carbamazepine on UiO-66 could be dissimilar to adsorption sites of tetracycline. The group conducted various characterization techniques to validate that physisorption of UiO-66 occurred for carbamazepine through cumulative roles played by hydrophobic effect along with electrostatic attraction combined with π–π electron donor-acceptor interaction. In contrast, the adsorption of tetracycline on UiO-66 was categorized predominantly as chemisorption, where strong electrostatic attraction plus π–π electron donor-acceptor interaction forces act in unison. It was an important and thorough investigation on the behavior of adsorption sites and pathways that was proved to be a guiding work for the future.

In quest of improved uptake capacities, Zhang et al., in 2018, synthesized two types of hierarchical pore metal–organic frameworks H-MOFs (H-UiO-66s) that possessed average mesopore sizes at 3.8 nm and 17.3 nm, respectively [96]. Both MOFs efficiently exhibited superior capacities to arrest tetracycline in aqueous solution. Notably, H-UiO-66 with pore size of 17.3 nm showed the highest adsorption intake of 666.67 mg/g with removal efficiency of more than 99% with the speculation to the interacting force involving pore structure also (Fig. 14.15).

Zhou et al. [97] in 2018 developed a highly stable luminescent zirconium-based MOF, PCN-128Y to remove tetracycline antibiotic from water as it possessed high adsorption capability by chelating metal-ligand bonding between Zr_6 nodes and tetracycline through solvent-assisted ligand incorporation (SALI) process [98]. In addition to removal of the tetracycline, another important highlight of this fluorescent MOF was its ability of detect the tetracycline in water. It seemed to be the first example of its kind, which opened a new prospect for treating wastewater by employing such luminescent MOFs.

Some other Zirconium based MOFs were studied for removal of pharmaceuticals from water system in addition to the family of UiO-66 MOFs. Zhao et al. [99] examined the affinity of chloramphenicol by PCN-222, in aqueous solution and reported 99.0% removal efficiency within 1 minute. Plentiful −OH sites on the node exhibited H-bonds to the hydroxyl, nitro, and carbonyl groups of chloramphenicol to aid the increased adsorption (Fig. 14.16).

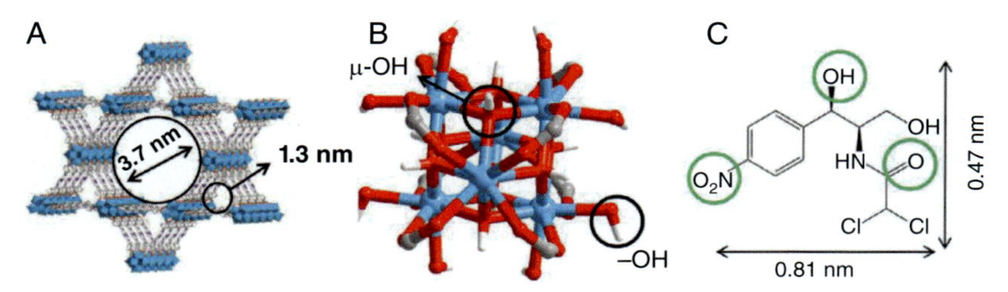

FIGURE 14.16 (A) Pore structure and (B) $Zr_6(l\text{-}OH)_8(OH)_8(CO_2)_8$ clusters of PCN-222, (C) structure and size of chloramphenicol (CAP). [Color of atoms: Zr (*blue*), C (*grey*), O (*red*), and H (*white*)]. (*Reproduced with permission from Elsevier [99]*).

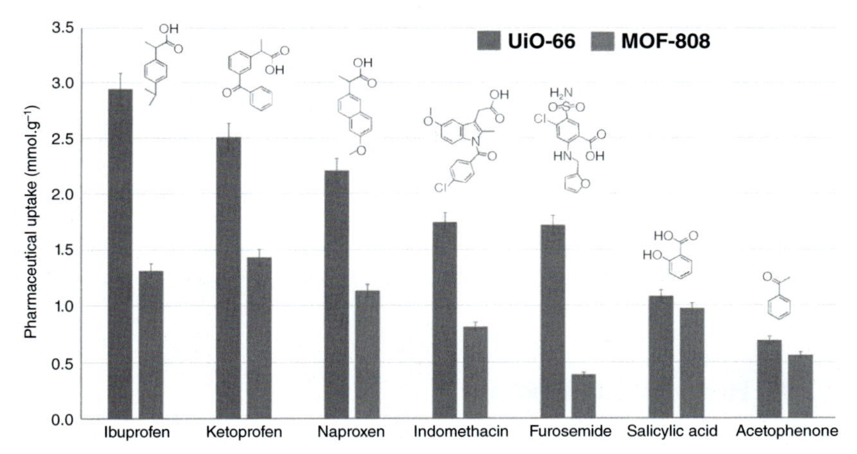

FIGURE 14.17 Study of saturated adsorption capacities of different pharmaceuticals in UiO-66 and MOF-808 [6.0 mg of MOF exposed to 1.0 mmol/L pharmaceutical at pH 6.0 ± 0.1] [100].

In another study, three porous MOFs, UiO-66, MOF-808, and MOF-802 were examined for the adsorption of a series of nonsteroidal anti-inflammatory (NSAID) pharmaceuticals – ibuprofen, ketoprofen, naproxen, indomethacin, furosemide, salicylic acid, and acetophenone – from water by Lin et al. in 2018. UiO-66 and MOF-808 showed outstanding adsorption due to affinity for the anionic pharmaceuticals of ibuprofen, ketoprofen, naproxen, indomethacin, and salicylic acid by cationic zirconium which is not fully coordinated in the cluster [100]. In addition, there was $\pi-\pi$ interactions between neutral NSAID molecules and the MOF linkers. The study demonstrated the importance of chemical functionalities in MOFs that could be used to remove pollutants with variety of structures effectively from wastewater (Fig. 14.17).

Wang et al. fabricated two fluorescent based MOFs, BUT-12 and BUT-13 containing truncated Zr_6-nodes and tritopic methyl-functionalized linkers in order to use for detection and removal of nitro-functionalized antibiotics [101]. Both MOFs exhibited good adsorption ability towards nitrofurazone and nitrofurantoin antibiotics as well as 2,4,6-trinitrophenol (TNP) and 4-nitrophenol

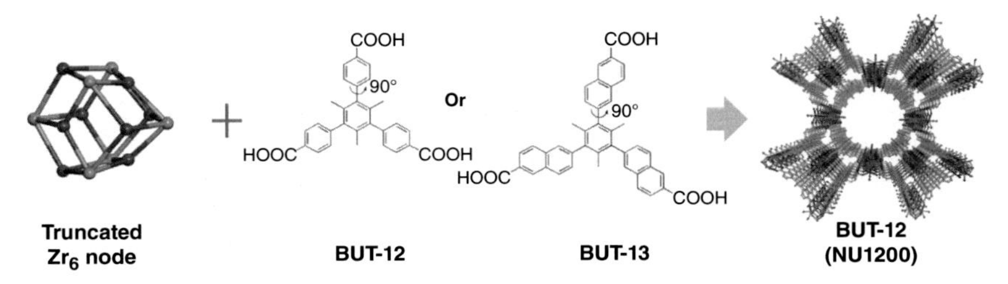

FIGURE 14.18 Zr-MOFs containing Zr_6 clusters with BUT-12 and BUT-13 linkers [80,101].

(4-NP) organic explosives in aqueous solution. The study provided a novel approach into the design of MOFs through tailoring of surface hydrophobicity plus incorporation of fluorescent moieties to detect as well as eliminate the pollutants simultaneously from water (Fig. 14.18).

Zhuang et al. [102] demonstrated a ZIF-67 MOF based alginate hydrogel for the removal of tetracycline. The process involved concurrent gelation of the alginate matrix accompanied by formation of the MOF through chelation involving metal Co^{2+} with alginate and 2-methylimidazole ligand leading to formation of MOF hydrogels. It could run for ten cycles resulting adsorption capacity of 364.89 mg/g and removal mechanism was reported to be based upon electrostatic interactions.

14.3.4 Heavy metals removal

In order to eliminate heavy metals from wastewater, the applications of MOFs have been a growing area of research [103–105]. In this viewpoint, prospective application of MOFs for removal of heavy metals are discussed in this section.

14.3.4.1 Arsenic removal

Zhu et al., in 2012, described the synthesis of tailored Fe-BTC—a porous MOF that contained iron nodes and 1,3,5-benzenetricarboxylic acid as linkers for arsenic (V) adsorption capabilities. The Fe−BTC polymer furthermore displayed improved dynamic adsorption capacity which was 6.5 times more than the adsorption ability of 50 nm Fe_2O_3 nanoparticles. Moreover, it was 37 times more efficient in comparison to powdered iron oxide as available commercially [106]. Jian et al. synthesized highly porous ZIF-8 nanaoparticles of 200–400 nm with a hierarchical structure possessing high surface area of 1063.5 m^2/g. Nanoparticles of ZIF-8 exhibited adsorptions of 49.49 mg/g and 60.03 mg/g, at neutral pH and at room temperature, for As(III) and As(V), respectively. The adsorptions occured through mechanism of electrostatic attraction in addition to the formation of arsenic complexes involving hydroxyl and amine groups on the adsorbent. Interference on adsorption was not found for co-existing SO_4^{2-}

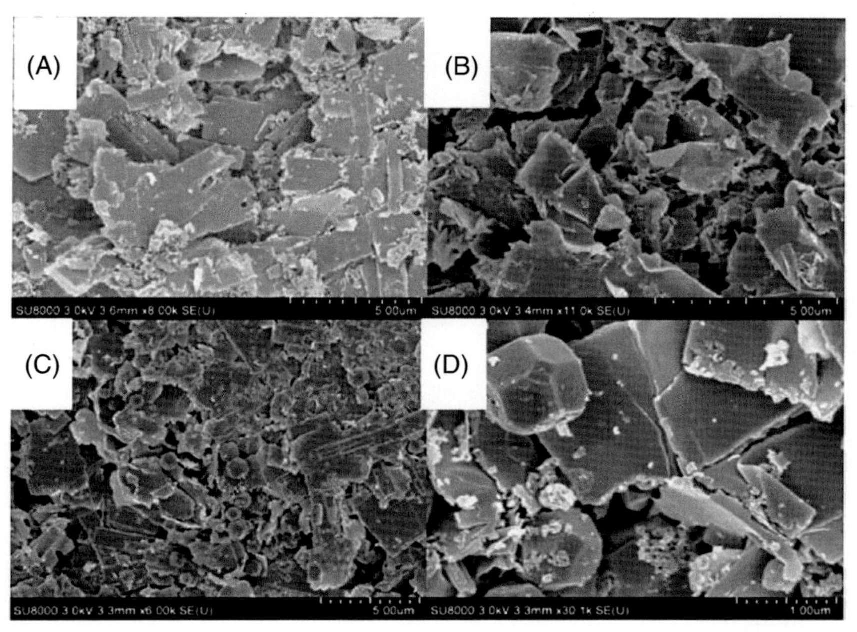

FIGURE 14.19 Scanning electron microscope images showing spent adsorbents: cubic (A), leaf-shaped (B) and dodecahedral (C), and (D). *(Reproduced with permission from Elsevier [108])*.

and NO_3^- anions, whereas adsorption was significantly affected by PO_4^{3-} and CO_3^{2-}[107]. Liu et al. fabricated three ZIFs having different morphologies—cubic, leaf-shaped, and dodecahedral ZIFs – as effective adsorbents for As(III) with a maximum adsorption value of 122.6 mg/g, 108.1 mg/g, and 117.5 mg/g for cubic, leaf-shaped and dodecahedral ZIFs, respectively [108]. The adsorption capacities were not associated with the surface area of MOFs as the leaf-shaped ZIF possessing approximately ten-fold lesser surface area exhibited adsorption capacity which was just a little less compared to the other two morphologies (Fig. 14.19).

Vu et al. in 2015, presented a detailed study on the adsorption of As(V) in MIL-53(Fe) that was prepared via HF-free solvothermal method. The MOF fashioned an adsorption of 21.27 mg/g of As(V) in aqueous solution through Lewis acid-base type interaction mechanism involving anionic $H_2AsO_4^-$ units and MOF nodes [109]. Li et al. prepared a similar MOF in 2014, MIL-53(Al), that showed highest adsorption intake (106 mg/g) of As(V) in the form of $HAsO_4^{2-}$ within 11 hours at an neutral pH range of 6–8. The mechanism of such adsorption for As(V) in MIL-53(Al) suggested that hydrogen bonding plus electrostatic interactions played a big role. (Fig. 14.20). Effectiveness of the MOF was also reported intact in the existence of other anions excluding PO_4^{3-} which reduced its removal ability by 14% [110].

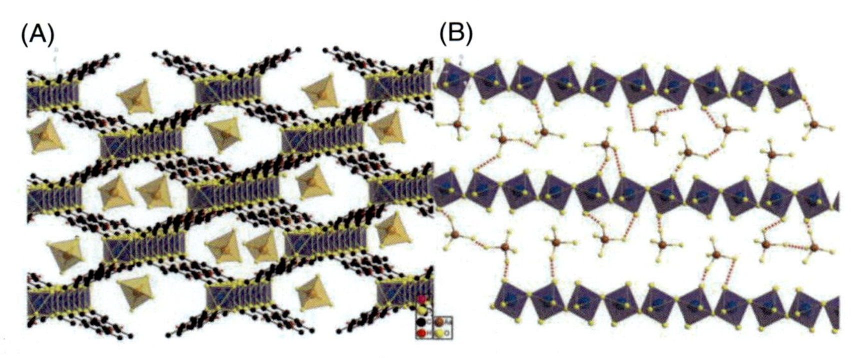

FIGURE 14.20 Simulated locations of arsenate in MIL-53(Al) as viewed (A) along the crystallographic *z* direction; all metal ions and arsenate as the polyhedral and (B) along the *x* direction; metal ions as the polyhedral and arsenate as the ball and stick model. Arsenate molecule linked through hydrogen-bonding interactions as represented by red dashed line. [Color of atoms: Al (*pink*); O (*yellow*); C (*dark*); H (*red*); As (brown)]. (*Reproduced with permission from Elsevier [110]*).

Li et al. in 2015 synthesized microwave assisted metal-organic framework, MOF-808 nanoparticles, as another As(V) adsorbent that exhibited excellent adsorbent for arsenic removal as well as possessed good chemical stability and reusability [111]. In 2015, Wang et al. first reported Zr-MOF, UiO-66, with micrometer order particle size for its adsorption of As(V) present in water [112]. The MOF functioned well in reasonably long range of pH exhibiting an exceptional adsorption of 303 mg/g at an optimal pH 2 and very little influence of some common anions was accounted. The study further reported that the mechanism of adsorption happened through coordination at hydroxyl groups on Zr-nodes or through substitution of BDC ligands in the framework (Fig. 14.21).

In 2016, Audu et al. [113] showed the use of the nodes and linkers to entrap anionic As(V) in $Na_2HAsO_4{\cdot}7H_2O$ and neutral As (III) in As_2O_3 "chemoselectively" in thiolated derivative of UiO-66, UiO-66-$(SH)_2$. Interestingly, both recognition motifs were found to be potentially incorporated into the identical framework aiming at dual-capture objectives and both the bindings appeared to be reversible through appropriate treatments.

14.3.4.2 Mercury removal

Fang et al. described their work in 2010 that PCN-100, a zinc-based framework utilizing TATAB (4,4′,4″-s-triazine-1,3,5-triyltri-*p*-aminobenzoate) linker with $Zn_4O(CO_2)_6$ SBU to show mercury(II) adsorption in aqueous solutions (Figs. 14.22 and 14.23) [114].

As observed through Inductively coupled plasma (ICP) analyses, PCN-100 adsorbed 1.38 Hg(II) per formula when the experiment was conducted in DMF solution. Ke et al., in 2011, demonstrated a novel way to get rid of Hg^{2+} through thiol-functionalized porous MOFs synthesized by thiol grafting on copper center, the unsaturated metal center in coordination viewpoint in a 3D porous MOF

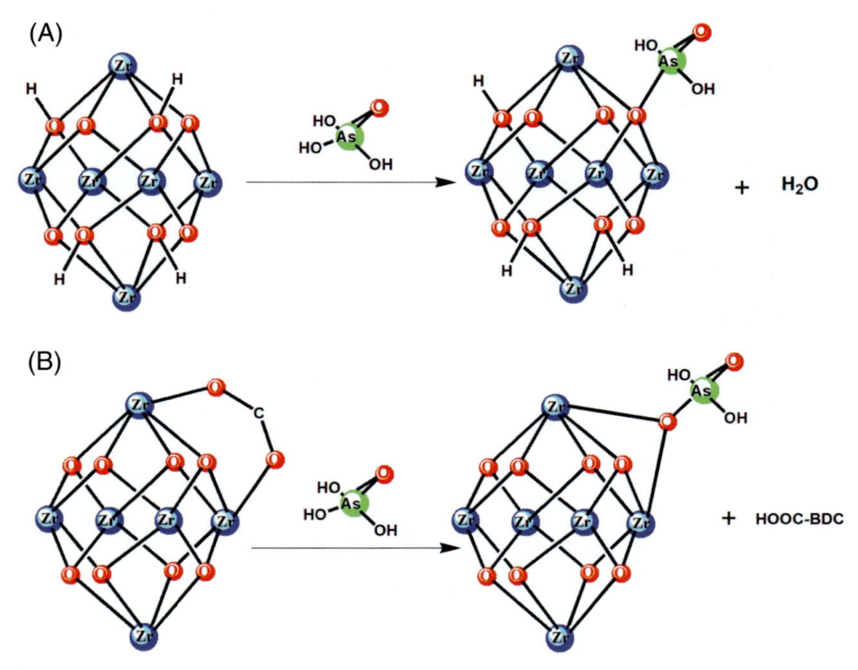

FIGURE 14.21 Proposed mechanism for adsorption of arsenate on UiO-66 via coordination at (A) hydroxyl group and (B) BDC ligand [112].

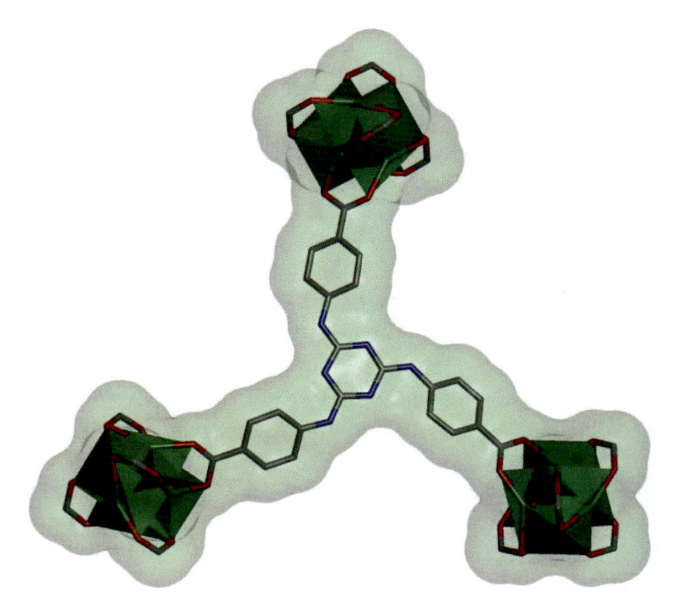

FIGURE 14.22 TATAB linked to three $Zn_4O(CO_2)_6$ SBUs in PCN-100 [114].

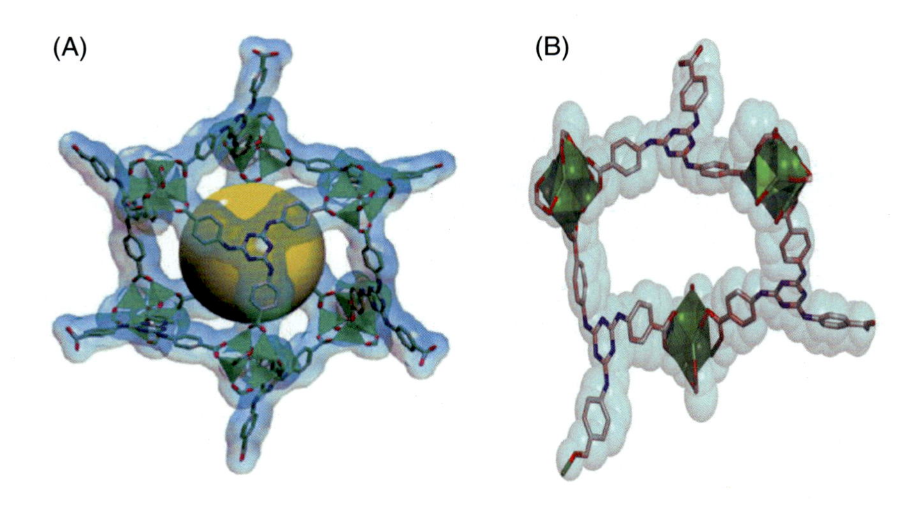

FIGURE 14.23 (A) PCN-100 with mesoporous cavity of 2.73 nm internal diameter and (B) a window of (A) with about 1.32 nm and 1.83 nm [114].

$[Cu_3(BTC)_2]$ [54]. Interestingly, the unfunctionalized $[Cu_3(BTC)_2]_n$ exhibited nil affinity to Hg^{2+} and the thiol-functionalized samples revealed significant adsorption capacity of 714.29 mg/g for Hg^{2+} separation from aqueous solution within the first 120 minutes. Large specific surface area as well as amount of functionalized adsorption sites inside pores of the MOF contributed the high uptake capacity of mercury in the removal process. Liu et al. reported, a group of novel, post-synthetically modified frameworks in 2014, where, Cr-MIL-101-AS was post-synthetically modified to contain densely packed thiol groups. It was effective in removal of Hg(II) from a 10 ppm solution with 99% adsorption within 6 hours and exhibited 93% efficiency in 0.1 ppm solution [115]. In 2013, Yee et al. reported mercury(II) adsorption by MOF analogous to UiO-66, Zr-DMBD with dimercapto-1,4-benzenedicarboxylic acid linkers (H_2DMBD) framework, in aqueous medium. As reported, it was a promising material for 100% Hg(II) uptake from starting solution of 10 ppm to 0.01 ppm within 12 hours. Detection of the adsorption process through naked-eye could be done on account of natural photoluminescence [116]. Sohrabi proposed, in 2013, removal of mercury through the use of $SH@SiO_2/Cu(BTC)_2$, a nanocomposite, which is based on HKUST-1 and thiol-modified silica nanoparticles immobilized within its structure. It showed maximum adsorption capacity of 210 mg/g selectively within 60 minutes at an optimal pH of 6 and was unaffected by other metal ions in solution [117]. Luo et al. developed a post-synthetically modified MIL-101 through thiol functional groups, MIL-101-Thymine, which showed mercury adsorption in aqueous solutions by coordinating with two thymine groups of the MOF (Fig. 14.24) [118].

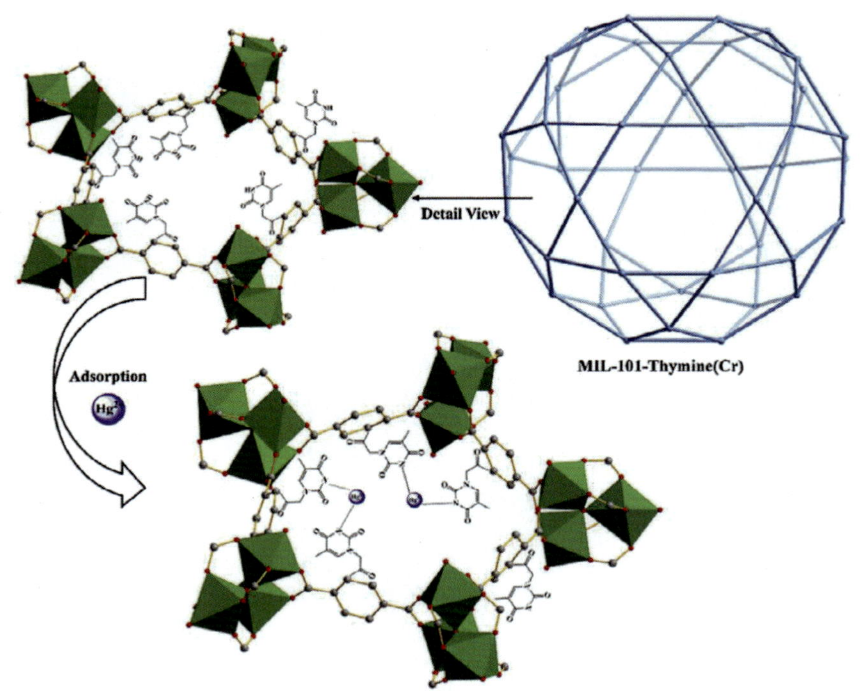

FIGURE 14.24 Illustration of two-coordination for mercury(II) with MIL-101-Thymine MOF. *(Reproduced with permission from Elsevier [118])*.

Subsequently, Saleem et al. in 2015 synthesized UiO-66-NHC(S)NHMe through covalent post-synthetic modifications on UiO-66-NH_2 to show the mercury(II) adsorption up to 99% after 240 minutes for a solution with 100 mg/L of target metal ion. On the contrary, only 4% mercury adsorption was displayed by UiO-66 only [119]. Abbasi et al. in 2015, evaluated a 3D Co(II) MOF for adsorption of mercury(II) in water. The MOF exhibited adsorption in solution up to 70% of the Hg(II), which reached equilibrium within 100 minutes at an optimal pH of 6 without any significant damage to the MOF [120]. A sulfur modified MOF, – FJI–H12, was developed in 2016 by Liang et al. to show mercury(II) uptake capabilities. Free NCS^- groups in the MOF were located within octahedral M_6L_4 cages which was built from Co(II) metal ions and 2,4,6-tri(1-imidazolyl)-1,3,5-triazine linkers [121]. Liang and his colleagues proposed that the uptake mechanism was correlated on both chemisorption and physisorption, as controlled by the SCN^- groups. The highest adsorption capacity was derived as 440 mg/g within 1 hour when optimum pH was fixed at 7.

Chakraborty et al., in 2015, synthesized a zinc metal ion and tetracarboxylate linkers based novel anionic MOF, AMOF-1, to employ it in separation of mercury from polluted water [122]. In one ppm level starting solution, adsorption

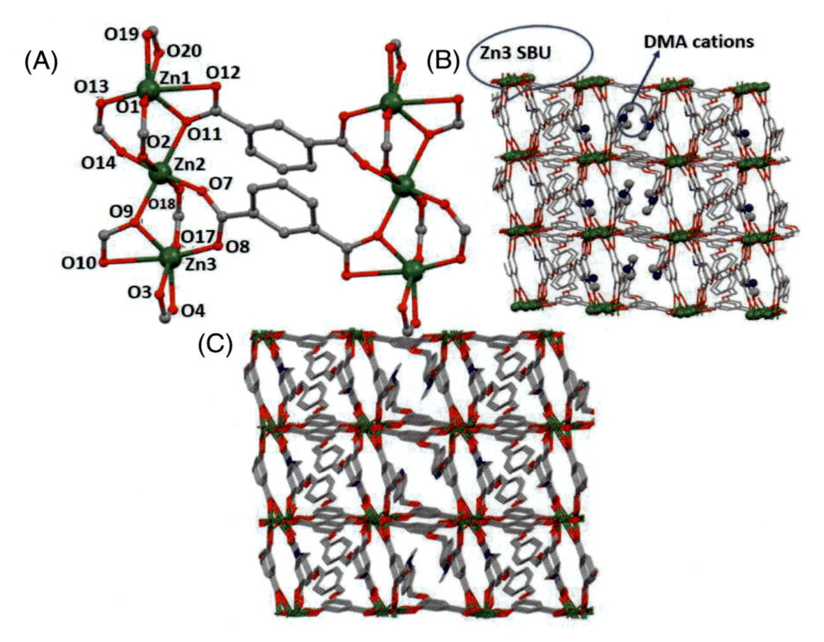

FIGURE 14.25 Views of (A) coordination environment around the metal center, Zn(II) and trinuclear SBU, (B) 3D framework with rectangular channel containing DMA cations, (C) biporous 3D framework with interconnected channels along crystallographic *a* direction [122].

of 94% of mercury(II) was found in first 18 hours, and adsorption reached 99% in doubled time. Adsorption mechanism as proposed was based on ion exchange with the dimethylammonium (DMA) ligands positioned inside the rectangular MOF channels (Fig. 14.25).

Huang et al., in 2015, synthesized a magnetic MOF, Fe_3O_4@-SiO_2 @HKUST-1, which was post-synthetic functionalized HKUST-1 with bismuthiol (Bi-I) for application in water-based adsorption of mercury(II) within 10 minutes resulting adsorption capacity of 264 mg/g with 99% adsorption in a solution having concentration of 20 mg/L [123]. Luo et al. designed MOF, Zn(hip)(L)·(DMF)(H_2O) (H_2hip=5-hydroxyisophthalic acid, L=N^4,$N^{4'}$-di(pyridine-4-yl)biphenyl4,4'-dicarboxamide) for adsorption of mercury(II) in aqueous solution [124]. Solvo (hydro)thermal process was followed as the synthesis route. The structure of the MOF displayed a single, hexagonal channel furnished with hydroxyl and acylamide active sites alongside the inner wall of the pores as depicted in Fig. 14.26. The MOF exhibited 333 mg/g uptake capacity after reaching equilibrium within 1 hour with a peak at pH 5 and worked at ultralow mercury(II) concentrations of ppb level.

One more example of MOF that was post-synthetically modified with thiol functionalization –ZIF-90-SH– was studied by Bhattacharjee et al. in 2015 for separation of mercury from polluted water at fair adsorption rates, notably, with increased uptake percentage with decreasing mercury concentration [125].

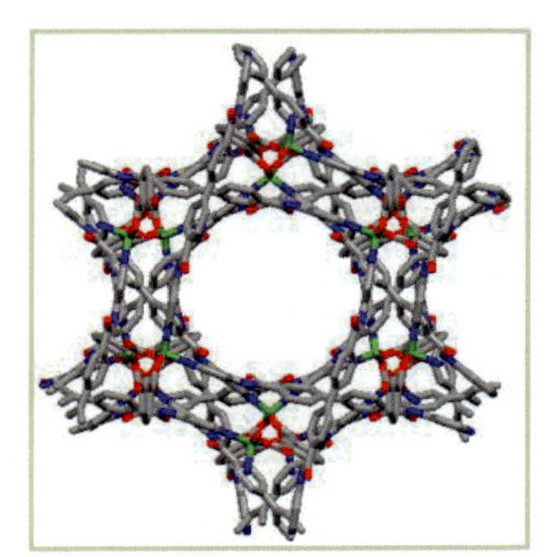

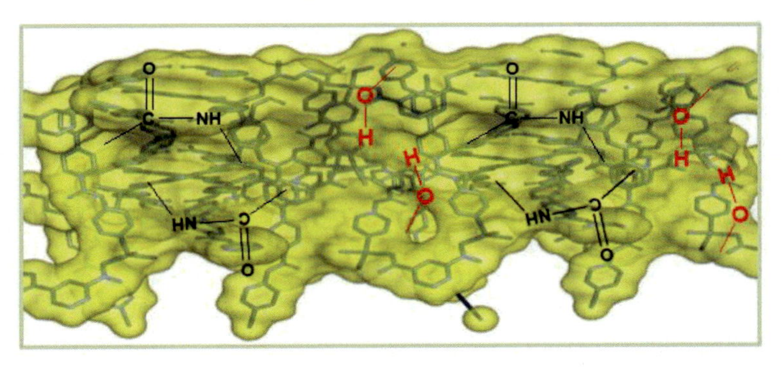

FIGURE 14.26 Structure of Zn(hip)(L)·(DMF)(H_2O) as perceived from above along the c-axis and as a labeled cross section along the length of the main channel [124].

Moreover, in 2016, a series of isoreticular and luminescent MOFs were designed by Rudd et al. in which one of the MOF, LMOF-263 with dicarboxylate-based ligands along with tetradentate chromophores, showed significant mercury uptake capacities with high selectivity [126]. Another stable mercury adsorbent MOF, indicated as BioMOF, was designed and represented by Mon et al. in 2016 with an exceptional uptake capacity. The ability of mercury adsorption was examined using $HgCl_2$ and CH_3HgCl in water and water/methanol media and 900 mg/g was found for $HgCl_2$ while 166 mg/g was reported for CH_3HgCl after 72 hours. The adsorption process could be inverted too by employing dimethyl sulfide making the MOF suitable for reuse [127]. Xiong et al. in 2017 tested MOF-74-Zn for mercury(II) adsorption in which the mechanism of adsorption was proposed as the combined effect of chemi- and physisorption, but more inclined to physisorption. Even, ultra-low Hg(II) concentrations of 50 ppb led to maximum 72% uptake at 45°C [128]. In 2017, Halder et al. synthesized a nickel based MOF, $[Ni(3\text{-bpd})_2(NCS)_2]_n$ (3-bpd = 1,4-bis(3-pyridyl)-2,3-diaza-1,3-butadiene), under mild conditions and employed it for selective detection and removal of Hg^{2+} ion from aqueous solution [129]. In this MOF, uncoordinated sulfur atoms of the bridging thiocyanato ligands were found to get involved with Hg^{2+} ions through bond formation as observed from change in color of the MOF from green to grey after treatment with Hg^{2+}. Maximum adsorption capacity was reported to be 713 mg/g with 94% removal capacity within 2 hours in aqueous solution of 10 mg/L [129].

14.3.4.3 Other heavy metals removal

Researchers paid due attention to other heavy metals like cadmium, lead and chromium by contributing detailed study for their removal from wastewater as those are highly toxic for the human body even their presence is very low in magnitude. Some reported MOFs are discussed here for suitable application in treatment of respective heavy metals from aqueous solution. It is further to note that some other less corrosive metals like silver, copper, iron, aluminum, etc. also pose threat to human health.

In 2011, Qin et al. reported MnO_2-MOF to adsorb lead(II) effectively in aqueous solutions showing uptake capacity of 917 mg/g in 1 hour [130]. Wang et al. reported $Cu_3(BTC)_2$-SO_3H, prepared through post-synthetic modification in 2015 to adsorb cadmium(II) at optimal 6 pH [131]. Previously discussed MOF, AMOF-1 synthesized by Chakraborty et al. was reported to be potential cadmium(II) as well as lead adsorbent in aqueous solution [122]. Another post-synthetically modified zinc-based MOF-5, HS-mSi@MOF-5 was reported by Zhang et al. in 2016 to remove cadmium(II) from water with maximum capacity of 98 mg/g. Moreover, it exhibited excellent removal property for Pb(II). In 2013, Zou et al. reported polyoxometalate-modified version – HKUST-1-MW@H3PW12O40 which was synthesized by employing microwave irradiation to remove cadmium(II) and lead(II) ions [132]. UiO-66-NHC(S)NHMe which has already been discussed for mercury removal in this section, was also reported for potential application as adsorbent for removal of cadmium and lead as well as chromium(III) [119]. Adsorption of chromium was also exhibited in a group of three azine- and imine functionalized MOFs, TMU-4,5, and 6, to adsorb chromium as reported by Tahmasebi et al. in 2014 [133]. TMU-5 was also reported to be an efficient cobalt(II) adsorbent as well as good adsorbent too for copper ions from water. Cobalt-based zeolitic imidazolate framework-67, ZIF-67, was studied by Li et al. to study the adsorption properties for chromium(VI) and ion exchange mechanism involving chromium(VI) anions ($Cr_2O_7^{2-}$) and the hydroxyl groups of the MOF was proposed [134].

In 2015, Cheng et al. reported MIL-53(Al) to exhibit adsorption of silver(I) ions in which an agglomeration effect adsorbs silver ions to accumulate together creating silver nanoparticles in order to get stabilized within the framework by the thiol groups [135]. Conde-Gonzalez et al. in 2016, reported silver(I) adsorbing HKUST-1 framework which was shown as silver nanoparticle adsorbent in the aqueous medium [136]. Abbasi et al. synthesized 3D Co(II) MOF which was potentially capable of removing aluminum(III) and iron(III) from water [120]. Rahimi and Mohaghegh synthesized a magnetic Cu-terephthalate MOF to assess

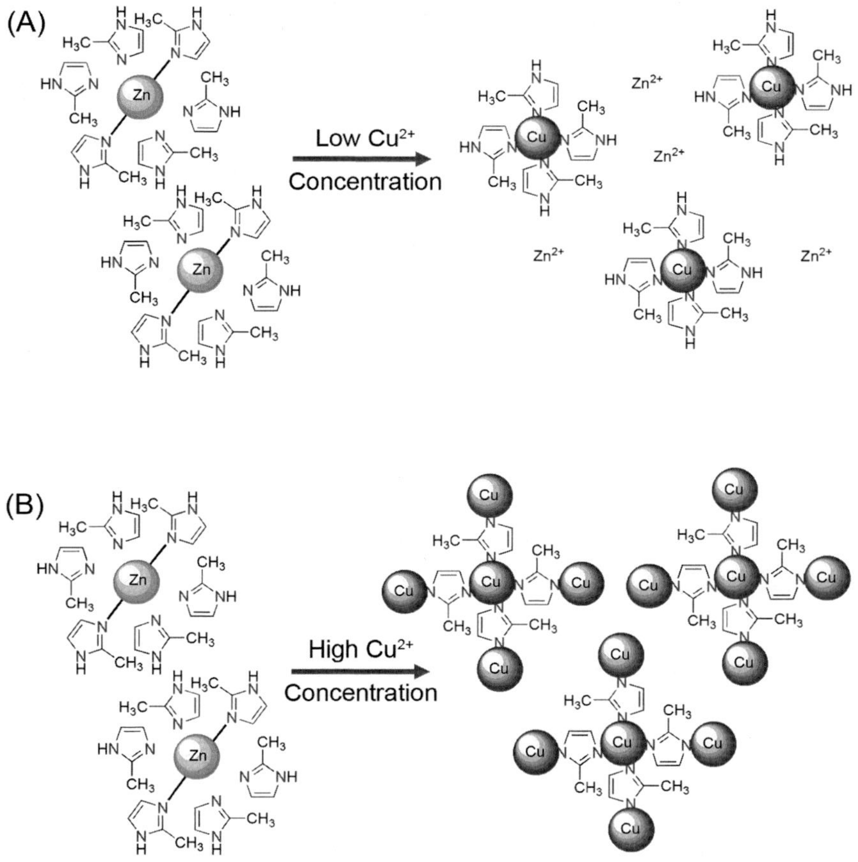

FIGURE 14.27 Mechanism for adsorption of Cu(II) ions by ZIF-8 at low (A) and high (B) concentrations of metal ion via ion exchange and coordination respectively [138].

adsorbing properties for iron(III), manganese(II) and zinc(II) in liquid phase. The MOF was also exhibited to adsorb copper(II) [137]. Wang et al. in 2016 introduced Chitosan-MOF to adsorb nickel(II) effectively in water as well as copper(II) with a maximum capacity of 55 mg/g. Zhang et al. reported ZIF-8 for Copper(II) adsorption via a transmetallation mechanism in ZIF-8 in which the toxicity of the metal ions, zinc, were replaced as in Fig. 14.27[138].

14.3.5 Nutrients removal

In view of environmental issues, removal of excess phosphates from water is highly noteworthy point of concern. Removal of phosphate anions from water was reported by Lin et al. by employing Zr-MOFs, UiO-66 and UiO-66-NH_2, for adsorptive exploration with 237 mg/g and 265 mg/g adsorption capacities, respectively [139]. The upward adsorption capacity indicated hydrogen bonding and electrostatic interactions between the amino functionality and phosphate anions. Lin and coworkers revealed that on exposure to high concentrations of phosphate like >25,000 ppm instead of 50 ppm, the BDC linkers of UiO-66 were replaced with PO_4^{3-} after getting extracted from the MOF leading to formation of porous, amorphous, and remarkably chemically robust ZrPhos and ZrOxyPhos with similar morphology as the parent UiO-66 [139].

14.3.6 Radioactive substances removal

Removal of radioactive substances from the aqueous environment is a major concern for wastewater remediation. There are important developments in MOF materials which can seize cationic and anionic radionuclides from aqueous solution [140,141].

Sequestration of fission products, ^{137}Cs and ^{90}Sr, were investigated by Y. L. Wang et al. through a rare example of a 3D uranyl organic framework material in which the structure was built through polycatenating three sets of graphene-like layers (Fig. 14.28) showing high adsorption with fast kinetics [142].

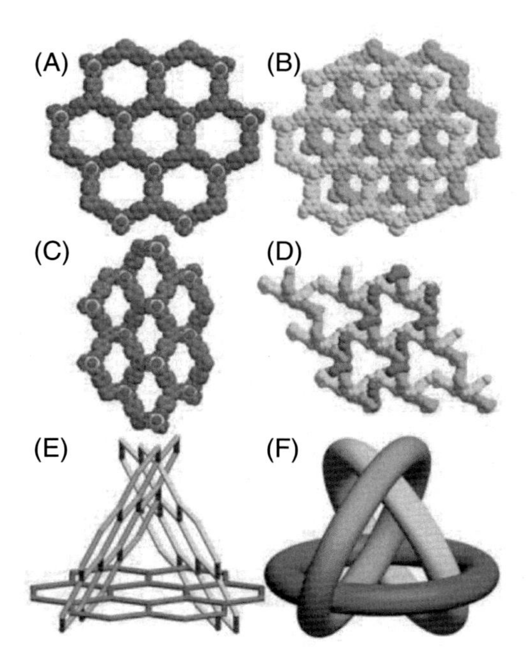

FIGURE 14.28 Structures of a rare polycatenated anionic uranyl organic framework material (A), (C) and (D), (B) $\pi-\pi$ stacking, (E) topological representation showing the three-set polycatenation, and (F) simplification of polycatenation. (*Reproduced with permission from American Chemical Society [142]*).

Wang et al. reported a Zr-based MOF, UiO-66-PYDC, constructed using pyridine-dicarboxylic acid (PYDC) for the efficient removal of I_2. The adsorption capacities of UiO-66-PYDC for I_2 was reported as 1250 mg/g due to the strong affinity of PYDC ligands to I_2 as well as high porosity and, additionally, it exhibited excellent renewable adsorption properties [143]. A novel magnetic MOF composite constituted of iron oxide magnetic nanoparticles and AMCA-MIL-53(Al) was employed for separation of Th(IV) and U(VI) metal ions from water. The process potentially could lead to practical applications in wastewater treatment as it was fast enough to reach the equilibrium that was acknowledged within 90 minutes for both the radioactive metals. As discussed previously, UiO-66 on exposure to high concentrations of phosphate like >25,000 ppm, the PO_4^{3-} linkers took the position of BDC to form ZrPhos and ZrOxyPhos with identical morphology of UiO-66 and those were found effective for removal of high-level nuclear waste like Sr, Pu, Np, and U [139]. In view of wastewater treatment, removal of radioactive metals using MOFs is a budding area of research and deserves more inputs.

14.4 Conclusion and future scope

Safe water is crucial to the survival of human health and ecosystem. Treatment of wastewater remains an uphill task in recent days owing to the ever-expanding manufacturing industries and agriculture sector. Different types of contaminants permeate to surface and groundwater sources. It poses a challenge to make clean water through numerous strategies. Among many approaches of purification, adsorption strategy has become a center of attraction on account of its user-friendly and energy-efficient method in removing hazardous contaminants from wastewater without much disturbing the environment. Reasonably, MOFs with high porosity and high surface areas have turned out as one of the most favored candidate to act as potential adsorbent for wide range of contaminants due to their remarkable structural features. MOFs are "tuned" prudently for selective adsorption as well as to increase the uptake capacity. In this book chapter, reports of many researchers for water-stable MOFs have been summarized including some interesting adsorption applications of zirconium-based MOFs having high porosity and large carbon content. Future scope of such target-specific MOFs greatly lies on effective sequestration of contaminants by employing viable and sustainable strategies.

However, special attention is needed in treating real samples which may not undergo preliminary treatment. In presence of possible interfering agents, the ability of MOFs in selective separation turns out to be a top priority. Performances of MOFs in this regard are paramount, which is often investigated in detail. Not only the structural stability of MOFs but also the number of cycles and the regeneration features of MOFs are also taken into consideration in treatment processes. The powder form of as-prepared MOFs seems a challenge for its practical applications in adsorption and catalysis. Efforts are to be further strenghtened to ovecome the limitations of poor processability and mechanical stability issues of MOFs. Improved stabilities of such MOFs by maintaining their crystalinity is also vital to avoid collapse when they are subjected in aqueous environments for wastewater treatment. In overcoming this issue, granulation technique aided by binders might lead to shaped pellets to improve physical stability of MOF-based materials without compromising surface area [144–148]. Another possible route to boost processability could be MOF–polymer composites [149]. As sufficiently long cycling life and mechanical properties are the requirements of such MOFs for effective implementation, it requires new insight to the durabale MOFs with considerable shelf-life to ensure the performance in repetitive use. At the same time the cost-effectiveness, environmental impacts and safety issues need to be addressed in designing frameworks through green approach. Selection of central metals is also important as metals would be ultimately released in the environment upon degradation. Ideal metal precursors like sulfates and acetates would be prefered to nitrates and chlorides as they might lead to potential toxic and corrosive effect [150,151]. To utilize oxides or hydroxides based metal precursors, alternate synthetic strategies are required owing to their poor solubilities. Harmless alkaline earth metals might be employed as central metal ions in place of transition metals after careful investigation of the features like lifecycle of MOFs, their biological fate, hazardous nature, etc. [151]. Biocompatible linkers would be preferred to linkers like carboxylic acids and imidazoles that eventually would lead to unsafe byproducts. Toxic behaviour of sodium salts of some linkers would also be taken into account as they could form mineral acid [152]. Such toxicological behaviour paved the way to construct MOFs involving muconic acid, fumaric acid [153], amino-acids [154], and γ-cyclodextrin [155], which are non-toxic linkers. Harmful solvents could be avoided too in the making of MOFs due to their toxicity concerns. For example, largely utilized solvent in the formation of of MOF structures, DMF (N,N′-dimethylformamide), could result corrosive and harmful products during solvothermal conditions [156]. Other flammable and volatile organic solvents which adversely affect human health and the environment could be avoided [157]. Accordingly, efficient green solvents need to be investigated involving simple synthetic methods with considerable recycling capability. Bigger attention would be drawn to water-based synthesis of MOFs keeping in view of its low cost with low reaction temperature and green approach that would eliminate the hazards associated with organic solvents [158,159]. Researchers are also exploring the development of thermochemical and mechanochemical synthesis of MOFs which could be based upon solvent-free reactions.

Tremendous advancements in MOFs have been noticed in recent times encompassing several fields of applications through effficient and rational designs. For example, dual-function MOF technologies have been employed in the design of composite materials to reap the advantages of both components. Likewise, the concept of single-crystal to single-crystal transformations are already being explored. It could be seen whether such concepts be possibly exploited in detection and adsorption or, even, in degradation of contaminants. Newly designed class of bio-MOFs deserve a thorough study in the aspect of future scope of their application in water remediation. Further investigations could also be carried out to develop MOFs in terms of introducing hydrophobic functionalities as well as to impart water-repellent properties which could potentially be effective strategies to expand the workability of MOFs in water remediation experiments in unfavourable working conditions like acidic or alkaline situations. Intrinsic magnetic properties of MOFs could also be used for wastewater treatment as well. In another challenge, accomodating large molecules in small pores could be addresed by modifications like carbonization in which the carbonization temperature would govern the pore size distribution and pore structure. Making good uses of competent properties of MOFs as well as by extending those properties through continuous developments, MOFs surely can deliver as a sustainable tool to trap wide variety of contaminants in different working condions in wastewater treatment.

References

[1] Website of National. Oceanic and Atmospheric Administration, USA. http://www.noaa.gov/education/resource-collections/freshwater/water-cycle.

[2] N. Malakar, Editor, Environmental Science, Licensed under the Creative Commons Attribution, License 4 (2019). http://cnx.org/content/col24970/1.1/.

[3] M. Kummu, J.H. Guillaume, H. de Moel, S. Eisner, M. Flörke, M. Porkka, S. Siebert, T.I. Veldkamp, P.J. Ward, The world's road to water scarcity: shortage and stress in the 20th century and pathways towards sustainability, Sci. Rep. 6 (2016) 1–6.

[4] United States Geological Survey, Scientific agency of the United States government website, https://www.usa.gov/federal-agencies/u-s-geological-survey.

[5] J.A. Cotruvo, 2017 WHO guidelines for drinking water quality: first addendum to the fourth edition, J. Am. Water Works Assoc. 109 (2017) 44–51.

[6] Sustainable Development Goal 6 on water and sanitation (SDG 6), UN-Water, United Nations, https://www.sdg6data.org/.

[7] N. Saha, M.S. Rahman, M.B. Ahmed, J.L. Zhou, H.H. Ngo, W. Guo, Industrial metal pollution in water and probabilistic assessment of human health risk, J. Environ. Manage. 185 (2017) 70–78.

[8] T. Tong, M. Elimelech, The global rise of zero liquid discharge for wastewater management: drivers, technologies, and future directions, Environ. Sci. Technol. 50 (2016) 6846–6855.

[9] S.A. Alrumman, A.F. El-kott, S.M. Keshk, Water pollution: source and treatment, Am. J. Environ. Sci. 6 (2016) 88–98.

[10] X. Liu, M. Wang, S. Zhang, B. Pan, Application potential of carbon nanotubes in water treatment: a review, J. Environ. Sci. 25 (2013) 1263–1280.

[11] M.N. Chong, B. Jin, C.W. Chow, C. Saint, Recent developments in photocatalytic water treatment technology: a review, Water Res. 44 (2010) 2997–3027.

[12] S.K. Gupta, K.Y. Chen, Arsenic removal by adsorption, J. Water Pollut. Control Fed. (1978) 493–506.

[13] L.S. Balistrieri, T.T. Chao, Selenium adsorption by goethite, Soil Sci. Soc. Am. J. 51 (1987) 1145–1151.

[14] L.S. Balistrieri, T.T. Chao, Adsorption of selenium by amorphous iron oxyhydroxide and manganese dioxide, Geochim. Cosmochim. Acta 54 (1990) 739–751.

[15] K. Mitchell, R.M. Couture, T.M. Johnson, P.R. Mason, P. Van Cappellen, Selenium sorption and isotope fractionation: iron(III) oxides versus iron(II) sulfides, Chem. Geol. 342 (2013) 21–28.

[16] M. Duc, G. Lefevre, M. Fedoroff, J. Jeanjean, J.C. Rouchaud, F. Monteil-Rivera, et al., Sorption of selenium anionic species on apatites and iron oxides from aqueous solutions, J. Environ. Radioact. 70 (2003) 61–72.

[17] H. Figueiredo, C. Quintelas, Tailored zeolites for the removal of metal oxyanions: overcoming intrinsic limitations of zeolites, J. Hazard. Mater. 274 (2014) 287–299.

[18] C.P. Huang, P.L. Fu, Treatment of arsenic(V)-containing water by the activated carbon process, J. Water Pollut. Control Fed. (1984) 233–242.

[19] R. Mahmudov, C.P. Huang, Selective adsorption of oxyanions on activated carbon exemplified by Filtrasorb 400 (F400), Sep. Purif. Technol. 77 (2011) 294–300.

[20] Y.T. Chan, W.H. Kuan, T.Y. Chen, M.K. Wang, Adsorption mechanism of selenate and selenite on the binary oxide systems, Water Res. 43 (2009) 4412–4420.

[21] C.P. Johnston, M. Chrysochoou, Mechanisms of chromate adsorption on boehmite, J. Hazard. Mater. 281 (2015) 56–63.

[22] R.M. Rego, G. Kuriya, M.D. Kurkuri, M. Kigga, MOF based engineered materials in water remediation: recent trends, J. Hazard. Mater. 403 (2021) 123605.

[23] Y.G. Chung, J. Camp, M. Haranczyk, B.J. Sikora, W. Bury, V. Krungleviciute, et al., Computation-ready, experimental metal–organic frameworks: a tool to enable high-throughput screening of nanoporous crystals, Chem. Mater. 26 (21) (2014) 6185–6192.

[24] H. Furukawa, K.E. Cordova, M. O'Keeffe, O.M. Yaghi, The chemistry and applications of metal-organic frameworks, Science 341 (2013) 1230444.

[25] N.S. Bobbitt, M.L. Mendonca, A.J. Howarth, T. Islamoglu, J.T. Hupp, O.K. Farha, et al., Metal–organic frameworks for the removal of toxic industrial chemicals and chemical warfare agents, Chem. Soc. Rev. 46 (2017) 3357–3385.

[26] O.K. Farha, A.M. Spokoyny, B.G. Hauser, Y.S. Bae, S.E. Brown, R.Q. Snurr, et al., Synthesis, properties, and gas separation studies of a robust diimide-based microporous organic polymer, Chem. Mater. 21 (2009) 3033–3035.

[27] V.I. Isaeva, L.M. Kustov, The application of metal-organic frameworks in catalysis, Pet. Chem. 50 (2010) 167–180.

[28] J. Gascon, A. Corma, F. Kapteijn, F.X. Llabres i Xamena, Metal organic framework catalysis: Quo vadis? ACS Catal. 4 (2014) 361–378.

[29] T. Zhang, W. Lin, Metal–organic frameworks for artificial photosynthesis and photocatalysis, Chem. Soc. Rev. 43 (2014) 5982–5993.

[30] M. Dincă, J.R. Long, Hydrogen storage in microporous metal–organic frameworks with exposed metal sites, Angew. Chem. Int. Ed. Engl. 47 (2008) 6766–6779.

[31] J.R. Li, R.J. Kuppler, H.C. Zhou, Selective gas adsorption and separation in metal–organic frameworks, Chem. Soc. Rev. 38 (2009) 1477–1504.

[32] O.K. Farha, A. Özgür Yazaydın, I. Eryazici, C.D. Malliakas, B.G. Hauser, M.G. Kanatzidis, et al., De novo synthesis of a metal–organic framework material featuring ultrahigh surface area and gas storage capacities, Nat. Chem. 2 (2010) 944–948.

[33] J.R. Li, J. Sculley, H.C. Zhou, Metal–organic frameworks for separations, Chem. Rev. 112 (2012) 869–932.

[34] P. Horcajada, C. Serre, M. Vallet-Regí, M. Sebban, F. Taulelle, G. Férey, Metal–organic frameworks as efficient materials for drug delivery, Angew. Chem. 118 (2006) 6120–6124.

[35] L.E. Kreno, K. Leong, O.K. Farha, M. Allendorf, R.P. Van Duyne, J.T. Hupp, Metal–organic framework materials as chemical sensors, Chem. Rev. 112 (2012) 1105–1125.

[36] M.J. Katz, J.E. Mondloch, R.K. Totten, J.K. Park, S.T. Nguyen, O.K. Farha, et al., Simple and compelling biomimetic metal–organic framework catalyst for the degradation of nerve agent simulants, Angew. Chem. 126 (2014) 507–511.

[37] M.J. Katz, S.Y. Moon, J.E. Mondloch, M.H. Beyzavi, C.J. Stephenson, J.T. Hupp, et al., Exploiting parameter space in MOFs: a 20-fold enhancement of phosphate-ester hydrolysis with UiO-66-NH_2, Chem. Sci. 6 (2015) 2286–2291.

[38] C.Y. Lee, O.K. Farha, B.J. Hong, A.A. Sarjeant, S.T. Nguyen, J.T. Hupp, Light-harvesting metal–organic frameworks (MOFs): efficient strut-to-strut energy transfer in bodipy and porphyrin-based MOFs, J. Am. Chem. Soc. 133 (2011) 15858–15861.

[39] M.C. So, G.P. Wiederrecht, J.E. Mondloch, J.T. Hupp, O.K. Farha, Metal–organic framework materials for light-harvesting and energy transfer, Chem. Commun. 51 (2015) 3501–3510.

[40] N.A. Khan, Z. Hasan, S.H. Jhung, Adsorptive removal of hazardous materials using metal-organic frameworks (MOFs): a review, J. Hazard. Mater. 244 (2013) 444–456.

[41] O.M. Yaghi, M. O'Keeffe, N.W. Ockwig, H.K. Chae, M. Eddaoudi, J. Kim, Reticular synthesis and the design of new materials, Nature 423 (2003) 705–714.

[42] O.K. Farha, I. Eryazici, N.C. Jeong, B.G. Hauser, C.E. Wilmer, A.A. Sarjeant, et al., Metal–organic framework materials with ultrahigh surface areas: is the sky the limit? J. Am. Chem. Soc. 134 (2012) 15016–15021.

[43] L. Zhu, L. Meng, J. Shi, J. Li, X. Zhang, M. Feng, Metal-organic frameworks/carbon-based materials for environmental remediation: a state-of-the-art mini-review, J. Environ. Manage. 232 (2019) 964–977.

[44] R. Grünker, V. Bon, P. Müller, U. Stoeck, S. Krause, U. Mueller, et al., A new metal–organic framework with ultra-high surface area, Chem. Commun. 50 (2014) 3450–3452.

[45] O.V. Gutov, W. Bury, D.A. Gomez-Gualdron, V. Krungleviciute, D. Fairen-Jimenez, J.E. Mondloch, et al., Water-stable zirconium-based metal–organic framework material with high-surface area and gas-storage capacities, Chem. Eur. J. 20 (2014) 12389–12393.

[46] T.C. Wang, W. Bury, D.A. Gómez-Gualdrón, N.A. Vermeulen, J.E. Mondloch, P. Deria, et al., Ultrahigh surface area zirconium MOFs and insights into the applicability of the BET theory, J. Am. Chem. Soc. 137 (2015) 3585–3591.

[47] H. Furukawa, N. Ko, Y.B. Go, N. Aratani, S.B. Choi, E. Choi, et al., Ultrahigh porosity in metal-organic frameworks, Science 329 (2010) 424–428.

[48] P. Li, N.A. Vermeulen, C.D. Malliakas, D.A. Gómez-Gualdrón, A.J. Howarth, B.L. Mehdi, et al., Bottom-up construction of a superstructure in a porous uranium-organic crystal, Science 356 (2017) 624–627.

[49] Y.K. Park, S.B. Choi, H. Kim, K. Kim, B.H. Won, K. Choi, et al., Crystal structure and guest uptake of a mesoporous metal–organic framework containing cages of 3.9 and 4.7 nm in diameter, Angew. Chem. Int. Ed. 46 (2007) 8230–8233.

[50] I. Senkovska, S. Kaskel, Ultrahigh porosity in mesoporous MOFs: promises and limitations, Chem. Commun. 50 (2014) 7089–7098.

[51] J. An, O.K. Farha, J.T. Hupp, E. Pohl, J.I. Yeh, N.L. Rosi, Metal-adeninate vertices for the construction of an exceptionally porous metal-organic framework, Nat. Commun. 3 (2012) 1–6.

[52] P.A. Kobielska, A.J. Howarth, O.K. Farha, S. Nayak, Metal–organic frameworks for heavy metal removal from water, Coord. Chem. Rev. 358 (2018) 92–107.
[53] S. Mukherjee, A. Kumar, M.J. Zaworotko, Metal-organic framework based carbon capture and purification technologies for clean environment, in: S.K. Ghosh (Ed.), Metal-Organic Frameworks (MOFs) for Environmental Applications, Elsevier, 2019, pp. 5–61.
[54] F. Ke, L.G. Qiu, Y.P. Yuan, F.M. Peng, X. Jiang, A.J. Xie, et al., Thiol-functionalization of metal-organic framework by a facile coordination-based postsynthetic strategy and enhanced removal of Hg^{2+} from water, J. Hazard. Mater. 196 (2011) 36–43.
[55] S. Horike, S. Shimomura, S. Kitagawa, Soft porous crystals, Nat. Chem. 1 (2009) 695–704.
[56] N.C. Burtch, H. Jasuja, K.S. Walton, Water stability and adsorption in metal–organic frameworks, Chem. Rev. 114 (2014) 10575–10612.
[57] A.J. Howarth, Y. Liu, P. Li, Z. Li, T.C. Wang, J.T. Hupp, et al., Chemical, thermal and mechanical stabilities of metal–organic frameworks, Nat. Rev. Mater. 1 (2016) 1–5.
[58] J. Canivet, A. Fateeva, Y. Guo, B. Coasne, D. Farrusseng, Water adsorption in MOFs: fundamentals and applications, Chem. Soc. Rev. 43 (2014) 5594–5617.
[59] M. Bosch, M. Zhang, H.C. Zhou, Increasing the stability of metal-organic frameworks, Adv. Chem. 2014 (2014) 1155.
[60] N. ul Qadir, S.A. Said, H.M. Bahaidarah, Structural stability of metal organic frameworks in aqueous media–controlling factors and methods to improve hydrostability and hydrothermal cyclic stability, Microporous Mesoporous Mater. 201 (2015) 61–90.
[61] C. Wang, X. Liu, N.K. Demir, J.P. Chen, K. Li, Applications of water stable metal–organic frameworks, Chem. Soc. Rev. 45 (2016) 5107–5134.
[62] R. Zhang, Z. Wang, Z. Zhou, D. Li, T. Wang, P. Su, et al., Highly effective removal of pharmaceutical compounds from aqueous solution by magnetic Zr-based MOFs composites, Ind. Eng. Chem. Res. 58 (2019) 3876–3884.
[63] P. Wu, Y. Liu, Y. Liu, J. Wang, Y. Li, W. Liu, et al., Cadmium-based metal–organic framework as a highly selective and sensitive ratiometric luminescent sensor for mercury(II), Inorg. Chem. 54 (2015) 11046–11048.
[64] B. Dutta, R. Jana, A.K. Bhanja, P.P. Ray, C. Sinha, M.H. Mir, Supramolecular aggregate of cadmium(II)-based one-dimensional coordination polymer for device fabrication and sensor application, Inorg. Chem. 58 (2019) 2686–2694.
[65] A. Karmakar, B. Joarder, A. Mallick, P. Samanta, A.V. Desai, S. Basu, et al., Aqueous phase sensing of cyanide ions using a hydrolytically stable metal–organic framework, Chem. Commun. 53 (2017) 1253–1256.
[66] M. Baghayeri, M. Ghanei-Motlagh, R. Tayebee, M. Fayazi, F. Narenji, Application of graphene/zinc-based metal-organic framework nanocomposite for electrochemical sensing of As(III) in water resources, Anal. Chim. Acta 1099 (2020) 60–67.
[67] M.U. Prathap, S. Gunasekaran, Rapid and scalable synthesis of zeolitic imidazole framework (ZIF-8) and its use for the detection of trace levels of nitroaromatic explosives, Adv. Sustain. Syst. 2 (2018) 1800053.
[68] S. Khan, A. Hazra, B. Dutta, R.M.J. Akhtaruzzaman, P. Banerjee, et al., Strategic design of anthracene-decorated highly luminescent coordination polymers for selective and rapid detection of TNP: an explosive nitro derivative and mutagenic pollutant, Cryst. Growth Des. 21 (2021) 3344–3354.
[69] E. Haque, J.W. Jun, S.H. Jhung, Adsorptive removal of methyl orange and methylene blue from aqueous solution with a metal-organic framework material, iron terephthalate (MOF-235), J. Hazard. Mater. 185 (2011) 507–511.

[70] S. Soni, P.K. Bajpai, J. Mittal, C. Arora, Utilisation of cobalt doped iron based MOF for enhanced removal and recovery of methylene blue dye from waste water, J. Mol. Liq. 314 (2020) 113642.

[71] A.R. Abbasi, M. Karimi, K. Daasbjerg, Efficient removal of crystal violet and methylene blue from wastewater by ultrasound nanoparticles Cu-MOF in comparison with mechanosynthesis method, Ultrason. Sonochem. 37 (2017) 182–191.

[72] H. Li, X. Cao, C. Zhang, Q. Yu, Z. Zhao, X. Niu, et al., Enhanced adsorptive removal of anionic and cationic dyes from single or mixed dye solutions using MOF PCN-222, RSC Adv. 7 (2017) 16273–16281.

[73] J.H. Cavka, S. Jakobsen, U. Olsbye, N. Guillou, C. Lamberti, S. Bordiga, et al., A new zirconium inorganic building brick forming metal organic frameworks with exceptional stability, J. Am. Chem. Soc. 130 (2008) 13850–13851.

[74] Q. Chen, Q. He, M. Lv, Y. Xu, H. Yang, X. Liu, et al., Selective adsorption of cationic dyes by UiO-66-NH_2, Appl. Surf. Sci. 327 (2015) 77–85.

[75] M.S. Embaby, S.D. Elwany, W. Setyaningsih, M.R. Saber, The adsorptive properties of UiO-66 towards organic dyes: a record adsorption capacity for the anionic dye Alizarin Red S, Chin. J. Chem. Eng. 26 (2018) 731–739.

[76] H. Molavi, A. Hakimian, A. Shojaei, M. Raeiszadeh, Selective dye adsorption by highly water stable metal-organic framework: long term stability analysis in aqueous media, Appl. Surf. Sci. 445 (2018) 424–436.

[77] A.A. Mohammadi, A. Alinejad, B. Kamarehie, S. Javan, A. Ghaderpoury, M. Ahmadpour, et al., Metal-organic framework Uio-66 for adsorption of methylene blue dye from aqueous solutions, Int. J. Environ. Sci. Technol. 14 (2017) 1959–1968.

[78] J.M. Yang, A facile approach to fabricate an immobilized-phosphate zirconium-based metal-organic framework composite (UiO-66-P) and its activity in the adsorption and separation of organic dyes, J. Colloid Interface Sci. 505 (2017) 178–185.

[79] Q. He, Q. Chen, M. Lü, X. Liu, Adsorption behavior of rhodamine B on UiO-66, Chin. J. Chem. Eng. 22 (2014) 1285–12890.

[80] R.J. Drout, L. Robison, Z. Chen, T. Islamoglu, O.K. Farha, Zirconium metal–organic frameworks for organic pollutant adsorption, Trends Chem. 1 (2019) 304–317.

[81] S. Tian, S. Xu, J. Liu, C. He, Y. Xiong, P. Feng, Highly efficient removal of both cationic and anionic dyes from wastewater with a water-stable and eco-friendly Fe-MOF via host-guest encapsulation, J. Clean Prod. 239 (2019) 117767.

[82] C. Zhang, H. Li, C. Li, Z. Li, Fe-loaded MOF-545 (Fe): peroxidase-like activity for dye degradation dyes and high adsorption for the removal of dyes from wastewater, Molecules 25 (2019) 168.

[83] J. Abdi, H. Abedini, MOF-based polymeric nanocomposite beads as an efficient adsorbent for wastewater treatment in batch and continuous systems: modelling and experiment, Chem. Eng. J. 400 (2020) 125862.

[84] Y. Su, Z. Li, H. Zhou, S. Kang, Y. Zhang, C. Yu, et al., Ni/carbon aerogels derived from water induced self-assembly of Ni-MOF for adsorption and catalytic conversion of oily wastewater, Chem. Eng. J. 402 (2020) 126205.

[85] H. Zhu, Q. Zhang, B.G. Li, S. Zhu, Engineering elastic ZIF-8-Sponges for oil–water separation, Adv. Mater. Interfaces 4 (2017) 1700560.

[86] T.B. Hayes, A. Collins, M. Lee, M. Mendoza, N. Noriega, A.A. Stuart, et al., Hermaphroditic, demasculinized frogs after exposure to the herbicide atrazine at low ecologically relevant doses, Proc. Natl. Acad. Sci. USA 99 (2002) 5476–5480.

[87] A. Ventura, G. Jacquet, A. Bermond, V. Camel, Electrochemical generation of the Fenton's reagent: application to atrazine degradation, Water Res. 36 (2002) 3517–3522.

[88] I. Akpinar, A.O. Yazaydin, Adsorption of atrazine from water in metal–organic framework materials, J. Chem. Eng. Data 63 (2018) 2368–2375.

[89] I. Akpinar, R.J. Drout, T. Islamoglu, S. Kato, J. Lyu, O.K. Farha, Exploiting π–π interactions to design an efficient sorbent for atrazine removal from water, ACS Appl. Mater. Interfaces 11 (2019) 6097–6103.

[90] Y.S. Seo, N.A. Khan, S.H. Jhung, Adsorptive removal of methylchlorophenoxypropionic acid from water with a metal-organic framework, Chem. Eng. J. 270 (2015) 22–27.

[91] X. Zhu, B. Li, J. Yang, Y. Li, W. Zhao, J. Shi, et al., Effective adsorption and enhanced removal of organophosphorus pesticides from aqueous solution by Zr-based MOFs of UiO-67, ACS Appl. Mater. Interfaces 7 (2015) 223–231.

[92] A. Pankajakshan, M. Sinha, A.A. Ojha, S. Mandal, Water-stable nanoscale zirconium-based metal–organic frameworks for the effective removal of glyphosate from aqueous media, ACS Omega 3 (2018) 7832–7839.

[93] K.L. Tan, K.Y. Foo, Preparation of MIL-100 via a novel water-based heatless synthesis technique for the effective remediation of phenoxyacetic acid-based pesticide, J. Environ. Chem. Eng. 9 (2021) 104923.

[94] M.R. Azhar, H.R. Abid, V. Periasamy, H. Sun, M.O. Tade, S. Wang, Adsorptive removal of antibiotic sulfonamide by UiO-66 and ZIF-67 for wastewater treatment, J. Colloid Interface Sci. 500 (2017) 88–95.

[95] C. Chen, D. Chen, S. Xie, H. Quan, X. Luo, L. Guo, Adsorption behaviors of organic micropollutants on zirconium metal–organic framework UiO-66: analysis of surface interactions, ACS Appl. Mater. Interfaces 9 (2017) 41043–41054.

[96] Y. Zhang, Q. Ruan, Y. Peng, G. Han, H. Huang, C. Zhong, Synthesis of hierarchical-pore metal-organic framework on liter scale for large organic pollutants capture in wastewater, J. Colloid Interface Sci. 525 (2018) 39–47.

[97] Y. Zhou, Q. Yang, D. Zhang, N. Gan, Q. Li, J. Cuan, Detection and removal of antibiotic tetracycline in water with a highly stable luminescent MOF, Sens. Actuators B Chem. 262 (2018) 137–143.

[98] P. Deria, J.E. Mondloch, E. Tylianakis, P. Ghosh, W. Bury, R.Q. Snurr, et al., Perfluoroalkane functionalization of NU-1000 via solvent-assisted ligand incorporation: synthesis and CO_2 adsorption studies, J. Am. Chem. Soc. 135 (2013) 16801–16804.

[99] X. Zhao, H. Zhao, W. Dai, Y. Wei, Y. Wang, Y. Zhang, et al., A metal-organic framework with large 1-D channels and rich OH sites for high-efficiency chloramphenicol removal from water, J. Colloid Interface Sci. 526 (2018) 28–34.

[100] S. Lin, Y. Zhao, Y.S. Yun, Highly effective removal of nonsteroidal anti-inflammatory pharmaceuticals from water by Zr(IV)-based metal–organic framework: adsorption performance and mechanisms, ACS Appl. Mater. Interfaces 10 (2018) 28076–28085.

[101] B. Wang, X.L. Lv, D. Feng, L.H. Xie, J. Zhang, M. Li, et al., Highly stable Zr (IV)-based metal–organic frameworks for the detection and removal of antibiotics and organic explosives in water, J. Am. Chem. Soc. 138 (2016) 6204–6216.

[102] Y. Zhuang, Y. Kong, X. Wang, B. Shi, Novel one step preparation of a 3D alginate based MOF hydrogel for water treatment, New J. Chem. 43 (2019) 7202–7208.

[103] S. Li, Y. Chen, X. Pei, S. Zhang, X. Feng, J. Zhou, et al., Water purification: adsorption over metal-organic frameworks, Chinese J. Chem. 34 (2016) 175–185.

[104] A.J. Howarth, Y. Liu, J.T. Hupp, O.K. Farha, Metal–organic frameworks for applications in remediation of oxyanion/cation-contaminated water, CrystEngComm 17 (2015) 7245–7253.

[105] Z. Hasan, S.H. Jhung, Removal of hazardous organics from water using metal-organic frameworks (MOFs): plausible mechanisms for selective adsorptions, J. Hazard. Mater. 283 (2015) 329–339.

[106] B.J. Zhu, X.Y. Yu, Y. Jia, F.M. Peng, B. Sun, M.Y. Zhang, et al., Iron and 1, 3, 5-benzenetricarboxylic metal–organic coordination polymers prepared by solvothermal method and their application in efficient As(V) removal from aqueous solutions, Phys. Chem. C 116 (2012) 8601–8607.

[107] M. Jian, B. Liu, G. Zhang, R. Liu, X. Zhang, Adsorptive removal of arsenic from aqueous solution by zeolitic imidazolate framework-8 (ZIF-8) nanoparticles, Colloids Surf. A Physicochem. Eng. Asp. 465 (2015) 67–76.

[108] B. Liu, M. Jian, R. Liu, J. Yao, X. Zhang, Highly efficient removal of arsenic(III) from aqueous solution by zeolitic imidazolate frameworks with different morphology, Colloids Surf. A Physicochem. Eng. Asp. 481 (2015) 358–366.

[109] T.A. Vu, G.H. Le, C.D. Dao, L.Q. Dang, K.T. Nguyen, Q.K. Nguyen, et al., Arsenic removal from aqueous solutions by adsorption using novel MIL-53 (Fe) as a highly efficient adsorbent, RSC Adv. 5 (2015) 5261–5268.

[110] J. Li, Y.N. Wu, Z. Li, M. Zhu, F. Li, Characteristics of arsenate removal from water by metal-organic frameworks (MOFs), Water Sci. Technol. 70 (2014) 1391–1397.

[111] Z.Q. Li, J.C. Yang, K.W. Sui, N. Yin, Facile synthesis of metal-organic framework MOF-808 for arsenic removal, Mater. Lett. 160 (2015) 412–414.

[112] C. Wang, X. Liu, J.P. Chen, K. Li, Superior removal of arsenic from water with zirconium metal-organic framework UiO-66, Sci. Rep. 5 (2015) 1–10.

[113] C.O. Audu, H.G. Nguyen, C.Y. Chang, M.J. Katz, L. Mao, O.K. Farha, et al., The dual capture of As V and As III by UiO-66 and analogues, Chem. Sci. 7 (2016) 6492–6498.

[114] Q.R. Fang, D.Q. Yuan, J. Sculley, J.R. Li, Z.B. Han, H.C. Zhou, Functional mesoporous metal−organic frameworks for the capture of heavy metal ions and size-selective catalysis, Inorg. Chem. 49 (2010) 11637–11642.

[115] T. Liu, J.X. Che, Y.Z. Hu, X.W. Dong, X.Y. Liu, C.M. Che, Alkenyl/thiol-derived metal–organic frameworks (MOFs) by means of postsynthetic modification for effective mercury adsorption, Chem. Eur. J. 20 (2014) 14090–14095.

[116] K.K. Yee, N. Reimer, J. Liu, S.Y. Cheng, S.M. Yiu, J. Weber, et al., Effective mercury sorption by thiol-laced metal–organic frameworks: in strong acid and the vapor phase, J. Am. Chem. Soc. 135 (2013) 7795–7798.

[117] M.R. Sohrabi, Preconcentration of mercury (II) using a thiol-functionalized metal-organic framework nanocomposite as a sorbent, Microchim. Acta 181 (2014) 435–444.

[118] X. Luo, T. Shen, L. Ding, W. Zhong, J. Luo, S. Luo, Novel thymine-functionalized MIL-101 prepared by post-synthesis and enhanced removal of Hg^{2+} from water, J. Hazard. Mater. 306 (2016) 313–322.

[119] H. Saleem, U. Rafique, R.P. Davies, Investigations on post-synthetically modified UiO-66-NH_2 for the adsorptive removal of heavy metal ions from aqueous solution, Microporous Mesoporous Mater. 221 (2016) 238–244.

[120] A. Abbasi, T. Moradpour, K. Van Hecke, A new 3D cobalt (II) metal–organic framework nanostructure for heavy metal adsorption, Inorg. Chim. Acta 430 (2015) 261–267.

[121] L. Liang, Q. Chen, F. Jiang, D. Yuan, J. Qian, G. Lv, et al., In situ large-scale construction of sulfur-functionalized metal–organic framework and its efficient removal of Hg(II) from water, J. Mater. Chem. A 4 (2016) 15370–15374.

[122] A. Chakraborty, S. Bhattacharyya, A. Hazra, A.C. Ghosh, T.K. Maji, Post-synthetic metalation in an anionic MOF for efficient catalytic activity and removal of heavy metal ions from aqueous solution, Chem. Commun. 52 (2016) 2831–2834.

[123] L. Huang, M. He, B. Chen, B. Hu, A designable magnetic MOF composite and facile coordination-based post-synthetic strategy for the enhanced removal of Hg^{2+} from water, J. Mater. Chem. A 3 (2015) 11587–11595.

[124] F. Luo, J.L. Chen, L.L. Dang, W.N. Zhou, H.L. Lin, J.Q. Li, et al., High-performance Hg^{2+} removal from ultra-low-concentration aqueous solution using both acylamide-and hydroxyl-functionalized metal–organic framework, J. Mater. Chem. A 3 (2015) 9616–9620.

[125] S. Bhattacharjee, Y.R. Lee, WS. Ahn, Post-synthesis functionalization of a zeolitic imidazolate structure ZIF-90: a study on removal of Hg(II) from water and epoxidation of alkenes, CrystEngComm 17 (2015) 2575–2582.

[126] N.D. Rudd, H. Wang, E.M. Fuentes-Fernandez, S.J. Teat, F. Chen, G. Hall, et al., Highly efficient luminescent metal–organic framework for the simultaneous detection and removal of heavy metals from water, ACS Appl. Mater. Interfaces 8 (2016) 30294–30303.

[127] M. Mon, X. Qu, J. Ferrando-Soria, I. Pellicer-Carreño, A. Sepúlveda-Escribano, E.V. Ramos-Fernandez, et al., Fine-tuning of the confined space in microporous metal–organic frameworks for efficient mercury removal, J. Mater. Chem. A 5 (2017) 20120–20125.

[128] Y.Y. Xiong, J.Q. Li, X.F. Feng, L.N. Meng, L. Zhang, P.P. Meng, et al., Using MOF-74 for Hg^{2+} removal from ultra-low concentration aqueous solution, J. Solid State Chem. 246 (2017) 16–22.

[129] S. Halder, J. Mondal, J. Ortega-Castro, A. Frontera, P. Roy, A Ni-based MOF for selective detection and removal of Hg^{2+} in aqueous medium: a facile strategy, Dalton Trans. 46 (2017) 1943–1950.

[130] Q. Qin, Q. Wang, D. Fu, J. Ma, An efficient approach for Pb(II) and Cd(II) removal using manganese dioxide formed in situ, Chem. Eng. J. 172 (2011) 68–74.

[131] Y. Wang, G. Ye, H. Chen, X. Hu, Z. Niu, S. Ma, Functionalized metal–organic framework as a new platform for efficient and selective removal of cadmium (II) from aqueous solution, J. Mater. Chem. A 3 (2015) 15292–15298.

[132] F. Zou, R. Yu, R. Li, W. Li, Microwave-assisted synthesis of HKUST-1 and functionalized HKUST-1-@ H3PW12O40: selective adsorption of heavy metal ions in water analyzed with synchrotron radiation, ChemPhysChem 14 (2013) 2825–2832.

[133] E. Tahmasebi, M.Y. Masoomi, Y. Yamini, A. Morsali, Application of mechanosynthesized azine-decorated zinc (II) metal–organic frameworks for highly efficient removal and extraction of some heavy-metal ions from aqueous samples: a comparative study, Inorg. Chem. 54 (2015) 425–433.

[134] X. Li, X. Gao, L. Ai, J. Jiang, Mechanistic insight into the interaction and adsorption of Cr (VI) with zeolitic imidazolate framework-67 microcrystals from aqueous solution, Chem. Eng. J. 274 (2015) 238–246.

[135] X. Cheng, M. Liu, A. Zhang, S. Hu, C. Song, G. Zhang, et al., Size-controlled silver nanoparticles stabilized on thiol-functionalized MIL-53 (Al) frameworks, Nanoscale 7 (2015) 9738–9745.

[136] J.E. Conde-González, E.M. Peña-Méndez, S. Rybáková, J. Pasán, C. Ruiz-Pérez, J. Havel, Adsorption of silver nanoparticles from aqueous solution on copper-based metal organic frameworks (HKUST-1), Chemosphere 150 (2016) 659–666.

[137] E. Rahimi, N. Mohaghegh, Removal of toxic metal ions from Sungun acid rock drainage using mordenite zeolite, graphene nanosheets, and a novel metal–organic framework, Mine Water Environ. 35 (2016) 18–28.

[138] Y. Zhang, Z. Xie, Z. Wang, X. Feng, Y. Wang, A. Wu, Unveiling the adsorption mechanism of zeolitic imidazolate framework-8 with high efficiency for removal of copper ions from aqueous solutions, Dalton Trans. 45 (2016) 12653–12660.
[139] K.Y. Lin, S.Y. Chen, A.P. Jochems, Zirconium-based metal organic frameworks: highly selective adsorbents for removal of phosphate from water and urine, Mater. Chem. Phys. 160 (2015) 168–176.
[140] C.W. Abney, K.M. Taylor-Pashow, S.R. Russell, Y. Chen, R. Samantaray, J.V. Lockard, et al., Topotactic transformations of metal–organic frameworks to highly porous and stable inorganic sorbents for efficient radionuclide sequestration, Chem. Mater. 26 (2014) 5231–5243.
[141] C. Xiao, M.A. Silver, S. Wang, Metal–organic frameworks for radionuclide sequestration from aqueous solution: a brief overview and outlook, Dalton Trans. 46 (2017) 16381–16386.
[142] Y. Wang, Z. Liu, Y. Li, Z. Bai, W. Liu, Y. Wang, et al., Umbellate distortions of the uranyl coordination environment result in a stable and porous polycatenated framework that can effectively remove cesium from aqueous solutions, J. Am. Chem. Soc. 137 (2015) 6144–6147.
[143] Z. Wang, Y. Huang, J. Yang, Y. Li, Q. Zhuang, J. Gu, The water-based synthesis of chemically stable Zr-based MOFs using pyridine-containing ligands and their exceptionally high adsorption capacity for iodine, Dalton Trans. 46 (2017) 7412–7420.
[144] A.H. Valekar, K.H. Cho, U.H. Lee, J.S. Lee, J.W. Yoon, Y.K. Hwang, et al., Shaping of porous metal–organic framework granules using mesoporous ρ-alumina as a binder, RSC Adv. 7 (2017) 55767–55777.
[145] J. Ren, N.M. Musyoka, H.W. Langmi, B.C. North, M. Mathe, X. Kang, et al., Hydrogen storage in Zr-fumarate MOF, Int. J. Hydrogen Energy 40 (2015) 10542–10546.
[146] S.A. Moggach, T.D. Bennett, A.K. Cheetham, The effect of pressure on ZIF-8: increasing pore size with pressure and the formation of a high-pressure phase at 1.47 GPa, Angew. Chem. Int. Ed. 48 (2009) 7087–7089.
[147] K.W. Chapman, G.J. Halder, P.J. Chupas, Pressure-induced amorphization and porosity modification in a metal−organic framework, J. Am. Chem. Soc. 131 (2009) 17546–17547.
[148] L.R. Redfern, L. Robison, M.C. Wasson, S. Goswami, J. Lyu, T. Islamoglu, et al., Porosity dependence of compression and lattice rigidity in metal–organic framework series, J. Am. Chem. Soc. 141 (2019) 4365–4371.
[149] Y. Zhang, X. Feng, S. Yuan, J. Zhou, B. Wang, Challenges and recent advances in MOF–polymer composite membranes for gas separation, Inorg Chem Front 3 (2016) 896–909.
[150] D. Farrusseng, Metal-Organic Frameworks: Applications From Catalysis to Gas Storage, John Wiley & Sons, 2011.
[151] P.A. Julien, C. Mottillo, T. Friščić, Metal–organic frameworks meet scalable and sustainable synthesis, Green Chem. 19 (2017) 2729–2747.
[152] M. Sánchez-Sánchez, N. Getachew, K. Díaz, M. Díaz-García, Y. Chebude, I. Díaz, Synthesis of metal–organic frameworks in water at room temperature: salts as linker sources, Green Chem. 17 (2015) 1500–1509.
[153] P. Horcajada, T. Chalati, C. Serre, B. Gillet, C. Sebrie, T. Baati, et al., Porous metal–organic-framework nanoscale carriers as a potential platform for drug delivery and imaging, Nat. Mater 9 (2010) 172–178.
[154] D. Sarma, K.V. Ramanujachary, S.E. Lofland, T. Magdaleno, S. Natarajan, Amino acid based MOFs: synthesis, structure, single crystal to single crystal transformation, magnetic and

related studies in a family of cobalt and nickel aminoisophthales, Inorg. Chem. 48 (2009) 11660–11676.

[155] R.A. Smaldone, R.S. Forgan, H. Furukawa, J.J. Gassensmith, A.M. Slawin, O.M. Yaghi, et al., Metal–organic frameworks from edible natural products, Angew. Chem. Int. Ed. 49 (2010) 8630–8634.

[156] S.M. Hawxwell, L. Brammer, Solvent hydrolysis leads to an unusual Cu(II) metal–organic framework, CrystEngComm 8 (2006) 473–476.

[157] N. Stock, S. Biswas, Synthesis of metal-organic frameworks (MOFs): routes to various MOF topologies, morphologies, and composites, Chem. Rev. 112 (2012) 933–969.

[158] J. Zhang, G.B. White, M.D. Ryan, A.J. Hunt, M.J. Katz, Dihydrolevoglucosenone (cyrene) as a green alternative to N, N-dimethylformamide (DMF) in MOF synthesis, ACS Sustain. Chem. Eng. 4 (2016) 7186–7192.

[159] I.A. Ibarra, P.A. Bayliss, E. Pérez, S. Yang, A.J. Blake, H. Nowell, et al., Near-critical water, a cleaner solvent for the synthesis of a metal–organic framework, Green Chem. 14 (2012) 117–122.

Chapter 15

Industrial aspects of water-based metal–organic frameworks

Atif Husain[a], Malik Nasibullah[a], Farrukh Aqil[b] and Abdul Rahman Khan[a]
[a] *Department of Chemistry, Integral University, Lucknow, Uttar Pradesh, India,* [b] *UofL Health-Brown Cancer Center and Department of Medicine, University of Louisville, Louisville, KY, United States*

15.1 Introduction

Metal–organic frameworks (MOFs) are porous crystalline solids or porous coordination polymers (PCPs) constructed by metal sites and organic or inorganic building blocks (metal ions or clusters) are some interesting representatives of coordination polymers. During the past several years they have received tremendous research attention from people of different domains due to their high surface area, permanent porosity, controllable morphology, tunable chemical properties, and flexible chemical structure [1]. There is still a major scope in structural optimization of their structure and properties which includes increasing the crystallinity and pore size to enhance their compatibility in accordance with various applications like gas storage, adsorption, and separation, large molecule encapsulation, supercapacitors, energy conversion, chemical sensors, biomedicine, catalysis, etc. Apart from the previously stated their potential applications also include acting as precursors and self-sacrificing templates for synthesizing metal oxides and heteroatom-doped carbons [2]. Hence, awareness and knowledge about MOFs and their potential applications with conceptual understanding are essential to explore and establishing new opportunities for versatile applications and new emerging fields. Albeit, the complete control of the reaction is complicated due to its unusual synthetic procedure. A flexibility in the structural orientation of these privileged structures can be attributed to the presence of a significant number of geometries formed between the metal ions or by incorporating oligo nuclear metal clusters, geometrical characteristics, and also the use of solvent [3].

Synthesis of Metal–Organic Frameworks via Water-Based Routes: A Green and Sustainable Approach.
DOI: https://doi.org/10.1016/B978-0-323-95939-1.00002-2

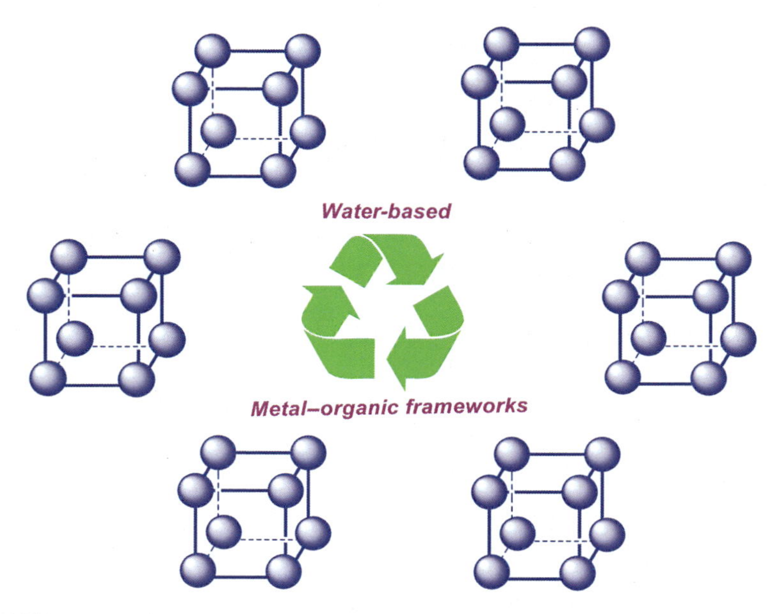

FIGURE 15.1 An illustration of water-based metal–organic frameworks.

The proportionality of water with MOFs to further enhance its applications in water utilization and efficient use of renewable or waste energy is still a matter of concern. To explore these domains the characteristic properties exhibited by MOFs like the uptake step of water, easy water release, and high mass-specific capacities, distinguish them from other most porous sorbents capacities, and make them attractive for various applications and processes where humidity control and energy reallocation using water as benign component is done are exploited according to the needs of industry [4]. Apart from their various benefits incorporation of water as an essential component is necessary as the traditional non-green synthetic pathways lead to deleterious environmental impacts and increased economic costs. The consequences arising from the same alignment towards a green preparation pathway could greatly reduce the environmental costs, energy, and the need for toxic organic solvents, and consequently reduce the production cost. These concepts of green chemistry need to be addressed in industrial production to comply with a sustainable approach [5–7]. This leads to the development of a green synthetic route to reduce the undesired effects that lead to improper utilization of resources. To date, strenuous efforts have been made to follow alternate pathways that lead to the minimal generation of hazardous organic solvents in the synthesis of MOFs like the solvent-free method, aerosol route, microwave radiation techniques, etc. [8].

This chapter presents the latest perspective of new emerging possibilities available by the apt use of water-based MOFs, their synthetic procedures in

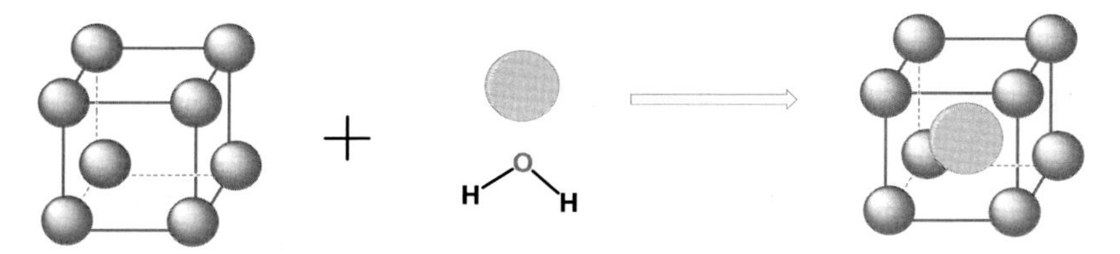

FIGURE 15.2 A basic synthetic route followed for the synthesis of water-based metal–organic frameworks.

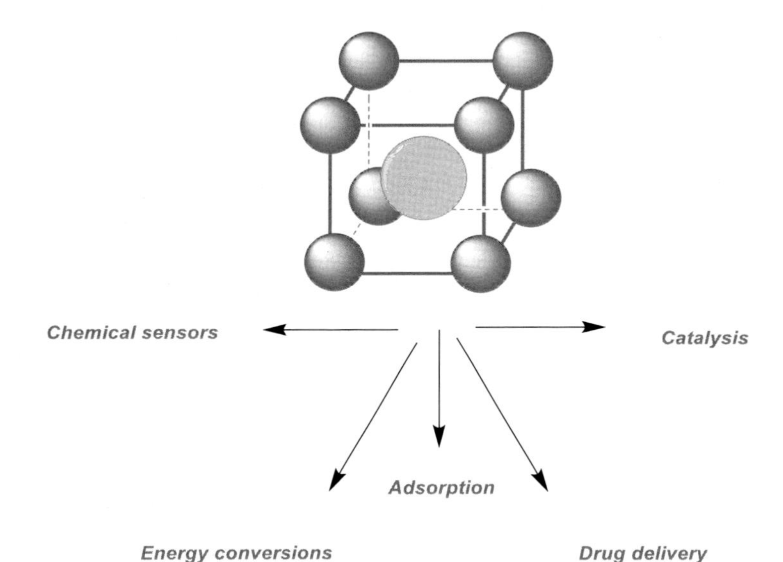

FIGURE 15.3 Potential applications of water-based metal–organic frameworks in various fields.

accordance with green chemistry, the role of water as a versatile component for its industrial production, and associated pros and cons for a wider industrial implementation of these water-based MOFs towards a greener and sustainable approach.

15.2 Synthetic procedures pertained to green chemistry

Another contrasting feature that adds to the popularity of MOFs is their easy synthetic procedures as compared to similar comparative structures like zeolites. Some of their intrinsic properties which help them achieve their task are their tendency to easily expand their surface area with tunable porosities and also their ability to function without breaking their bonds. Hence, they can be easily synthesized with a broad array of diverse structures and skeletons [9]. Certain parameters which are pre-requisite for the greener and sustainable industrial production of these MOFs are the first and foremost selection of cheaper, safer, and biocompatible building units minimizing the energy input, easy activation, and continuous chain manufacturing. Finding and exploring a green method for MOF synthesis is a major aspect of research in the same field employing harmless reactants, mild solvent conditions, and lesser byproducts are some key factors that play a prominent role in the green synthesis of MOFs [10]. Selection of metal salts containing benign anions or linker salts with innocuous metal cations is done in such a way as to avoid possible byproducts such as corrosive acids generated during the course of the reaction ultimately not only securing that water is the only byproduct, but also to achieve high atomic efficiency. The use of an innocuous solvent is another important parameter for green synthesis by replacing some commonly used solvents like DMF which apart from being toxic also decomposes into undesirable components on exposure to high temperatures [11].

A crucial factor in achieving the goal of sustainability and environmental friendliness, solvent-free conditions are the best choice for MOF synthetic procedures as they involve safe and cleaner pathways. The established routes which help in achieving the condition of the solvent-free condition are mechanochemical, thermochemical, and diffusion-controlled reactions. One representative of this class named zeolite imidazole framework (ZIF) can be synthesized by the last diffusion-controlled method also known as the accelerated aging method employing water as the best alternative solvent [12]. In the due course of time, a large number of MOFs have been synthesized in water due to their environment-friendly nature facilitating subsequent purification and recycling and a scope for further developments is much awaited. Due to these characteristically potent properties, the dominance of synthetic procedures pertaining to green chemistry is effectively selected by researchers of various domains. These synthesized MOFs have proven to exhibit superiority over traditional materials and procedures in solving various problems associated with academic research and industrial applications. With the industrial demand for a green and sustainable approach in the near future, the discovery of newer advanced materials for efficient industrial uses needs to be focused on.

15.3 Advantages of water-based synthetic routes as a green method

Water-based synthetic routes are exclusively selected routes in contrast to various other established routes over the past few decades due to the privileged space occupied by them. During the same course of time, the synthesis of water-based frameworks is selectively oriented towards the water-based synthetic pathways. Albeit these pathways also possess some demerits nevertheless the dominance of merits of the same is considered. The significance of this water-based synthesis is not only taken into account as they not only reduce manufacturing and address environmental aspects but also provide a direction to synthesize new materials with outstanding structures and properties [13]. At a larger scale only, a few of them have been successfully exploited for practical applications, the use of clean and renewable water as solvents to save costs, while avoiding contamination of water, can be considered as a key benefit for industrial-scale production [14]. Some representatives of these MOFs like UiO-66, MIL-160, and CAU-10 that can be synthesized in water are stable in water and can also lead to enhanced material properties [15]. Among the different classes of MOFs which have been synthesized in water some of the influential classes like zeolitic imidazole frameworks (ZIFs), MIL, UiO, IRMOFs, CPL, etc. have shown some interesting and exploring features that pave the way for more explorations in green synthesis of MOFs in aqueous systems [16].

A brief study of the synthetic procedures following water-based routes has shown some extraordinary features of them exhibiting distinct advantages in terms of their simpler, versatile, and environment-friendly nature; tendency to reduce the generation of harmful byproducts and easy scale-up procedures leading towards a green and sustainable approach [2]. Based on the detailed evaluative criteria proposed by Reinsch et al. [14] and Julian et al. [17] to illustrate how MOF industrial synthesis and application is orienting towards a greener and sustainable approach are (1) using water or other non-toxicity or low-toxicity solvents (e.g., ionic liquids) to replace toxic solvents in the synthesis and activation/purification process; (2) minimizing energy input and preferably the room temperature and pressures; (3) improving product yields and STYs, the maximization of incorporating raw material or feedstock to the resultant product and the minimum of the synthesis time; (4) avoiding the generation of additional by-product or waste (except water) and synthesis steps; and (5) continuous manufacturing routes.

15.4 Versatility of water leading to a sustainable approach and new domains of industry

Water plays a key role in day-to-day activities of mankind like as drinking water, indoor air conditioning, and food preparation ultimately in sustaining and survival of human life and also in large-scale industrial procedures and agriculture. MOFs play a key role in better utilization of these at a larger industrial scale ranging from phase transformation to efficient usage of renewable or waste energy resources [18]. Due to the innate nature of specific properties of water and further improvements in the intrinsic properties related to water absorption there lies a scope in the enhancement of thermodynamically associated parameters. These parameters pave the way for the exclusive study of water adsorption in MOFs by categorizing the prescribed mechanism into different categories which are (1) chemisorption on open metal sites, (2) capillary condensation (physisorption), and (3) layer/cluster adsorption (physisorption) [19]. Further a detailed study for elucidation of different criteria to determine the suitable MOFs in the same previously described water adsorption-related applications are: (1) hydrothermal stability, (2) a steep uptake isotherm at a specific relative pressure for pour filling or condensation, (3) a large water working capacity for the requisite maximum delivery of water or energy effect, (4) minimal or no hysteresis in desorption, (5) high cycling durability, and (6) facile adsorption–desorption, fast mass and heat kinetics for the desired energy efficiency [20].

Due to those above-described features of water as a versatile media, the selection of water as the most suitable solvent is made despite of various demerits associated with the same. On the basis of different comparative studies made during the past few years, the selection of water is efficiently made keeping in mind the environmental risks posed by different other organic solvents. As a result, the use of water along with other similar green solvents like ethanol, acetone, and ethyl acetate is abruptly increasing at a constant pace as compared to similar fewer green solvents like THF, DMSO, toluene, acetic acid, etc. In this regard, different preparation strategies are followed for the preparation of water-based MOFs that could reduce as much as possible the consumption of non-environment-friendly substances [21].

15.5 Emerging industrial applications of water-based MOFs

The past few decades have witnessed a swift change in the construction of these MOFs encompassing research to material, energy, environmental research, and green chemical engineering varying from a small standard laboratory scale. The depletion of natural resources and the growing scarcity of fossil fuels which is accompanied by increasingly severe environmental problems and climate change that is a result of rapid civilization and industrialization has played a key role in the development of alternative green fuels and pollutant management methods. As a result, the development of adsorption-based technology has successfully illustrated a wide range of modifiable functional groups ultimately leading to the development of an emerging attractive category of materials for adsorption applications [10]. This has led to the development of newer technologies based on MOF adsorption alternatives to compression and liquefaction to provide an everlasting challenge to the safe storing and efficient usage of gases like H_2 and CH_4 providing a new face to the development of devices for gas storage [22]. The development of newer and more advanced energy-related applications enclosed within the boundaries of a greener, safer, environmentally sustainable approach has become an urgent focus in contemporary green engineering. Due to this among various energy storage and conversion systems, photochemical and electrochemical water splitting, and CO_2 transformation are prime approaches to translate solar and electrical energy into the chemical bonds of simple species such as H_2, O_2, and CH_4 that are easy to reserve and transport [23].

A new array of development of newer potential applications on a green production scale is also emerging at a faster pace in fields like adsorption, separation, sensing, catalysis, and proton conduction. Among these, some are not yet established but are progressing at a constant rate. Over the last decade, the emergence of new potential developments and diverse unconventional synthetic approaches for MOFs have been introduced and developed including reflux under ambient pressure [24], electrochemical synthesis [25], microwave-assisted synthesis [26], mechanochemical synthesis [27], spray-drying synthesis [28], and flow chemistry [29]. These new domains of emerging applications could be undoubtedly regarded as a significant leap in the practical production of MOFs toward actual applications.

Pelletization or extrusion of MOFs has emerged as another instrumental technique in the production of actual protective equipment requiring shape particles to maintain the intrinsic properties of materials [30]. A concrete development in the study of the mechanical stability of MOFs by potentially supporting the new developments in the same field through theoretical and experimental works has led to new promising trends in the same field [31]. The selection of promising MOFs for various industrial implications by molecular simulation has also proved effective in an efficient selection process for further development and analysis. Hence, the development of newer MOFs has new domains for their industrial applications providing a safer, greener, and sustainable approach towards the synthesis of water-based MOFs.

15.6 Present and future challenges for a wider industrial implementation of water-based MOFs

As we are moving towards a brighter path by effective and efficient synthesis of MOFs via water-based routes there lies some obstacles of cost, scale-up preparation, processability and stability issues which need to resolved to further promote their practical utilization on industrial scale. The generalized methods leading to the continuous production of MOFs have however low generality, and involve some complicated synthetic procedures and equipment [32]. This leads to the demand of urgent development of a simple and general method for green synthesis overcoming these obstacles by an efficient green and environment strategy. Some other challenges and limitations which hinder the full-fledged large-scale industrial production of MOFs are poor crystalline structure, decreased porosity, and low yields however cannot be fully neglected [33]. Their needs a development of an advanced strategy by minimizing the consumption of energy and maximizing the production rate. As MOF synthesis is largely conducted in solvothermal or hydrothermal conditions requiring a high temperature, pressure, and long reaction time an efficient strategy overcoming these drawbacks still needs to be developed [34]. The production of these MOFs at a larger industrial scale requires a low manufacturing cost keeping in mind the related safety hazards and environmental impacts. Thus, the application of the principles of "green synthesis" in the preparation of MOFs is the key to their implementation in industrial productions and commercial applications.

The development of a readily available method to achieve the desired goal of green synthesis is the need of an hour. Due to this reason the utmost task of "synthesis of MOFs via water-based routes" is still developing and requires full-fledged attention from various research groups and manufacturing agencies.

Thus, a need for exploring and operating advanced synthetic routes for green synthesis and their true implementation leading to MOF commercialization needs to be urgently overcome. Different MOFs have specific challenges associated with them due to their unique composition, structure, and properties ultimately presenting their synthetic procedures much more complicated than that of zeolites. Different strategies like an electrochemical, mechanochemical, and microwave-assisted synthesis of MOFs via an electrochemical process for the industrial preparation of one such representative named HKUST-1 by using metal electrodes directly as a metal source for the generation of metal anions are also recorded [35].

15.7 Conclusion and future prospects

This privileged class of MOF with the progressive scope of controlled morphology and modifiable modular structure has attracted much attention due to its specific structural features and porous nature leading to wide variety of established and emerging applications. The presence of selective properties like good thermal stability, specific pore size, large surface area help in establishing the prominent role of MOF-based materials in functional applications. There are various methods from which MOF-based materials can be synthesized by applying the ambient conditions or high temperature and pressure. In this chapter, we have concisely drawn the aspects highlighting the industrial possibilities of water-based MOFs and the synthetic protocols followed during their preparation pertained with green chemistry. The advantages and versatility of water as an efficient and green media and in aqueous systems toward a sustainable, greener, environment-friendly approach with the potential of creating new dimensions in the industry. The emerging industrial applications of these water-based MOFs in adsorption, energy storage and utilization, catalysis, sensing, electrochemical synthesis, microwave and mechanochemical synthesis, flow chemistry, etc. In the end, the present and future challenges which hinder effective and efficient processing and synthesis at a larger industrial scale are also discussed.

Keeping in mind the previously discussed parameters future trends will involve the preparation of robust MOFs and MOF-based devices using a facile, green, and economical approach ranging from micro-scale to a larger industrial scale capable of meeting industrial requirements. In the present scenario, a wide class of many novel functional MOFs have been synthesized and will soon be ready for practical implementation. This helps us in drawing a conclusion that these MOFs will be able to compete in the near future with today's well-known industrial materials consequently minimizing the global pressure on energy and the environment and establishing their indispensability to the modern world. Therefore, the exploration of advanced synthetic routes to reach the level of "green synthesis" is the need of the hour for the future industrial production of MOFs. Ultimately this task led by academia and industry will lead to a better, more effective, and efficient implementation of these water-based MOFs consistent with the demands of "green synthesis and green industrial utilization" leading to their wider commercialization.

References

[1] P. Szuromi, Mesoporous metal-organic frameworks, Science 359 (6372) (2018) 172–174.

[2] C. Duan, Y. Yu, J. Xiao, X. Zhang, L. Li, P. Yang, J. Wu, H. Xi, Water-based routes for synthesis of metal-organic frameworks: a review, Sci. China Mater. 63 (5) (2020) 667–685.

[3] M.S. Khan, M. Khalid, M. Shahid, Engineered Fe_3 triangle for the rapid and selective removal of aromatic cationic pollutants: complexity is not a necessity, RSC Adv. 11 (2021) 2630.

[4] X. Liu, X. Wang, F. Kapteijn, Water and metal−organic frameworks: from interaction toward utilization, Chem. Rev. 120 (2020) 8303–8377, doi:10.1021/acs.chemrev.9b00746.

[5] P.T. Anastas, Introduction: green chemistry, Chem. Rev. 107 (6) (2007) 2167–2168.

[6] P. Anastas, N. Eghbali, Green chemistry: principles and practice, Chem. Soc. Rev. 39 (1) (2010) 301–312.

[7] C.-J. Li, P.T. Anastas, Green chemistry: present and future, Chem. Soc. Rev. 41 (4) (2012) 1413–1414.

[8] H. Bux, F. Liang, Y. Li, et al., Zeolitic imidazolate framework membrane with molecular sieving properties by microwave-assisted solvothermal synthesis, J. Am. Chem. Soc. 131 (2009) 16000–16001.

[9] M.S Khan, M. Khalid, M. Shahid, What triggers dye adsorption by metal organic frameworks? The current perspectives, Mater. Adv. 1 (2020) 1575.

[10] X.-J. Kong, J.-R. Li, An overview of metal–organic frameworks for green chemical engineering, Engineering 7 (2021) 1115–1139. https://doi.org/10.1016/j.eng.2021.07.001.

[11] J. Chen, K. Shen, Y. Li, Greening the processes of metal–organic framework synthesis and their use in sustainable catalysis, Chem. Sus. Chem. 10 (16) (2017) 3165–3187.

[12] M.J. Cliffe, C. Mottillo, R.S. Stein, D.K. Bučar, T. Friščić, Accelerated aging: a low energy, solvent-free alternative to solvothermal and mechanochemical synthesis of metal–organic materials, Chem. Sci. (Camb.) 3 (8) (2012) 2495–2500.

[13] J. Ren, X. Dyosiba, N.M. Musyoka, et al., Review on the current practices and efforts towards pilot-scale production of metal-organic frameworks (MOFs), Coord. Chem. Rev. 352 (2017) 187–219.

[14] H. Reinsch, "Green" synthesis of metal-organic frameworks, Eur. J. Inorg. Chem. 2016 (27) (2016) 4290–4299.

[15] M.J. Katz, Z.J. Brown, Y.J. Colón, et al., A facile synthesis of UiO-66, UiO-67 and their derivatives, Chem. 49 (2013) 9449–9451.

[16] H.C. Zhou, S. Kitagawa, Metal-organic frameworks (MOFs), Chem. Soc. Rev. 43 (2014) 5415–5418.

[17] P.A. Julien, C. Mottillo, T. Friščić, Metal-organic frameworks meet scalable and sustainable synthesis, Green Chem. 19 (2017) 2729–2747.

[18] C. Forman, I.K. Muritala, R. Pardemann, B. Meyer, Estimating the global waste heat potential, Renew. Sustain. Energy Rev. 57 (2016) 1568−1579.

[19] J. Canivet, A. Fateeva, Y. Guo, B. Coasne, D. Farrusseng, Water adsorption in MOFs: fundamentals and applications, Chem. Soc. Rev. 43 (2014) 5594−617.

[20] H. Furukawa, F. Gándara, Y.-B. Zhang, J. Jiang, W.L. Queen, M.R. Hudson, O.M. Yaghi, Water adsorption in porous metalorganic frameworks and related materials, J. Am. Chem. Soc. 136 (2014) 4369−4381.

[21] S. Wang, C. Serre, Toward green production of water-stable metal–organic frameworks based on high-valence metals with low toxicities, ACS Sustain. Chem. Eng. 7 (14) (2019) 11911–11927, doi:10.1021/acssuschemeng.9b01022.

[22] Y. He, F. Chen, B. Li, G. Qian, W. Zhou, B. Chen, Porous metal–organic frameworks for fuel storage, Coord. Chem. Rev. 373 (2018) 167–198.

[23] J. Zhou, B. Wang, Emerging crystalline porous materials as a multifunctional platform for electrochemical energy storage, Chem. Soc. Rev. 46 (22) (2017) 6927–6945.

[24] A.H. Valekar, K.-H. Cho, U.H. Lee, J.S. Lee, J.W. Yoon, Y.K. Hwang, S.G. Lee, S.J. Cho, J.-S. Chang, Shaping of porous metal–organic framework granules using mesoporous ρ-alumina as a binder, RSC Adv. 7 (88) (2017) 55767–55777.

[25] M. Li, M. Dincă, Reductive electrosynthesis of crystalline metal–organic frameworks, J. Am. Chem. Soc. 133 (33) (2011) 12926–12929.

[26] K.M.L. Taylor-Pashow, J. Della Rocca, Z. Xie, S. Tran, W. Lin, Postsynthetic modifications of iron-carboxylate nanoscale metal−organic frameworks for imaging and drug delivery, J. Am. Chem. Soc. 131 (40) (2009) 14261–14263.

[27] P.A. Julien, K. Užarević, A.D. Katsenis, S.A.J. Kimber, T. Wang, O.K. Farha, Y. Zhang, J. Casaban, L.S. Germann, M. Etter, R.E. Dinnebier, S.L. James, I. Halasz, T. Friščić, In situ monitoring and mechanism of the mechanochemical formation of a microporous MOF-74 framework, J. Am. Chem. Soc. 138 (9) (2016) 2929–2932.

[28] L. Garzón-Tovar, M. Cano-Sarabia, A. Carné-Sánchez, C. Carbonell, I. Imaz, D. Maspoch, A spray-drying continuous-flow method for simultaneous synthesis and shaping of microspherical high nuclearity MOF beads, React. Chem. Eng. 1 (5) (2016) 533–539.

[29] M. Faustini, J. Kim, G.-Y. Jeong, J.Y. Kim, H.R. Moon, W.-S. Ahn, D.-P. Kim, Microfluidic approach toward continuous and ultrafast synthesis of metal–organic framework crystals and hetero structures in confined microdroplets, J. Am. Chem. Soc. 135 (39) (2013) 14619–14626.

[30] M.I. Nandasiri, S.R. Jambovane, B.P. McGrail, H.T. Schaef, S.K. Nune, Adsorption, separation, and catalytic properties of densified metal-organic frameworks, Coord. Chem. Rev. 311 (2016) 38−52.

[31] P.Z. Moghadam, S.M. Rogge, A. Li, C.-M. Chow, J. Wieme, N. Moharrami, M. Aragones-Anglada, G. Conduit, D.A. Gomez-Gualdron, V. Van Speybroeck, Structure-mechanical stability relations of metal-organic frameworks via machine learning, Matter 1 (2019) 219.

[32] P. Baláž, M. Achimovičová, M. Baláž, et al., Hallmarks of mechanochemistry: from nanoparticles to technology, Chem. Soc. Rev. 42 (2013) 7571.

[33] J.M. Woodley, New opportunities for biocatalysis: making pharmaceutical processes greener, Trends Biotech. 26 (2008) 321–327.

[34] C. Duan, H. Zhang, A. Peng, et al., Synthesis of hierarchically structured metal-organic frameworks by a dual-functional surfactant, ChemistrySelect 3 (2018) 5313–5320.

[35] P. Silva, S.M.F. Vilela, J.P.C. Tomé, F.A. Almeida Paz, Multifunctional metal organic frameworks: from academia to industrial applications, Chem. Soc. Rev. 44 (19) (2015) 6774–6803.

Index

Page numbers followed by "*f*" and "*t*" indicate, figures and tables respectively.

X

Z

Printed in the United States
by Baker & Taylor Publisher Services